# Advances in Atom and Single Molecule Machines

Advances in Atom and Single Molecule Machines is the first comprehensive series of books dealing specifically with single atom and molecule machines. Derived from a number of long-term European Commission Future and Emerging Technologies (FET) projects including AtMol, Elfos, Focus, Diamant, Artist, PAMS and MEMO, volumes in this series comprise topical reviews, lecture-course derived textbooks and re-worked proceedings of workshops.

More information about this series at http://www.springer.com/series/10425

Masakazu Aono
Editor

# Atomic Switch

## From Invention to Practical Use and Future Prospects

Springer

*Editor*
Masakazu Aono
International Center for Materials Nanoarchitectonics (MANA)
National Institute for Materials Science (NIMS)
Tsukuba, Ibaraki, Japan

ISSN 2193-9691 ISSN 2193-9705 (electronic)
Advances in Atom and Single Molecule Machines
ISBN 978-3-030-34874-8 ISBN 978-3-030-34875-5 (eBook)
https://doi.org/10.1007/978-3-030-34875-5



This Springer imprint is published by the registered company Springer Nature Switzerland AG.
The registered company address is: Gewerbestrasse 11, 6330 Cham, Switzerland

*Dedicated to the memories of our late*
*Dr. Heinrich Rohrer*

# Preface

The volume editor and co-workers invented a novel nanometer-scale electronic switch at the beginning of the 2000s and named it "atomic switch". This switch is critically different from the conventional semiconductor-based transistor switch in that on/off switching is caused by electrochemically controlled movement of metal atoms (ions) in a nanometer-scale gap between two electrodes. The atomic switch is therefore a two-terminal nonvolatile switch basically, being compact in structure and small in power consumption.

The invention of the atomic switch was made in RIKEN (Institute of Physical and Chemical Research, Japan) and NIMS (National Institute for Materials Science, Japan). Soon later, we started collaborative research and development of the atomic switch with NEC Corp., and after more than a decade, in 2016, the atomic switch was put into practical use in the form of FPGA (field-programmable gate array). As compared to conventional semiconductor-based FPGA (SRAM-FPGA), NEC's new FPGA (AtomSW-FPGA or NanoBridge-FPGA) is characterized not only by nonvolatility, small size (~1/3) and low power consumption (~1/10) but also by high tolerance to electromagnetic noise and radiation including cosmic rays (~100 times). We also revealed that the atomic switch has close similarity to the synapse in human brain and that random networks of such synaptic atomic switches exhibit fascinating novel characteristics.

Recently, we opened an international symposium entitled "Atomic Switch; Invention, Practical Use and Future Prospects" (Tsukuba, Japan, 27–28 March 2017) in commemoration of the practical use of the atomic switch. This book consists of 12 chapters relating to representative papers presented at the symposium and a chapter showing an almost comprehensive list of papers related to the atomic switch published so far. The volume editor is happy and honored to have led the invention and development of the atomic switch over two decades. I hope this book will give new inspirations to scientists and engineers in various fields including ICT, AI and IoT.

Finally, I would like to express my sincere thanks to Prof. Christian Joachim (CNRS, France) as well as to Ms. Charlotte Hollingworth and Ms. Stephanie Kolb (Springer Nature Publishing) for their patient encouragement in publishing this book. Also, I thank the associate editors of this book, Dr. Kazuya Terabe, Prof. Tsuyoshi Hasegawa, Dr. Tomonobu Nakayama, Prof. James Gimzewski, Dr. Ilia Valov, Dr. Tsuruoka Tohru and Dr. Takashi Tsuchiya, for their support in editing this book.

Tsukuba, Ibaraki, Japan Masakazu Aono

# Contents

# Editor and Associate Editors

## Editor

**Masakazu Aono** International Center for Materials Nanoarchitectonics (MANA), National Institute for Materials Science (NIMS), Tsukuba, Ibaraki, Japan

## Associate Editors

**Kazuya Terabe** International Center for Materials Nanoarchitectonics (MANA), National Institute for Materials Science (NIMS), Tsukuba, Ibaraki, Japan

**Tsuyoshi Hasegawa** School of Advanced Science and Engineering, Waseda University, Tokyo, Japan

**Tomonobu Nakayama** International Center for Materials Nanoarchitectonics (MANA), National Institute for Materials Science (NIMS), Tsukuba, Ibaraki, Japan

**Toshitsugu Sakamoto** System Platform Research Laboratories, NEC Corporation, Ibaraki, Japan

**James K. Gimzewski** California NanoSystem Institute (CNSI), UCLA, Los Angeles, CA, USA

# Invention and Development of the Atomic Switch

**Kazuya Terabe, Tsuyoshi Hasegawa, Tomonobu Nakayama, and Masakazu Aono**

**Abstract** Our invention of atomic switches with novel operating principles, using movement of atoms (ions) within solids, are serendipitous in a sense. In earlier study, we were trying to work on an experiment where atoms were arranged into lines in order to draw on a substrate by generating high electric field at an apex of a scanning tunneling microscope's (STM) tip made of an electron and metal-ion mixed conductor material, and depositing metal atoms one by one from the tip. In doing so, without forethought, we found that by controlling the voltage applied to the STM tip, a protrusion at that apex, consisting of a small number of metallic atoms, could be reversibly grown and shrunk. We immediately came up with the idea of using these reversible processes for atomic-scale electrical switching. Thereafter, we have created the atomic switches, and found various unique physical properties such as quantized conductance, and valuable functions such as logic-gate operation in developed atomic switches.

## 1 Introduction

Since a great number of electronics switch devices typified by semiconductor transistors are used in the information and communications equipment, upgrading of that equipment largely depends on improving performance of transistors, which are operated by controlling the migration of electrons within semiconductors. Though semiconductor transistors have seen remarkable progress with technological development in miniaturization and integration, it is currently feared that the progress is beginning to slow. Thus, it is becoming important to create switch devices that operate on a completely new set of principles, together with further development

K. Terabe (✉) · T. Nakayama · M. Aono
International Center for Materials Nanoarchitectonics (MANA), National Institute for Materials Science (NIMS), Tsukuba, Ibaraki, Japan
e-mail: terabe.kazuya@nims.go.jp

T. Hasegawa
Department of Applied Physics, Waseda University, Tokyo, Japan

 
M. Aono (ed.), *Atomic Switch*, Advances in Atom and Single Molecule Machines,
https://doi.org/10.1007/978-3-030-34875-5_1

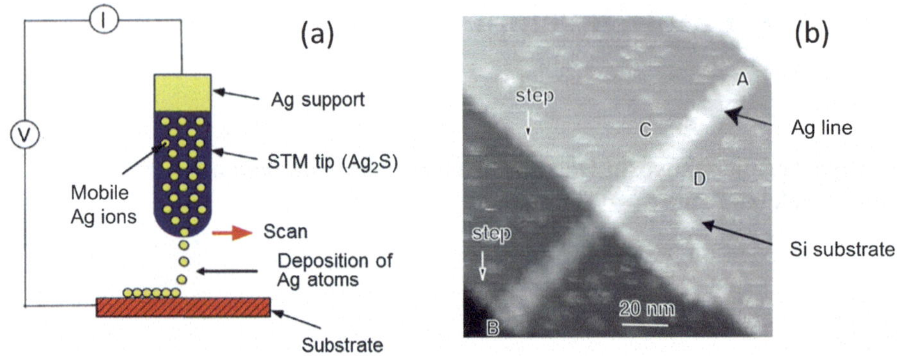


**Fig. 1** (**a**) Schematic diagram of a method for constructing nanostructures by applying Ag atoms onto the substrate from the STM tip of the $Ag^+$ ion and electron mixed conductor $Ag_2S$. (**b**) Ag thin line formed by applying Ag atoms onto the substrate while scanning the $Ag_2S$ tip. [5] © (2002), AIP publishing LLC

of semiconductor transistors. If a revolutionary novel switch device is developed, it will be possible to innovatively advance information and communication equipment.

Our development of innovative atomic switches, whose operating principles are completely different from that of the semiconductor transistor, are due to the discovery of unexpected solid electrochemical nanophenomenon, namely by serendipity [1–4]. In our initial research, we were developing nanotechnology to deposit metal atoms one by one on a substrate, by utilizing metal ion migration within the solid material of the STM tip. The new method illustrated in Fig. 1a was examined. Silver (Ag) atoms were arranged into lines in order to draw on a substrate by generating a high electric field at an apex of the STM tip made of a silver sulfide ($Ag_2S$) crystal, and depositing Ag atoms from the tip. Since the $Ag_2S$ crystal is an electron and silver ($Ag^+$) ion mixed conductor [5, 6], $Ag^+$ ions of the $Ag_2S$ crystal are moving around in a rigid lattice made of sulfur ($S^-$) ion as if in a liquid. When negative bias voltage is applied to the substrate between the STM tip and the substrate electrodes, positively charged $Ag^+$ ions in the $Ag_2S$ tip are expected to move to the apex and be extracted from the apex and deposited onto the counter substrate one by one along the scan of the STM tip parallel to the substrate. Figure 1b shows the fabricated nanoscale line structure made of Ag atoms on the substrate by scanning the STM tip under applying appropriate bias voltage and tunneling current [7]. This fabrication process appears as if the line is drawn on substrates by using a "nano-fountain pen "containing ink of $Ag^+$ ions.

In doing related researches, without forethought, we found that by controlling the bias voltage and tunneling current applied to the STM tip, a cluster at the apex of the tip, consisting of a small amount of Ag atoms, could be reversibly grew and shrank [8, 9]. Figure 2 shows the growth and shrinkage behavior of the Ag atom cluster at the apex of $Ag_2S$ tip. The behaviors were examined by monitoring the distance of the displacement of the $Ag_2S$ tip in the direction perpendicular to the substrate surface under a constant-current mode. The Ag cluster grow and shrank reversibly when the

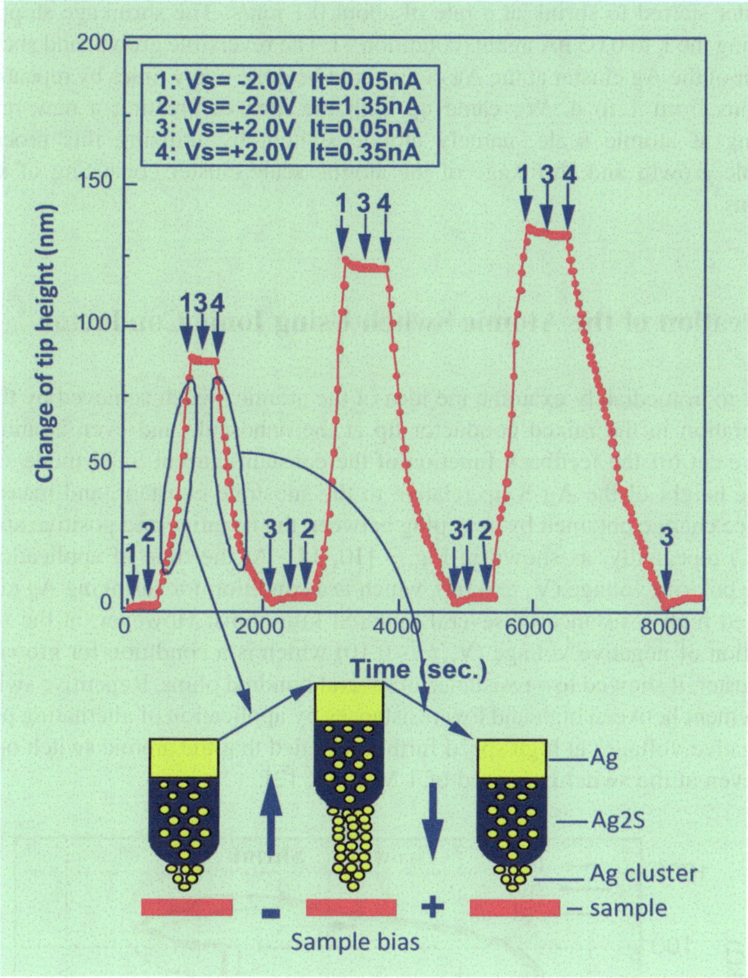


**Fig. 2** Reversible growth and shrinkage of Ag cluster on the $Ag_2S$-STM tip by changing the condition of applied bias voltage. [9] © (2002), AIP publishing LLC

polarity of $V_s$ and magnitude of $I_t$ were changed, in which $V_s$ and $I_t$ are the bias voltage applying to the substrate and the tunneling current, respectively. The conditions of $V_s$ and $I_t$ were changed in the order of 1, 2, 1, 3, 4, and 3 conditions, and this series of operation was repeated three times in Fig. 2. No Ag cluster growth was observed at $V_s = -2.0$ V and $I_t = 0.05$ nA (condition 1). However, when $I_t$ was increased to 1.35 nA with $V_s$ kept constant (condition 2), the Ag cluster started to grow at the apex of the $Ag_2S$ tip at a growth rate of about 0.1–0.2 nm/s. The growth stopped by returning the condition from 2 to 1. No change was observed in the grown cluster by the polarity switching of $V_s$ from negative to positive, with keeping $I_t$ at the same (condition 3). However, when $I_t$ was increased to 0.35 nA (condition 4), the cluster started to shrink at a rate of about 0.1 nm/s. The shrinkage stopped by decreasing the $I_t$ to 0.05 nA again (condition 3). The reversible growth and shrinkage behavior of the Ag cluster at the $Ag_2S$ tip could be done many times by repeating the conditions from 1 to 4. We came up with the idea of creating a new resistive switching at atomic scale, namely atomic switch, by utilizing this process of reversible growth and shrinkage of the atomic scale cluster consisting of several Ag atoms.

## 2 Creation of the Atomic Switch Using Ionic Conductor

In order to immediately examine the idea of the atomic switch achieved by the $Ag^+$ ion migration in the mixed conductor tip at the nanoscale and even atomic scale level, we cut off the feedback function of the constant-current STM mode so as to keep the height of the $Ag_2S$ tip relative to the substrate constant, and measured a resistance change obtained by sweeping between the negative and positive applying bias ($V_s$) repeatedly, as shown in Fig. 3 [10, 11]. At the time of application of a positive polarity voltage ($V_s \simeq 0.05$), which is a condition for shrinking Ag clusters, it showed high resistance of several hundred kilo-ohms. However, at the time of application of negative voltage ($V_s \simeq -0.10$) which is a condition for growing Ag atom cluster, it showed low resistance of several hundred ohms. Repetitive switching measurement between high and low resistances by application of alternating positive and negative voltages at high speed further revealed that the atomic switch operates stably even at the switching speed of 1 MHz [2, 12].

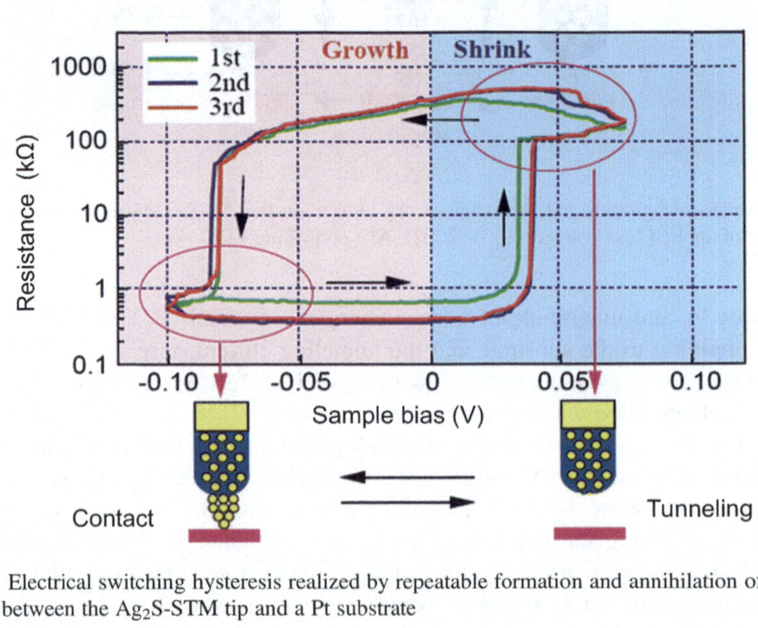


**Fig. 3** Electrical switching hysteresis realized by repeatable formation and annihilation of the Ag bridge between the $Ag_2S$-STM tip and a Pt substrate

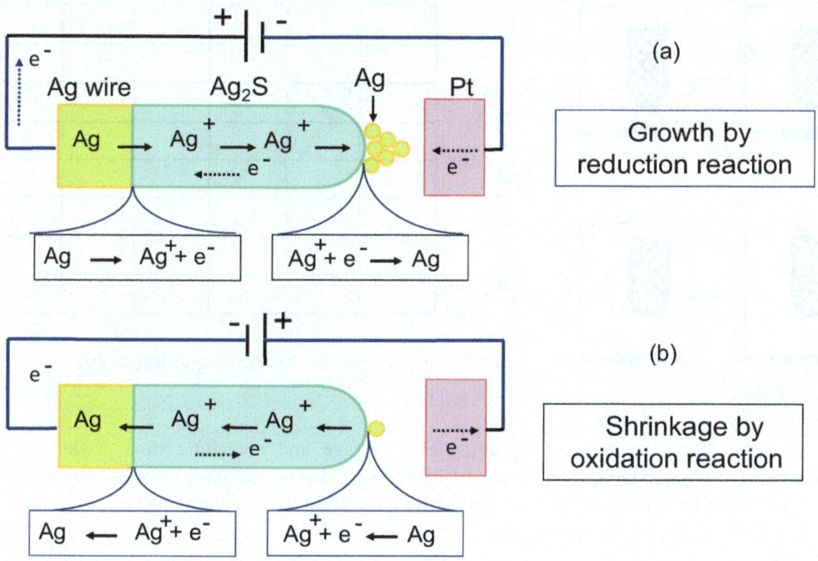


**Fig. 4** Mechanism of growth and shrinkage of silver atom cluster achieved by solid electrochemical reaction using tunnel electron. The growth (**a**) and shrinkage (**b**) are caused by the reduction and oxidation reactions, respectively

## *2.1 Basic Operation Principle*

Figure 4a, b show schema diagrams of the electrochemical behaviors in STM tip made of the $Ag_2S$ needle-like crystal, which is electron and $Ag^+$ ion mixed conductor, under the application of negative and positive bias voltage, respectively. A Ag wire serving as supply sources of $Ag^+$ ions to the $Ag_2S$ crystal, in addition to the role of the electrode, is bonded to ends of the $Ag_2S$ crystal. The other platinum (Pt) electrode is located about 1 nm away from the $Ag_2S$ tip so that tunneling current flows between them. In Fig. 4a, when negative bias voltage is applied to the Pt electrode with respect to the Ag electrode/$Ag_2S$ tip, $Ag^+$ ions in the $Ag_2S$ are moved to the apex and are neutralized to Ag atoms by reduce reaction ($Ag^+_{(Ag2S)} + e^- \rightarrow Ag_{(cluster)}$). Electrons are supplied by flowing from the Pt electrode. As a result, The Ag cluster grows on the apex of the $Ag_2S$ tip, forming the bridge between the tip and the Pt electrode. As the cluster grows, the chemical potential of $Ag^+$ ion in the $Ag_2S$ crystal decreases, but its potential is kept constant by supplying $Ag^+$ ions from the Ag electrode. Thereafter, when a positive voltage with opposite polarity is applied to the Pt substrate in Fig. 4b, the Ag cluster is oxidized to $Ag^+$ ions, and dissolved into the $Ag_2S$ crystal ($Ag_{(cluster)} \rightarrow Ag^+_{(Ag2S)} + e^-$). As results, the bridge between the $Ag_2S$ tip and the Pt electrode is annihilated.

In this manner, the atomic switch with the tunneling gap structure, namely "gap-type atomic switch" can be operated by the growth and shrinkage of the metal-atom

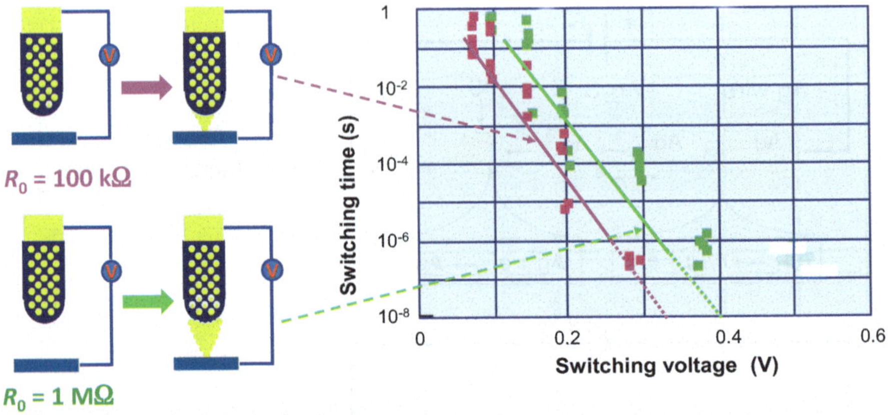


**Fig. 5** The relationship between applying bias voltage and switching time. Switching time behaviors for initial tunneling resistances of 1 MΩ and 100 kΩ are shown by two lines, respectively. Both switching times decrease exponentially with increasing bias voltages [12]

cluster with controlling the metal ion migration and the solid electrochemical reaction in a mixed conductor material from the nanoscale to atomic scale level [10].

The switching time ($t_{sw}$) dependence on switching voltage was examined as shown in Fig. 5 [12]. The switching time was defined as the time required for the resistance to decrease from initial Off states (1 MΩ and 100 kΩ) to the ON state (12.9 kΩ) after applying voltage. The value of 12.9 kΩ is considered to be the resistance of a single atomic contact with a substrate. The switching times decreased exponentially as switching voltages increased in both atomic switches with the OFF state values of 1 MΩ and 100 kΩ. These behaviors are caused by an activation process of the solid electrochemical reaction ($Ag^{+}_{(Ag2S)} + e^{-} \rightarrow Ag_{(cluster)}$). As expected from the operating mechanism based on the solid electrochemical reaction causing the growth and shrinkage of the Ag cluster, the switching rate depends on the activation process of the reaction, which is characterized by $t_{sw} \propto \exp(E_a/k_BT)$, where $E_a$ is the activation energy for generating the electrochemical reaction, $k_B$ is Boltzmann constant and T is temperature.

## 2.2 *Fabrication of the Gap-Type Atomic Switch*

After fundamental studies of atomic switches using STM, a fabricating method for devices of gap-type atomic switches that were formed at each crossing points of crossbar structure was developed using conventional technique for semiconductor device miniaturization [1, 2]. The typical cross-bar structure with nanogaps was fabricated utilizing a $Ag_{(\sim 1nm\ thickness)}/Ag_2S/Ag$ wire and two Pt wires (Fig. 6a). Nanogaps of about 1 nm between the $Ag_2S/Ag$ wire and Pt wires were formed by solid electrochemical reaction after the deposition of these wires. The overall process

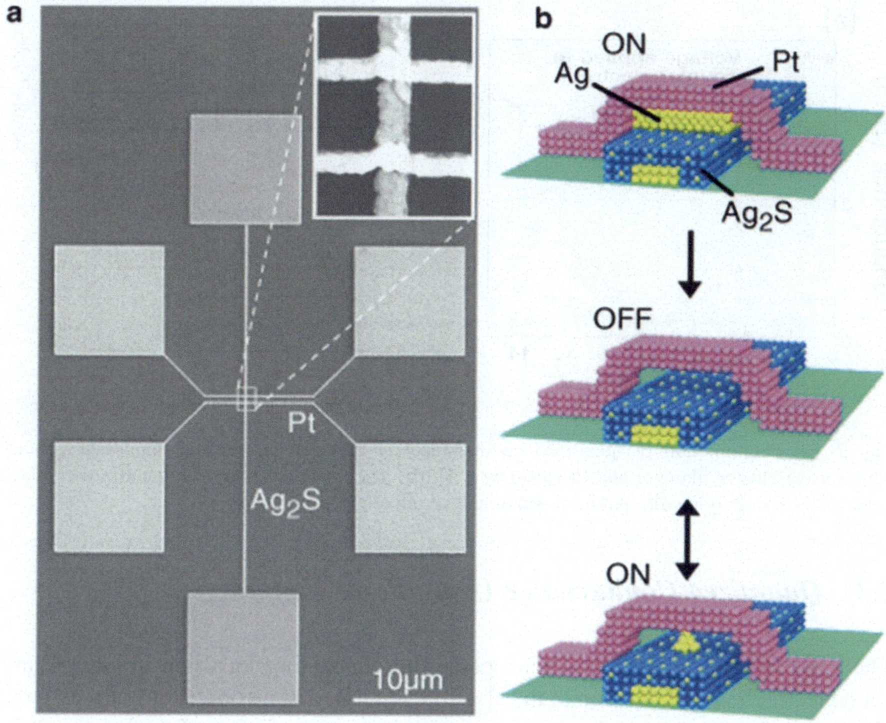


**Fig. 6** (**a**) Scanning electron microscope image of the gap-type atomic switch fabricated with the cross-bar structure of Ag/$Ag_2S$/Ag wire and Pt wires. (**b**) Schematic diagram of the gap-type atomic switch with the cross-bar structure of the $Ag_{(\sim 1nm\ thickness)}$/$Ag_2S$/Ag wire and the Pt wire. As-formed structure of ON state without the nanogap (top), switched OFF state with the nanogap (middle) and switched ON state with a Ag filament bridge after the initial switching OFF state process (bottom) [1]

is as follows. First, a Ag nanowire was deposited on an insulator substrate, and the Ag nanowire was partially sulfurized to form the $Ag_2S$ layer by introducing sulfur vapor in the deposition chamber. The Ag thin film of about 1 nm was then deposited on the $Ag_2S$/Ag wire. Continuously, Pt two wires were deposited so as to form the cross-bar structure. After fabricating the crossbar structure consisting of the $Ag_{(\sim 1nm\ thickness)}$/$Ag_2S$/Ag wire and two Pt wires, a certain voltage was applied between these wires so as to form the nanogaps using the solid electrochemical reaction ($Ag_{(\sim 1nm\ thickness)} \rightarrow Ag^{+}_{(Ag2S)} + e^{-}$). By this solid electrochemical reaction, $Ag_{(\sim 1nm\ thickness)}$ layers were dissolved into the $Ag_2S$ layers, as shown in Fig. 6b.

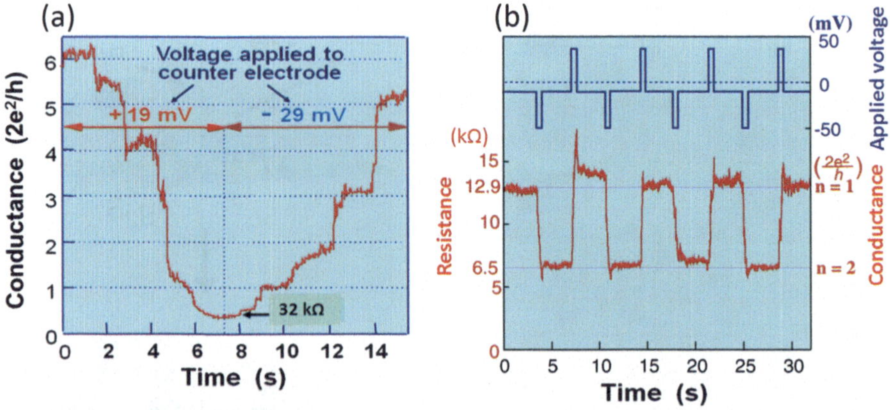


**Fig. 7** (**a**) Phenomenon of quantized conductance by precisely controlling the point contact between Ag cluster filament and Pt substrate [11]. (**b**) The switching between quantized conductance states by applying alternating positive and negative pulse voltages

## *2.3 Quantized Conductance Using Point Contact*

The gap-type atomic switches that operate by controlling local ion migration and solid electrochemical reaction in the electron and ion mixed conductor not only have resistance switching but also interesting physical properties and functions such as quantized conduction at room temperature [1, 11]. The quantized conductance was observed when Ag atom cluster is slowly grown and shrunk between the $Ag_2S$-STM tip and the Pt substrate by applying relatively small voltage near the threshold voltage to start the electrochemical reactions ($Ag^+_{(Ag2S)} + e^- \leftrightarrows Ag_{(cluster)}$). The experimental situation is the same as that in Fig. 3. Quantized conductance appeared, as shown in Fig. 7a, by controlling the point contact between the Ag cluster on the $Ag_2S$ tip and the Pt substrate. The time on the horizontal axis, in this figure, starts just before the disappearance of the point contact between the Ag cluster and the Pt substrate. The quantized conductance of the vertical axis is indicated by the number of quantized unit, as calculated by dividing the electrical conductance by $2e^2/h$ (where $e$ is the charge of an electron and $h$ is Plank's constant). By continuing to apply a positive voltage of 19 mV to the Pt substrate, the appeared quantum conductance decreased from 6 to 1 in steps, while decreasing the point contact area between the Ag cluster and the Pt substrate. Subsequently the polarity of applying voltage was switched to negative, and −29 mV was applied. The Ag cluster regrew and formed the point contact. By increasing the area of the point contact, the quantum number increased from 1 to 5 in steps again. Interestingly, it was found that this quantized state can be controlled by applying a pulsed voltage to the $Ag_2S$-STM tip. In Fig. 7b, the state of the point contact whose quantum number is 1 was formed between the STM tip and the substrate in advance. Next, when voltages of 50 mV and −50 mV were applied to the substrate in pulses, the quantum

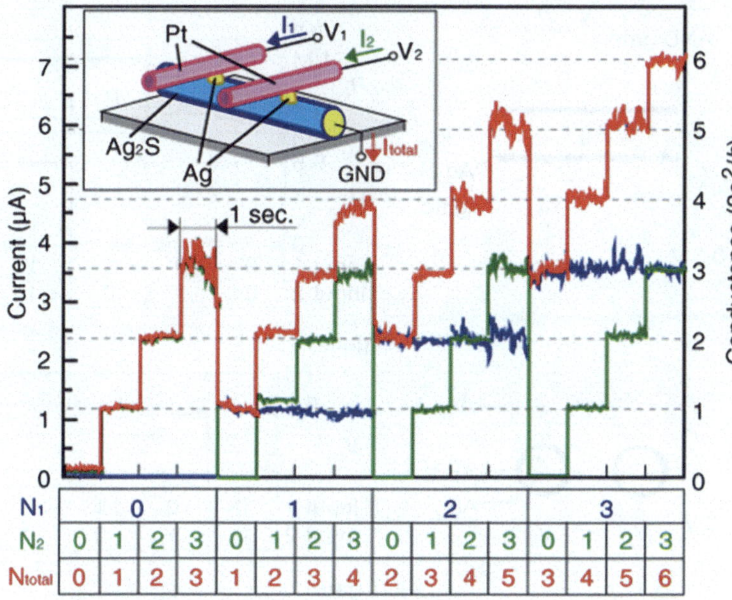


**Fig. 8** 1 × 2 array of gap-type atomic switches formed at cross-points of $Ag/Ag_2S$ wire and Pt wires. States of the quantized conductance of these atomic switches can be changed independently from N = 0 to N = 3 [1]

number can be alternately controlled between 1 and 2 by precisely controlling the growth and shrinkage of the Ag cluster on the STM tip.

To apply the repeatable switching between these quantum number to interesting device function, 1 × 2 array atomic switches with the cross-bar structure consisting $Ag/Ag_2S$ nanowire and two Pt nanowires were fabricated [1]. The structure is the same as that in Fig. 6. Each atomic switch exhibits quantized conductance by precisely controlling the applied positive and negative pulse voltages. Quantum numbers can be arbitrarily switched accordingly by applying voltages of appropriate polarity and magnitude in a pulsed manner. Figure 8 shows that 4 × 4 = 16 states can be obtained by connecting two atomic switches capable of producing four kinds of quantum number, from 0 to 3. This result shows that multi-valued memory and adder circuit can be developed by using this function.

## 2.4 *Logic–Gate Operation*

By using one or two gap-type atomic switches, it is possible to make basic logic gates for logic operation of the computer [1]. Figure 9a–c show the structures of the logic gates AND, OR, NOT and their experimental operations, respectively. Here, a value of 1 in these figures indicates a state where voltage is applied to the input

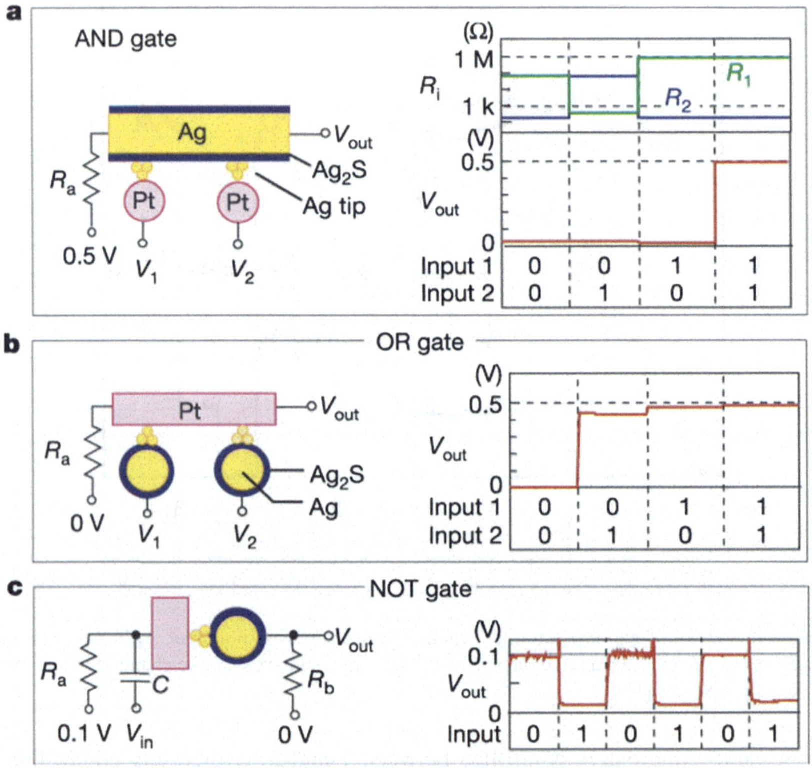


**Fig. 9** Gap-type atomic switch based logic gates. (**a**) Schema of AND gate fabricated using two atomic switches and its operating result. (**b**) Schema of OR gate fabricated using two atomic switches and its operating result. (**c**) Schema of NOT gate fabricated using the atomic switch and its operating result [1]

1 ($V_1$) or input 2 ($V_2$), and the value of 0 indicates a state where no voltage is applied to them. Depending on the voltage application conditions of input 1 and input 2, bridges of the Ag filaments between the Ag/$Ag_2S$ wire and the Pt wires form or disappear at each atomic switch. For instance, an AND gate was consisted of two atomic switches with one Ag/$Ag_2S$ wire and two Pt wires (see Fig. 9a). The output voltage ($V_{out}$) was detected only when voltages are applied to both input 1 and input 2. Also, in an OR gate consisted of two atomic switches with two Ag/$Ag_2S$ wires and one Pt wire, no voltage of $V_{out}$ was detected only when no voltage was applied to either of the two inputs of 1 and 2 (see Fig. 9b). In the NOT gate consisted of one atomic switch with one Ag/$Ag_2S$ wire and one Pt wire, the output voltage of $V_{out}$ was detected only when no voltage was applied to the input 1 (see Fig. 9c). By combining these basic logic gates of the AND gate, the OR gate and the NOT gate, it is possible to construct logic circuits for the computer.

## 3 Gapless (Junction)-Type Atomic Switch

Gapless-type atomic switches operated by controlling the migration of metal ions and the solid electrochemical reaction from the nanoscale to the atomic scale in the ion conductor have been also developed [11]. In the aforementioned gap-type atomic switches (Fig. 10a), local ion migration and electrochemical reaction are controlled by using tunnel electrons in the gap. As another control method, gapless atomic switches use the nanospaces of ionic conductors or mixed conductors such as nanowires and thin films (Fig. 10b). Ion migration and electrochemical reactions occur in the nanospaces within the ionic conductors or mixed conductors, thereby controlling the growth and shrinkage of metal nanofilament. Figure 11 illustrates the growth process of the Ag filament (cluster) inside the $Ag_2S$ nanowire of an electron and $Ag^+$ ion mixed conductor. When an appropriate negative voltage is applied to the Pt electrode, $Ag^+$ ions migrate to the Pt electrode side, and Ag atoms are precipitated to form the Ag cluster by the reduction reaction. The Ag cluster grows to form the bridge between two electrodes by the continuation of the reduce reaction. Subsequently, when the polarity of the voltage applied to the Pt electrode is switched to positive, the Ag filament is oxidized to $Ag^+$ ions and shrinks by solid-dissolving in $Ag_2S$. In this way, the gapless-type atomic switch operates by utilizing the growth and shrinkage of the metal filament in the nanospace of the mixed conductor or ion conductor. Gapless atomic switches using sulfide-based mixed conductors such as $Ag_2S$ and cupper sulfide ($Cu_2S$) [13], have been fabricated, which are operated by controlling migration of $Ag^+$ or $Cu^+$ ions and their solid electrochemical reactions. Furthermore, gap-type atomic switches utilizing the migration and electrochemical reaction of $Ag^+$ ions or $Cu^+$ ions in transition metal oxides such as tantalum oxide ($Ta_2O_5$) have been developed [14]. Although $Ta_2O_5$ is known as an insulator, has small ion and electron conductivities. Thus, an atomic switch based on a $Ta_2O_5$ thin film can be operated only by local moving of small amount of $Cu^+$ ions and electrons

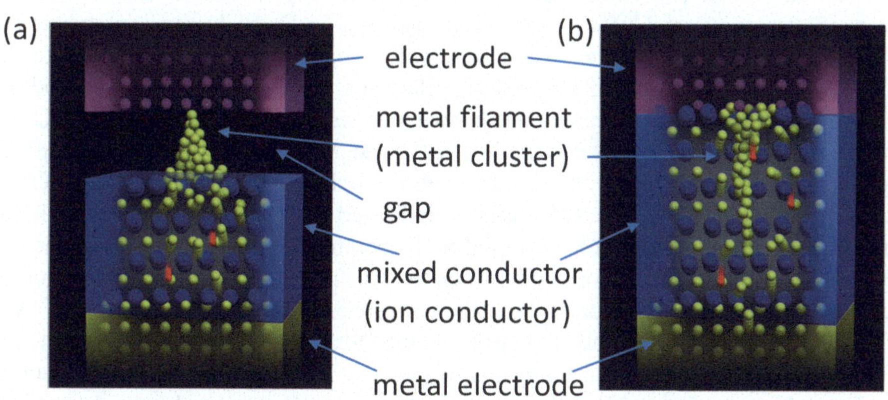


**Fig. 10** Schematic diagrams (**a**), (**b**) of the gap-type atomic switch and the gap-less type atomic switch, respectively

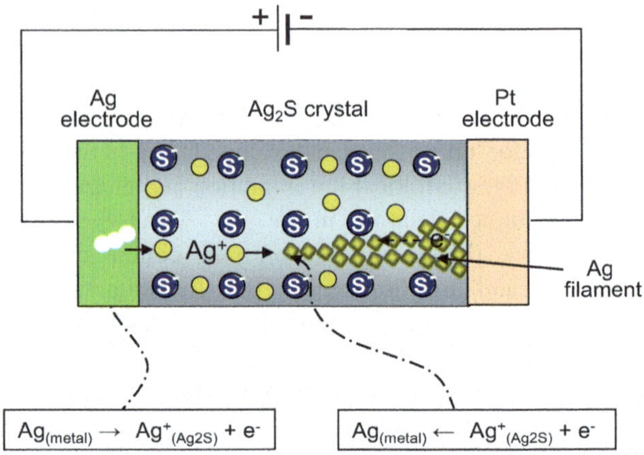


**Fig. 11** Switching mechanism of the gap-less type atomic switch using a silver atom filament grown by solid electrochemical reaction [11]

in the nanospace to form the nanoscale Cu filament. Gap-type atomic switches using not only inorganic materials but also polymer materials have been developed. Switching behavior of the atomic switch consisting of solid-polymer-electrolyte film, such as polyethylene oxide and Ag perchlorate complexes, has been demonstrated [15].

## *3.1 Switch Operation Principle*

It is described in detail here that although both-type atomic switches operate utilizing local ion migration and its electrochemical reaction, the gapless-type atomic switch operates with a mechanism different from that of the gap-type atomic switch [16].

Current-voltage (I-V) characteristics of the gapless-type atomic switches made of stacked structures of Cu/$Ta_2O_5$/Pt layers and Ag/$Ta_2O_5$/Pt layers are shown in Fig. 12a, b, respectively. The switches are achieved by formation and shrinkage of bridges, which are made of Cu and Ag filaments in $Ta_2O_5$, between electrodes, respectively. Both as-fabricated atomic switches showed resistances of 10 GΩ or more. When the applied bias voltages to the Cu and Ag electrodes were first swept to the positive polarity, the elements indicate the switches from high resistances (OFF states) to the low resistances (ON states) as indicated by the blue (I-V) curves, which are called SET operation. In these initial SET operations, bridges of metal filaments (Cu and Ag filaments) are formed between the Cu, Ag and Pt electrodes, which are called the "forming process". Since the energies required to form the Cu, Ag filaments are different, the voltages at the SET operation are different in Fig. 12a, b. However, the mechanisms of these switches caused by the formation and shrinkage of Cu and Ag filaments are the same. For example, the formation mechanism of the Cu filament in $Ta_2O_5$ at this SET operation is shown in Fig. 12c. When a positive bias is applied to the Cu electrode, an anodic oxidation

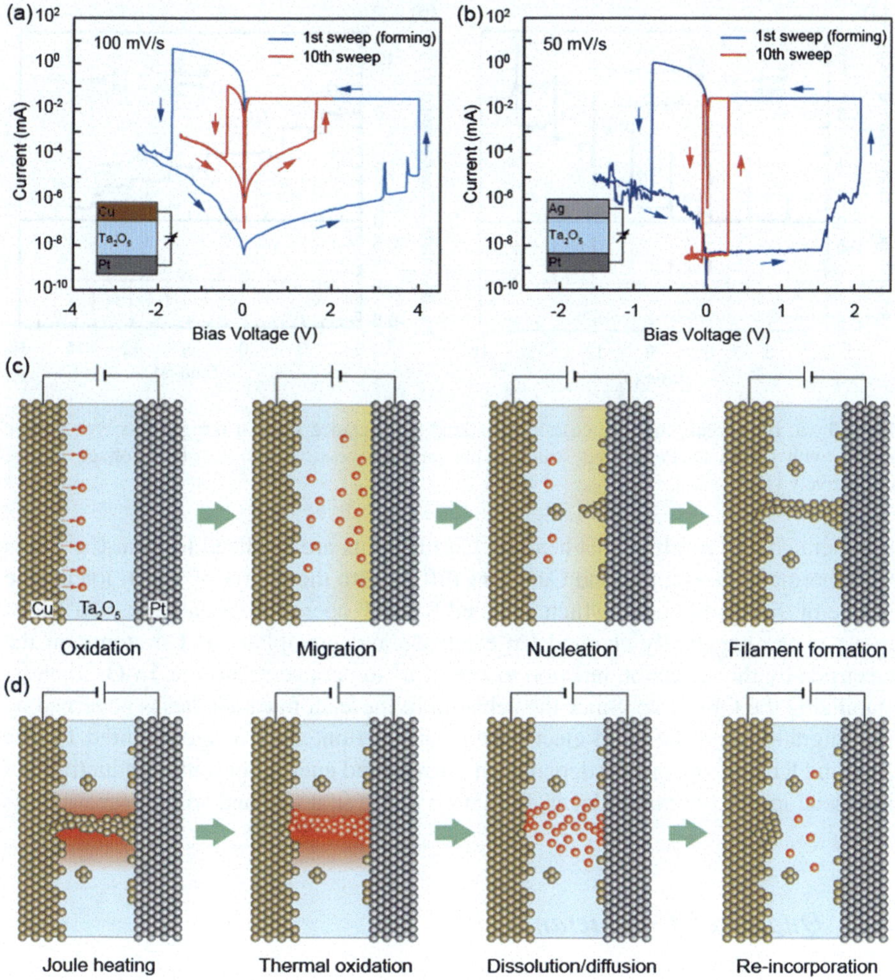


**Fig. 12** (**a, b**) Typical forming behaviors (1st sweep) and subsequent switching behaviors (10th sweep) for gap-less type atomic switches with element structures of $Cu/Ta_2O_5/Pt$ and $Ag/Ta_2O_5/Pt$, respectively. (**c, d**) The SET and RESET mechanisms of the $Cu/Ta_2O_5/Pt$ cell under application of positive and negative bias to the Cu electrode, respectively [16]

($Cu \rightarrow Cu^{z+} + ze^-$, where z is either 1 or 2) occurs at the $Cu/Ta_2O_5$ interface, and then positively charged $Cu^{z+}$ ions in $Ta_2O_5$ migrate toward the Pt negative electrode. The concentration of $Cu^{z+}$ ions near the $Ta_2O_5/Pt$ interface gradually increases and reaches supersaturation, resulting in heterogeneous Cu nucleation on the Pt electrode. Subsequently, as the Cu nucleus continues to grow by continuous voltage application, the bridge of Cu filaments is formed between the Cu electrode and the Pt electrode, and the element shows the switching from the OFF state to the ON state.

Subsequently, when a negative voltage is swept to the Cu electrode, high density current flows in the formed Cu nanofilament due to the low resistivity and Joule heat

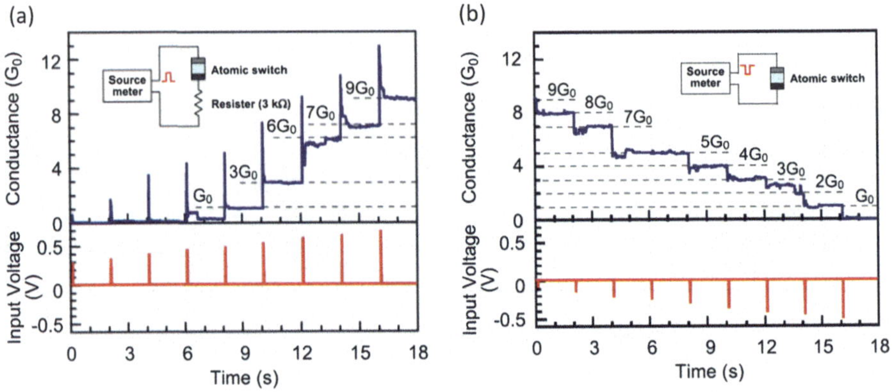


**Fig. 13** (**a**, **b**) Increase and decrease in quantized conductance state of the gap-less type atomic switch with $Cu/Ta_2O_5/Pt$ layered structure by applying positive and negative voltage pulses, respectively [17]

is generated. As result of this heating, Cu filaments are oxidized to form $Cu^{z+}$ ions with thermally assistance, and $Cu^{z+}$ ions diffuse into the matrix of $Ta_2O_5$ to cuts the bridge of the Cu filament, which is called RESET operation. Finally, the $Cu^{z+}$ ions move to the negatively charged Cu electrode and precipitate as Cu atoms on the electrode by the reduction reaction to keep $Cu^+$ ion concentration in $Ta_2O_5$, thereby stabilizing the OFF state. Since the behavior of these atomic switches is governed by ion migration and the solid electrochemical reaction, the voltage required for the SET and RESET operations depends on the required energy for ionic conductivity of the metal ion of the matrix layer, formation of the filament and so on.

## 3.2 *Quantized Conductance*

The gapless-type atomic switches also show quantized conductance by controlling the point contact between metal filaments and electrodes, as well as the gap-type atomic switches [17]. As an example, Fig. 13a, b show the quantum conduction characteristics of the gapless-type switch consisted of $Ag/Ta_2O_5/Pt$ layers. When positive voltage pulses with a width of 20 ms are applied to the Ag electrode at intervals of 2 s, the conductance increase stepwise with increasing the pulse amplitude of positive voltage, as shown in Fig. 13a. Each step change corresponds to the quantized conductance behavior. The appeared states of the quantum conductance increased from 1 to 9 in steps. Subsequently, when the negative voltage pulses were applied to the Ag electrode, the change of the quantized conductance was changed as shown in Fig. 13b. The conductance also decreased stepwise with increasing the pulse amplitude of negative voltage. The states of the quantum conductance decreased from 9 to 1 in steps. It reveals that the states of quantized conduction can be also reversibly controlled by precisely manipulating the point contact between the electrode and metal filament in the gapless atomic switch.

## References


1. Terabe, K., Hasegawa, T., Nakayama, T., Aono, M.: Quantized conductance atomic switch. Nature. **433**, 47 (2005)
2. Hasegawa, T., Tearbe, K., Sakamoto, T., Aono, M.: Nanoionics switching device: atomic switches. MRS Bull. **34**, 929 (2009)
3. Hino, T., Haegawa, T., Terabe, K., Tsuruoka, T., Nayak, A., Ohno, T., Aono, M.: Atomic switches: atomic-movement-controlled nanodevices for new type of computing. Sci. Technol. Adv. Mater. **12**, 013003 (2011)
4. Hasegawa, T., Terabe, K., Tsuruoka, T., Aono, M.: Atomic switches: atom/ion movement controlled devices for beyond Von-Neumann computers. Adv. Mater. **24**, 252 (2012)
5. Rickert, H.: Electrochemistry of Solids – An Introduction. Springer, Berlin (1982)
6. Ohachi, T., Taniguchi, I.: Controlled filamentary growth of silver from silver compounds. J. Cryst. Growth. **13–14**, 191 (1972)
7. Terabe, K., Nakayama, T., Hasegawa, H., Aono, M.: Ionic/electronic mixed conductor tip of a scanning tunneling microscope as a metal atom source for nanostructuring. Appl. Phys. Lett. **80**, 4009 (2002)
8. Terabe, K., Nakayama, T., Iyi, N., Aono, M.: In: Furukawa, Y., Mori, Y., Kataoka, T. (eds.) Proceedings of 9th International Conference on Production Engineering, p 711. The Japan Society for Precision Engineering, Osaka (1999)
9. Terabe, K., Nakayama, T., Hasegawa, T., Aono, M.: Formation and disappearance of a nanoscale silver cluster realized by solid electrochemical reaction. J. Appl. Phys. **91**, 10110 (2002)
10. Terabe, K., Hasegawa, T., Nakayama, T., Aono, M.: Quantum point contact switch realized by solid electrochemical reaction. RIKEN Rev. **37**, 7 (2001)
11. Terabe, K., Hasegawa, T., Liang, C.H., Aono, M.: Control of local ion transport to create unique functional nanodevice based ion ionic conductors. Sci. Technol. Adv. Mater. **8**, 536 (2007)
12. Tamura, T., Hasegawa, T., Terabe, K., Nakayama, T., Sakamoto, T., Sunamura, H., Kawaura, H., Hosaka, S., Aono, M.: Switching property of atomic switch controlled by solid electrochemical reaction. Jpn. J. Appl. Phys. **45**, L364 (2006)
13. Sakamoto, T., Sunamura, H., Kawaura, H., Hasegawa, T., Nakayama, T., Aono, M.: Nanometer-scale switches using copper sulfide. Appl. Phys. Lett. **82**, 18 (2003)
14. Sakamoto, T., Lister, K., Banno, N., Hasegawa, T., Terabe, K., Aono, M.: Electronic transport in $Ta_2O_5$ resistive switch. Appl. Phys. Lett. **91**, 092110 (2007)
15. Wu, S., Tsuruoka, T., Terabe, K., Hasegawa, T., Hill, J.P., Ariga, K., Aono, M.: A polymer-electrolyte-based atomic switch. Adv. Funct. Mater. **21**(93), (2011)
16. Tsuruoka, T., Terabe, K., Hasegawa, T., Aono, M.: Forming and switching mechanisms of a cation-migration-based oxide resistive memory. Nanotechnology. **21**, 425205 (2010)
17. Tsuruoka, T., Hasegawa, T., Terabe, K., Aono, M.: Conductance quantization and synaptic behavior in $Ta_2O_5$-based atomic switch. Nanotechnology. **23**, 4357058 (2012)

# Pathway to Atomic-Switch Based Programmable Logic

**T. Sakamoto, M. Miyamura, Y. Tsuji, X. Bai, A. Morioka, R. Nebashi, M. Tada, N. Banno, K. Okamoto, N. Iguchi, H. Hada, and T. Sugibayashi**

**Abstract** Atomic switch featuring non-volatility, large on/off conductance ratio, and small foot print advantageously applies to programmable logic such as field programmable gate array (FPGA). FPGA is a reconfigurable hardware with a high energy efficiency but has a drawback of area penalty for programmability. Conventional FPGA utilizes a large number of programmable switches composed of SRAM and pass transistor. Atomic switch replaces the programmable switch, resulting in reducing the area and power consumption and enhancing the performance.

## 1 Introduction

Energy efficient computing has been required for power-constrained apparatus due to the ceiling of power supply or the limited cooling power in electrical devices. Power of micro-processors in server rack or in offices is limited to about 100 W because of the air cooling power of available heat sink. For the processors in the automobile, the cooling system also determines its maximum power, which is about 10 W. For the mobile apparatus and for Internet-of-Things (IoT) applications, power consumption is the most critical issue, since power supply or lifetime of battery determines the usable power. At the same time, higher performance of processors is expected. Hence, enhancement in performance without increasing the power consumption requires highly energy efficient computing.

Up to now, the higher energy efficiency has been obtained by scaling the feature size or the supply voltage of Complementary metal-oxide-semiconductor (CMOS) transistor. But, recently, the scaling has been challenging because of the increased electrical fields in the transistor. And the operation frequency of the CMOS circuit has been saturated due to the heat density issue [1]. We need the other technologies to boost the energy efficiency. Figure 1 shows the position of System-on-chip (SoC) platforms in terms of programmability and energy efficiency. Application specific

T. Sakamoto (✉) · M. Miyamura · Y. Tsuji · X. Bai · A. Morioka · R. Nebashi · M. Tada · N. Banno · K. Okamoto · N. Iguchi · H. Hada · T. Sugibayashi
System Platform Research Laboratories, NEC Corporation, Ibaraki, Japan
e-mail: t-sakamoto@dp.jp.nec.com



M. Aono (ed.), *Atomic Switch*, Advances in Atom and Single Molecule Machines,
https://doi.org/10.1007/978-3-030-34875-5_2

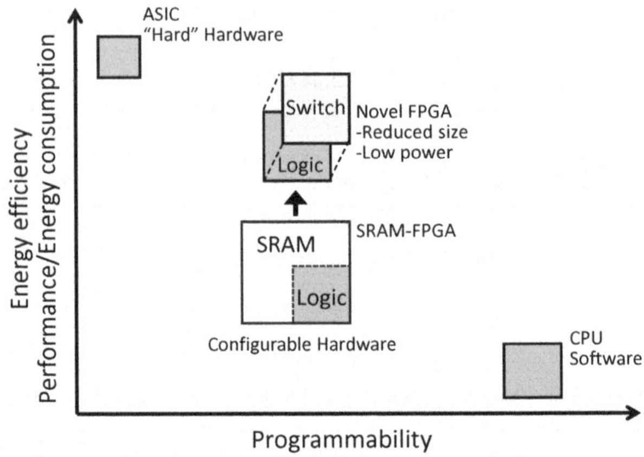


**Fig. 1** System-on-Chip (SoC) platforms applications in terms of programmability and energy efficiency. Concept of "switch over the logic" of field programmable gate array (FPGA) is realized by atomic switch, resulting in higher energy efficiency. ASIC stands for application specific integrated circuit

integrated circuit (ASIC) shows the highest performance and lowest power consumption but it has low flexibility for various applications. It also needs large development cost and long time to market. A microprocessor or CPU is widely used due to its good programmability. But, the processing performance of CPU is restrained by memory access (that is "von Neumann bottleneck"). For the real time and intelligent processing in Internet-of-Things (IoT) applications, a low latency and a higher performance is desired. Compared with ASIC, FPGA has better flexibility and low development cost. Compared with CPU, FPGA has better performance due to massive parallel processing and hard wired logic (that is non von-Neumann architecture). FPGA is utilized as an energy-efficient computation.

FPGA has better energy efficiency than CPU. But, FPGA still consumes large power, which is arising from the area penalty for programmability. To achieve a hardware programmability, FPGA equips a large number of programmable switches composed of a static random access memory (SRAM) as a configuration memory cell and a pass transistor as a switch. The SRAM and the pass transistor occupy a large area in FPGA. Kuon reported that the area difference between AISC and FPGA is about thirty, resulting in a large performance gap. Long wire length in FPGA induces the dynamic power or signal delay [2]. The SRAM also induces the static leakage power. As it is the common issue for SoC platforms, a leakage current of transistors exponentially increases with ambient temperatures. For example, the leakage current of n-channel transistor increases by two orders of magnitude from room temperature to 150 °C. The soft error in SRAM in FPGA is also challenging. Single event effect (SEE) in SRAM may change the logic function and cause failure in the apparatus.

Many research groups try to replace SRAM in FPGA with emerging nonvolatile memories, such as magnetic-RAM (MRAM) [3], resistive-RAM (ReRAM) [4], and phase change memory (PCM) [5], which are integrated in the interconnect, so as to reduce the chip size, power consumption, and soft errors. For these novel FPGAs, the external read only memory (ROM) for storing the configuration memory required for SRAM-based FPGA is not needed and the instant-on is possible.

However, performance evaluations based on the benchmark circuit have not been done for these novel devices. Furthermore, the pass transistor is still used in the novel FPGAs for routing signals.

This paper describes atomic-switch based FPGA. Atomic switch, which features non-volatility, high on/off conductance ratio, and small foot print, replaces both SRAM and pass transistor, resulting in reducing chip size and enhancing energy efficiency. In the following Sect. 2, concept of "switch over the logic" is discussed. And in Sect. 3, atomic switch technology is disclosed. Its operation principle and features are described. And then, atomic-switch based FPGA will be described in Sect. 4. Circuit architecture and performance evaluation using benchmark circuit will be disclosed. Operation in harsh environments is discussed in Sect. 5. Finally, in Sect. 6, an atomic-switch based FPGA as an accelerator of CPU is demonstrated, where FPGA core is integrated with CPU core monolithically.

## 2 Concept of "Switch Over Logic"

Vertical structure of integrated circuit (IC) is divided into two parts: a transistor layer and a Cu interconnect layer. In FPGA, both logic block and programming circuitry of SRAM and pass transistors share the transistor layer. Stacking the programming circuitry on top of the logic blocks closes the performance gap between FPGA and ASIC (see Fig. 1). The concept of "switch over the logic" minimizes the area of programmable logic [6]. Reduced length, parasitic capacitance, and resistance of the interconnect improve a signal delay and dynamic power consumption. Implementing configuration memory into the interconnect layer is achieved by using MRAM [3], ReRAM [4] or PCM [5]. Further, stacking both the configuration memory and switch over the logic block significantly improves logic density and performance. Atomic switch is a promising candidate for this purpose. The non-volatility also conserves stand-by power, while an SRAM-based FPGA is suffering from gate tunneling current and source-drain leakage in SRAM.

## 3 Atomic Switch Technology

The atomic switch is one of ReRAMs, but it has a high on/off conductance ratio. The metal filament formed between two electrodes results in a very low resistive state. Atomic switch was originally developed by Terabe [7]. The developed atomic switch has a nanometer-scale air-gap where the metal filament is formed electrically. The silver electrode coated with silver sulfide is separated from Pt electrode by a small air gap. The silver sulfide is an ionic conductor. At this small gap, a silver filament forms via electrochemical reaction. By applying a voltage pulse, the conductance increases step by step, of which conductance is quantized in units of $2e^2/h$. This indicates that the switching can be done by a few atoms.

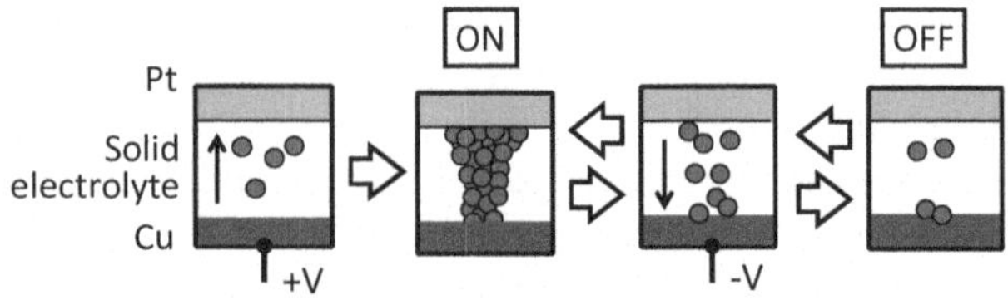


**Fig. 2** Schematic of atomic switch and operation principle. Atomic switch is composed of solid electrolyte sandwiched between two electrodes of Pt and Cu. When positive voltage is applied to Cu, Cu is ionized and precipitated at Pt electrode, and then Cu filament is formed. When negative voltage is applied to Cu, Cu filament is dissolved and switch is turned off. Transition to low resistive state is called as set operation and that to high resistive state as reset

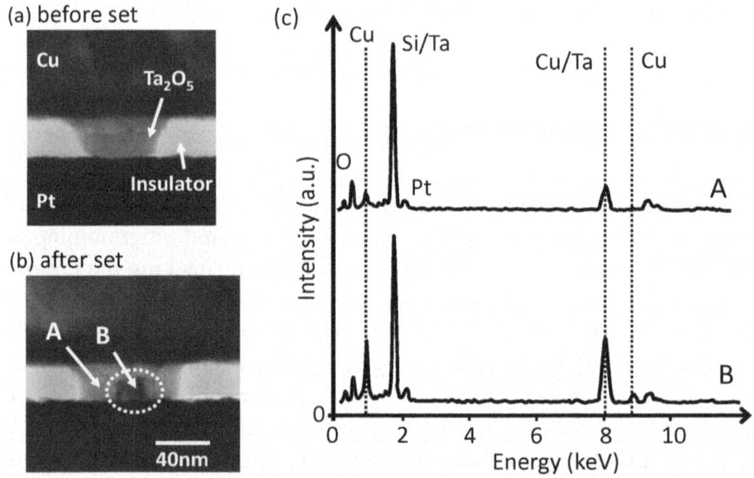


**Fig. 3** Observation of Cu filament. Transmission electron microscopy (TEM) images of atomic switch (**a**) before set operation or (**b**) after set operation is also shown. Atomic switch is composed of $Ta_2O_5$ sandwiched between Pt and Cu. (**c**) Energy dispersive X-ray (EDX) element analysis is performed at points A and B in TEM image after set operation. Signals from Cu are shown by dotted lines

Differently from the above atomic switch, a metal filament is formed in the solid electrolyte for the atomic switch in this article, as shown in Fig. 2 [8]. The atomic switch is composed of the solid electrolyte sandwiched between Pt and Cu. When a positive voltage is applied to Cu, Cu is ionized and precipitated at Pt electrode, and then Cu filament is formed. The conductance of the switch changes to high (or on state). This operation is named by "set". When a negative voltage is applied to the on-state atomic switch, the Cu filament is broken and dissolved into the solid electrolyte and the switch turns off (namely, "reset" operation). Figure 3 shows photo images of the atomic switch which is composed of the tantalum-oxide ($Ta_2O_5$) solid-electrolyte sandwiched by two electrodes of Cu and Pt electrodes. Each image corresponds to that before the set operation (Fig. 3a) or after the set operation

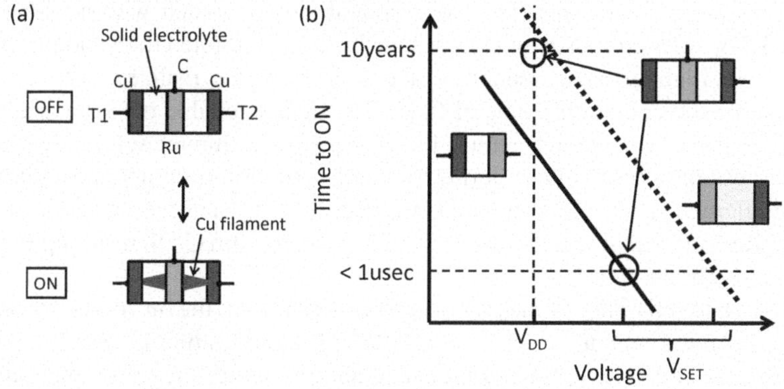


**Fig. 4** Concept of complementary atom switch (CAS). (**a**) Structure of CAS. CAS is composed of two atomic switches connected in an opposite direction. To turn on the switch, both elements are set to on by applying voltage between T1 and C and T2 and C, sequentially. For off-state, both elements are set to off. During the logic operation, signal passes through between T1 and T2. (**b**) Schematic of time-to-on versus voltage on atomic switches with different solid electrolyte thickness. Time-to-on of 10 years is required when operation voltage ($V_{DD}$) is applied to switch. At the same time, a shorter time-to-on is also required for set operation. Two circles indicates operation points for complementary atom switch (CAS). Set voltage ($V_{SET}$) is as small as the voltage of two-terminal atomic switch with thinner solid electrolyte. And time-to-on is 10 years for logic operation

(Fig. 3b). After the set operation, the dark region is observed between the two electrodes. Element analysis shows that the dark region is composed of Cu [9]. The thickness of the Cu filament is around 30 nm. The thickness depends on the on the conductance of the switch, which is controlled by the current during set operation. Here we intentionally form a thick filament for observation and the on resistance is a several tens ohm. This experiment shows that the low conductance is originated from the Cu filament in the solid electrolyte.

To apply the atomic switch to the programmable switch in FPGA, a high off-state reliability is required. During a logic operation, a constant operation voltage ($V_{DD}$) is applied to the off-state switch. Thus, the switch needs to persist its off state for ten years while $V_{DD}$ (~1 V) is applied. For this purpose, we have developed a complementary atom switch (CAS) [10].

The CAS is composed of two atomic switches connected in an opposite direction (Fig. 4a). To turn on CAS, both elements are programmed to be on state by applying a voltage between T1 and C, and T2 and C, sequentially. During the logic operation, signal passes through between T1 and T2. For an off state, both elements are programmed to be off. Figure 4b shows the time-to-on versus an applied voltage of atomic switches with different $V_{SET}$ (or different solid-electrolyte thickness). For off-state reliability, the time-to-on should be more than 10 years for $V_{DD}$ of ~1 V. Figure 4b shows the trade-off between high off-state reliability and low $V_{SET}$. Atomic switch with a thick solid electrolyte has a high off-state reliability but a high $V_{SET}$ (shown in dotted line in Fig. 4b). Use of high-voltage-tolerant transistor results in a large circuit size. The two-terminal atomic switch with a higher $V_{SET}$ needs a 3.3-V-tolerant transistor [11]. When the CAS is applied to the programmable switch, both high off-state reliability and low $V_{SET}$ are realized. The circles in Fig. 4b show the operation points of CAS. The highly reliable off-state is achieved even when the atomic switch with a lower $V_{SET}$ (or a thinner solid electrolyte) is used, because the stress voltage during the logic operation is complementally shared by two elements. A 1.8-V-tolerant transistor with small foot print is used [12, 13]. Recently, a logic transistor is available to program CAS, resulting in further reduction of circuit area [14].

The on-state reliability of the atomic switch depends on the on conductance or the thickness of the filament, which is controlled by a programming current. To get a higher on-state reliability, lower on-conductance or larger programming current is needed. The reliability also depends on the fabrication process or materials of the atomic switch. We have explored various Ru alloy materials instead of simple Ru, which is the inert electrode of atomic switch [15]. Then, the on-state reliability is improved by two orders of magnitudes. It is noted that Ru is used for implementing atomic switch in IC to avoid a metal contamination due to Pt. Moreover, when the Cu filament is induced to form at the edge of the Cu electrode, the on-state reliability is improved further. We also optimize the process steps of switch stack, fail bit count can be eliminated for millions of switches. Highly reliable on-state and low programming current can be obtained by optimizing the material, device structure and process conditions.

## 4 Atomic-Switch Based FPGA

This section describes the circuit architecture and the performance evaluation results of atomic-switch based FPGA [11–13, 16]. Figure 5 shows the schematic circuit diagram of atomic-switch based FPGA. FPGA is composed of array of configurable logic block (CLB), Input-output (IO) cell, and programming drivers for atomic switch (Fig. 5a). The routing tracks spread on the CLB arrays (Fig. 5b). Each of 4 routing tracks with 4 segment length has 3 lanes in all 4 directions (vertical or horizontal direction). Since the tracks are unidirectional, they include the tracks for positive and negative directions. Each signal from the CLB can then reach the CLB at 4 units away in either direction by using only one wire. This contributes to the routability and the reduction in the signal delay. The segment wires are buffered by AND gate at the section between the second and third CLBs. The AND gate also terminates unused wires to reduce the parasitic wire capacitance. The segment length and the lane number are chosen to have the capability of routing all of the MCNC "golden 20" benchmark circuits [17]. CLB has a crossbar switch for signal routing and two basic logic-elements (BLEs) for logic operation, which is composed of 4-input look-up-table (LUT) and D-flip flop (DFF) (Fig. 5c). The crossbar has 51 inputs comprising of 48 routing tracks, 2 feedback lines, and the one fixed-low

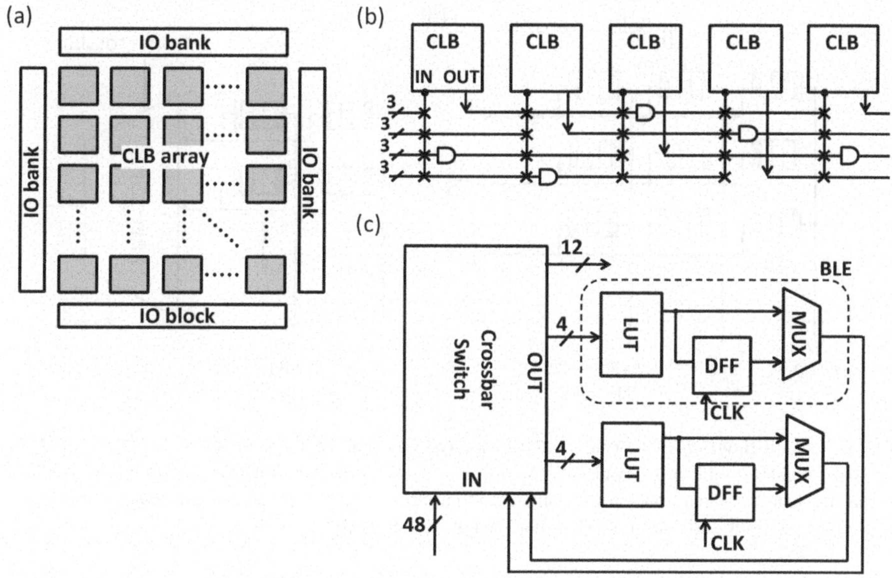


**Fig. 5** Architecture of atomic switch based FPGA. (**a**) FPGA fabric composed of IO cells and array of clustered logic block (CLB). (**b**) Routing between CLBs. There are 4 tracks with 3 lanes for one direction (e.g. +x direction). Each track spreads 4 segments of CLB and is inserted by buffer at middle. CLB has 12 inputs and 3 outputs for each direction, and this means that CLB has 48 inputs and 12 outputs totally. (**c**) Schematic of CLB. CLB has crossbar for routing signals and two basic logic elements (BLEs) composed of 4-input look-up table (LUT) and D-flip flop (DFF). Output of BLE selected by multiplexer (MUX) enters into crossbar switch

line preventing floating node. The crossbar also has 20 outputs composing 8 outputs to LUT and 12 outputs to the neighboring CLBs via buffers.

CAS is utilized in crossbar switch and memory bit of LUT (Fig. 6). Hence, the circuit configuration maintains even when the power is turned off, and the external non-volatile memory is not required. For routing signals, we use the fully populated crossbar switch with CAS, which is placed at each cross point (Fig. 6a). Output line is connected to one of input lines via CAS after circuit configuration. When the output line is not selected, it is connected to GND by using the fixed-low line. One row of crossbar switch (shown by dotted lines in Fig. 6a) corresponds to 51-to-1 multiplexer (MUX). MUX in the conventional FPGA is composed of dozens of SRAMs and pass transistors. Since the area of SRAM (200–300$F^2$) is much larger than that of pass transistor (~30$F^2$), the usage of SRAM is controlled. "F" is the feature size of transistors and corresponds to the technology node of CMOS. When considering 8-to-1 MUX, the three-level structure is the most area efficient. But, the signal passes through 3 pass transistors to be selected and the delay becomes longer, compared to a single-level structure. For the compromise deal between area and delay, typical 8-input routing MUX, realized as a two-level structure comprised of 12 pass transistors sharing 6 SRAMs. For the CAS based MUX, the output is selected by turning on one of the CAS in the row of crossbar switch, which is a

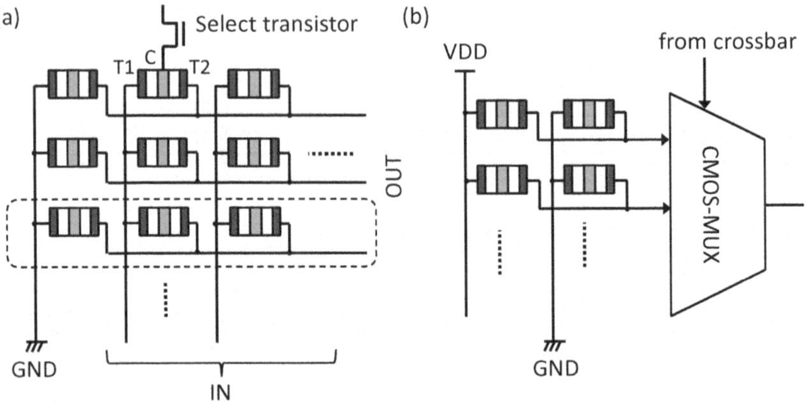


**Fig. 6** Circuit diagram of atomic-switch based (**a**) crossbar switch and (**b**) look-up table (LUT). Crossbar has 50 inputs and 12 outputs, including 51 $\times$ 12 complementary atom switches (CAS) and select transistors. Select transistor is connected to middle node (C) of CAS. Terminals T1 and T2 are connected to input and output of crossbar switch, respectively

single level structure (shown by dotted line in Fig. 6a). The CAS based MUX exhibits clear advantage in signal delay due to the single level structure and small capacitance. The capacitance of the off-state two-terminal atomic switch ($<$ 0.2 fF) is smaller than that of pass transistor (typically $\sim$1 fF). And advantageously, the parasitic capacitance of the select transistor connected to the middle node of CAS is shielded by the off-state atomic switch.

The CAS is also used for the data memory of LUT and configuration bits in CLB. The 4 input LUT with 16 memory bits is used in BLE. Each of memory bits is composed of two CAS connected to either $V_{DD}$ or GND (Fig. 6b). The voltage at the middle node represents the LUT data and configuration bit. When one of CASs is set to on state and the other is set to off state alternately, the middle node provides a logic "1" or "0". According to 4 inputs of LUT, only one data memory is selected in CMOS-based MUX.

Figure 7 shows the physical implementation of CAS in FPGA. The chip includes a 64 $\times$ 64 logic cell array, programming driver and decorder circuits for programming atomic switch. Figure 7b and c show plane views of layout of CLB. Figure 7b indicates the interconnect layer and Fig. 7c the transistor layer. The chip is fabricated by a 65 nm node CMOS process with 7 layered Cu back-end of line. The CAS is integrated between Metal 4 and 5 (Fig. 7d). Totally 4.6-Mbit CAS are integrated in the chip. The embedded process of CAS is fully compatible with CMOS fabrication process, since no degradations in the CMOS and the Cu interconnects have been confirmed. Two atomic-switches are formed on the edge of two Cu lines, so as to improve on-state reliability. We use a polymer-solid electrolyte (PSE) for CAS, which contributes to low programming voltage [18]. A high free volume in the PSE is supposed to result in the smooth formation of the Cu filament without destroying

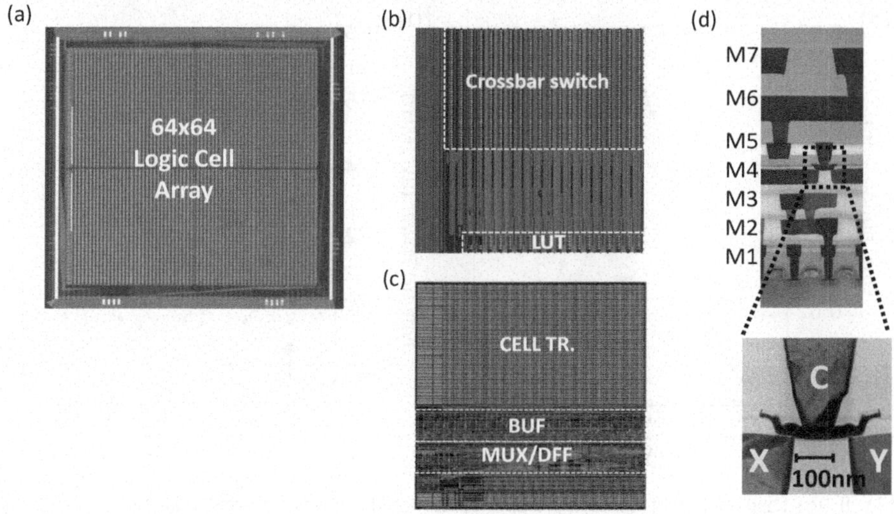


**Fig. 7** Physical implementation of atomic switch in integrated circuit. (**a**) Atomic-switch based FPGA. (**b**) Interconnect layer of clustered logic block (CLB). (**c**) Transistor layer of CLB. (**d**) Cross sectional view of chip. Atomic switch is integrated between Metal 4 and 5. 65 nm-node CMOS with 7 layered Cu back-end of line is used

the electrolyte, thereby resulting in forming-less programming, which is required to ReRAM with metal oxide resistive layer.

Logic function is demonstrated by mapping the application on the fabricated programmable logic. The configuration data for CAS is generated by in-house design-automation tools. The tool chain is composed of LUT-mapping, CLB-clustering, and placement and routing (P&R) tools. The tool chain generates the configuration data from a resistor-transfer-level (RTL) code. First, the gate level netlist is prepared by logic synthesis. Then, the gate level netlist are converted into the format comprising of 4-input LUT and DFF. Up to 2 pairs of LUT and DFF are configurable corresponding to the CLB. Then, a P&R is carried out and the binary data to program CAS is generated. The developed static timing analysis (STA) tool evaluates a signal delay. The difference from the existing STA tool for SRAM-based FPGA is that the tool is necessary to prepare a variety of primitives in accordance with the on/off state of CAS in these primitives. The propagation delay primarily depends on the number of the on-state CAS, because the capacitive load of output line of crossbar is coupled by the on-state CAS. The dynamic power consumption is also calculated by the power analyzer tool, so as to optimize the configuration data before programming. Figure 8 shows the results of STA and power analyzer on the configuration data generated by the tool chain. A 16-bit arithmetic logic unit (ALU) circuit is used for application circuit. The obtained signal delay in a critical path has various values for each configuration data, ranging from 40 to 65 ns. The variation is originated from the nature of simulated annealing used in the tool chain. The dynamic

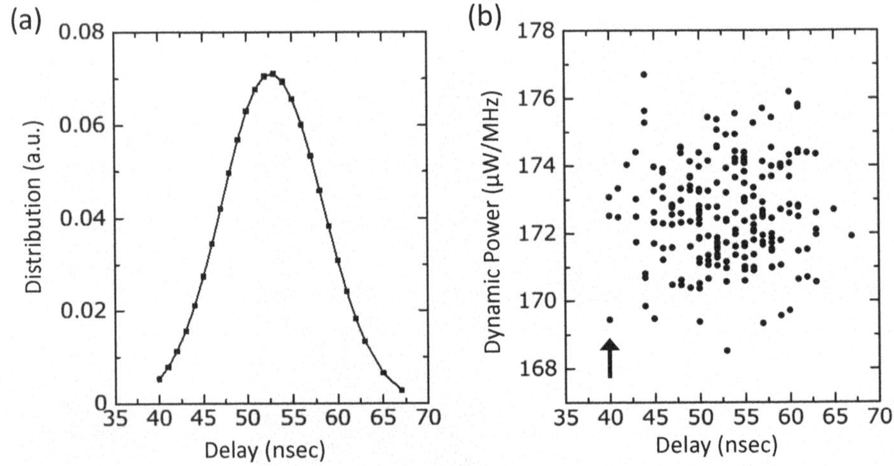


**Fig. 8** (**a**) Distribution of simulated signal delay based on configuration data by static timing analysis (STA). (**b**) Simulated dynamic power consumption versus signal delay

power also has a variation (Fig. 8b). The best data to be configured on the chip can be selected among a lot of trials.

The operational speed and power consumption are measured on 64 × 64 programmable logic cell arrays and compared with those of a commercial low-power oriented FPGA [19]. For the comparison, the same RTL code of 16-bit ALU is used. The 16-bit ALU circuit with a 1 k-gate scale includes an instruction decoder and input signal generator. 332 LUTs and 73 DFFs are used. A count-up signal generated by the mapped 16-bit counter is the input as the operand of ALU. All 28 instructions are cyclically asserted using a one-hot signal generator. The appropriate configuration data is selected by using STA tool. Each logic cell has 2240-bit atomic switches (or 1120 CAS) for signal routing and configuration memory. The configuration is done by setting the designated 140-bit atomic switches (or 70-bit CASs) to the on state in each cell. The verification pattern is generated from the Verilog test bench, and both devices are evaluated using a logic tester.

Figure 9a shows 2-dimensional shmoo plots in terms of clock period and $V_{DD}$. The circuit mapped on atomic-switch based FPGA cell arrays operates at 2.5 times faster clock frequency when $V_{DD} = 0.8$ V, compared to that on the state-of-the-art low power FPGA shown in Fig. 9b [19]. The novel FPGA also operated down to $V_{DD}$ as low as 0.5 V. The power consumption is compared at the minimum $V_{DD}$ ($V_{DDmin}$) for 15-MHz. The CAS-based FPGA operates at 0.73 V and 15 MHz with the active power of 550 μW (Table 1). Contrarily, the $V_{DDmin}$ of the reference chip is 0.94 V, and the active power 630 μW. The dynamic power of the programmable logic cell arrays (= 28.0 μW/MHz) is also lower than that of the reference (= 39.5 μW/MHz). These improvements are mainly originated from the small capacitance of the CAS and a single level structure in CAS-based MUX. High on/off conductance ratio contributes to suppress static power.

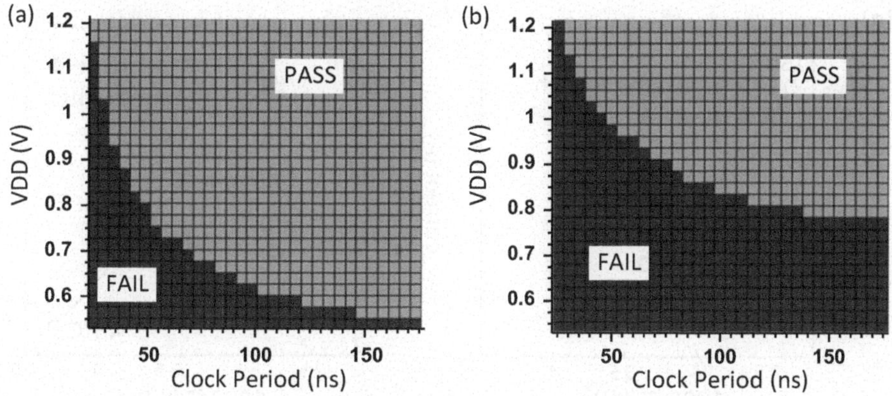


**Fig. 9** Operational region in terms of clock period and operation voltage ($V_{DD}$) for (**a**) atomic-switch based FPGA and (**b**) SRAM-based FPGA. 16bit-ALU is mapped on both FPGAs

**Table 1** Performance comparison between atomic-switch based FPGA and commercial low power FPGA

| | AtomSW FPGA | SRAM FPGA |
|---|---|---|
| Routing switch | Atomic switch | SRAM + Pass Tr. |
| Process node | 65 nm | 40 nm |
| Number of LUTs | 8192 | 1280 |
| Max. speed@0.8 V | 18.2 MHz | 7.1 MHz |
| VDDmin@15 MHz | 0.73 V | 0.94 V |
| Active power@VDDmin | 550 μW | 630 μW |

## 5 Operation in Harsh Environment

The operation of atomic-switch based FPGA in harsh environments is discussed. Here the high temperature or low temperature ambient is considered. As shown in Fig. 10, temperature effect both on the on resistance and the off current of atomic-switch is smaller than those of nMOSFET [20]. Small variations in the on or off conductance of atomic-switch contributes to prevent from degrading an operation margin of FPGA. Moreover, at the high temperatures, the increase in the leakage current is suppressed, which is the one of the issues of SRAM-based FPGA. We have also confirmed that atomic-switch based FPGA operates at the temperatures ranging from −55 to 150 °C.

The SRAM has a chance to suffer from soft errors due to the radiation from space. Especially, the satellites in the space are suffering from the high energy particles. While SRAM used in the satellites are required to be extremely reliable, they are susceptible to circuitry failures (that is SEE). Even on the ground, the neutron generated in the atmosphere by primary cosmic rays is the main source of SEE. The occurrence of failures due to radiation should be considered for highly reliable system like a commutation apparatus and automobile. SEE of atomic switch has

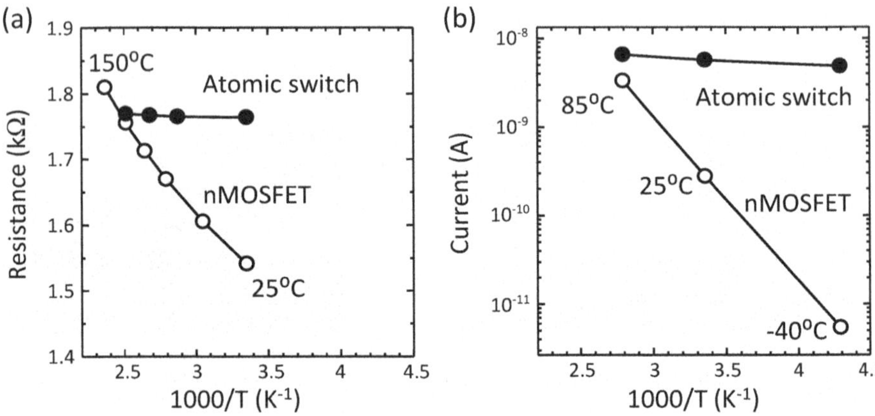


**Fig. 10** (**a**) Temperature dependence of on resistance and (**b**) that of off current for atomic switch and nMOSFET

been check by Takeuchi of JAXA. For this evaluation, 128 k-bit atomic-switch ROM was used. We have not observed SEE, showing that SEE cross section is at least 100 times lower than that of NAND flash. And this result is much lower than that of SRAM.

## 6 FPGA Accelerator

Combination of CPU and nonvolatile FPGA accelerator (that is SoC FPGA) provides both the programmability and the high energy efficiency, especially for Internet-of-Things (IoT) applications or wireless sensor nodes [21]. Figure 11 shows the block diagram of CPU and atomic-switch based FPGA for accelerator [22]. Since the data transfer in the chip is one of the performance bottlenecks, three kinds of data paths (DPs) are implemented: DP(1) through the register connected to data bus for peripheral modules, DP(2) to directly read/write SRAM by FPGA, and DP(3) for connecting modules outside of chip. Isolation cells are placed on the edge of FPGA core. System operation with CPU and FPGA is done by vectored interrupt scheme and specific 64 Byte mail-box register. Atomic switch constitutes both code-ROM [23] and nonvolatile FPGA by the same process. The chip is fabricated by a 65 nm CMOS process (shown in Fig. 7). Since the voltage stress during the operation is different for each other, one transistor and one resistor (1T1R) cell structure is used for ROM and 1T2R (or CAS) for FPGA.

A 32-bit cyclic redundancy check (CRC32) is used to evaluate the energy efficiency of MCU with no help of FPGA. The software process in the CPU operates up to 25 MHz when the operation voltage ($V_{DD}$) is 0.41 V. At this condition the highest power efficiency for CPU of 33 μW/MHz is achieved. The CRC32 execution is off-loaded from CPU to the configured FPGA by waking up when necessary. Data

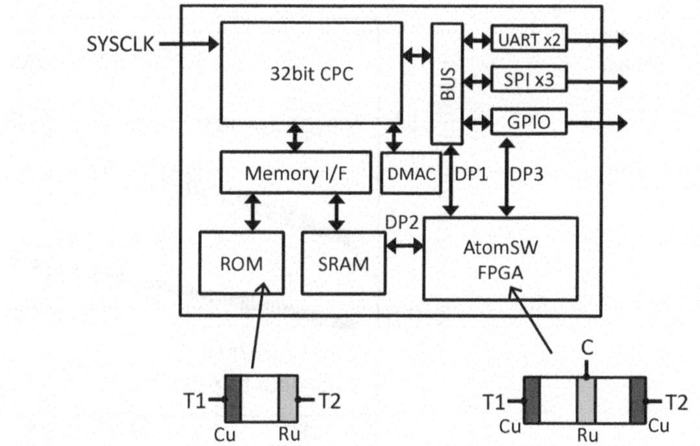


**Fig. 11** Circuit diagram of micro-controller unit with atomic-switch based FPGA as accelerator. Two-terminal atomic switch is used of code ROM and complementary atom switch (CAS) for programmable switch in FPGA

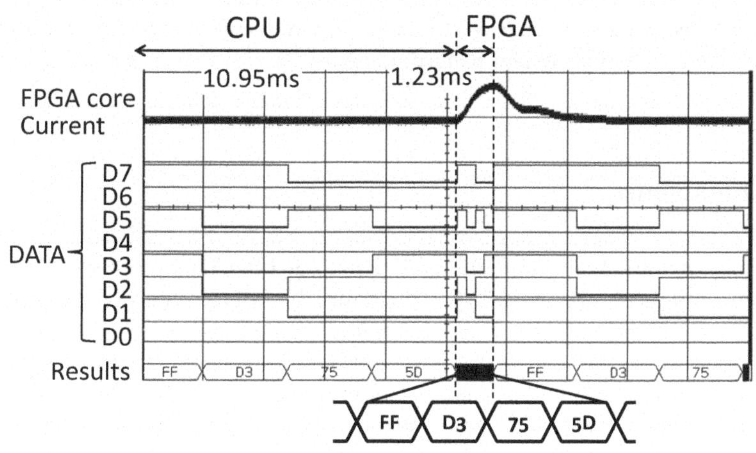


**Fig. 12** Observed signals of 32-bit cyclic redundancy check (CRC32) when switching CPU (software) processing mode to FPGA-accelerating mode. Measured current of FPGA core is also shown. Execution time for CPU or FPGA is 10.95 or 1.23 ms, respectively

transfer from SRAM or the others macros is continuously controlled by CPU using data-path DP(1). With each CRC32 execution for 32 bit - 256 word, the lowest byte of the processing result "checksum" is transferred to I/O ports (GPIO). As shown in Fig. 12, the executing speed is nine times faster when switching software-processing mode to FPGA accelerating mode. A 57.9 μW/MHz @25 MHz energy efficiency is obtained for accelerating mode. Considering the power consumption for data

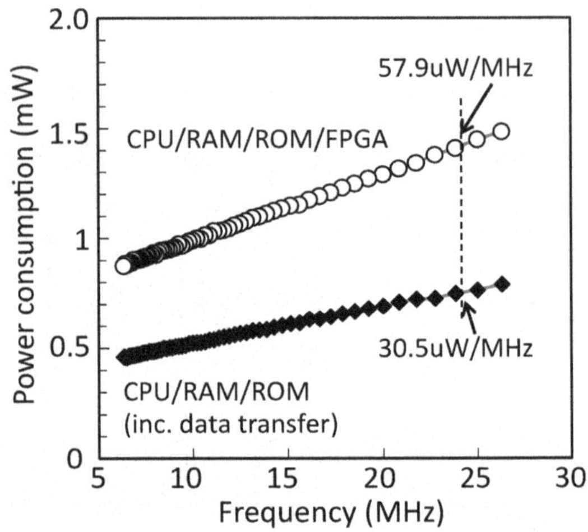


**Fig. 13** Active power consumption of with FPGA in open circle and without FPGA in solid diamond

transfer (30.5 μW/MHz @25 MHz), the total active power is 88.4 μW/MHz @25 MHz (Fig. 13). Hence, the total energy of the CRC32 execution by using FPGA accelerator became one third of that without FPGA. The demonstrated result of CRC32 ensures a high energy efficiency in applications including bit and logic calculation, e.g. encryption and pattern matching.

## 7 Conclusions

A low-power nonvolatile programmable logic (or FPGA) is proposed for energy-constrained applications. A 64 × 64 programmable-logic cell array includes a 4.6-Mbit CAS for the routing switch and configuration memory. A 16-bit ALU, which is a building block of the micro-controller unit, is implemented to compare the speed and power consumption with a state-of-the-art low power FPGA. The atomic-switch based FPGA exhibited 30% dynamic power reduction and 2.5 times faster operation in the low-voltage region. The operations in harsh environments are also demonstrated. By using atomic-switch based FPGA as CPU accelerator, both programmability and energy efficiency can be achieved. When a software process in the CPU is offloaded to NPL, the nine times faster processing speed and three times higher energy efficiency are realized.

**Acknowledgments** A part of this work was supported by NEDO. A part of the device processing was operated by AIST, Japan.

## References


1. Horowitz, M.: Computing's energy problem (and what we can do about it). IEEE ISSCC Dig. Tech. Papers. 10–14 (Feb. 2014)
2. Kuon, I., Rose, J.: Measuring the gap between FPGAs and ASICs. IEEE Trans. Comput.-Aided Des. Integr. Circuits Syst. **26**(2), 203–215 (Feb. 2007)
3. Suzuki, D., Natsui, M., Ikeda, S., Hasegawa, H., Miura, K., Hayakawa, J., Endoh, T., Ohno, H., Hanyu, T.: Fabrication of a nonvolatile lookup-table circuit chip using magneto/semiconductor-hybrid structure for an immediate-power-up field programmable gate array. Symposium on VLSI Circuits Dig. Tech. Papers, 80–81 (June 2009)
4. Liauw, Y.Y., Zhang, Z., Kim, W., El Gamal, A., Wong, S.S.: Nonvolatile 3D-FPGA with monolithically stacked RRAM-based configuration memory. IEEE ISSCC Dig. Tech. Papers, 406–407 (Feb. 2012)
5. Wen, C.-Y., Li, J., Kim, S., Breitwisch, M., Lam, C., Paramesh, J., Pileggi, L.T.: A non-volatile look-up table design using PCM (phase-change memory) cells. Symp VLSI Circuits Dig. Tech. Papers, 302–303 (June 2011)
6. Sakamoto, T., Kaeriyama, S., Sunamura, H., Mizuno, M., Kawaura, H., Hasegawa, T., Terabe, K., Nakayama, T., Aono, M.: A nonvolatile programmable solid-electrolyte nanometer switch. IEEE ISSCC Dig. Tech. Papers. 290–291 (Feb. 2004)
7. Terabe, K., Hasegawa, T., Nakayama, T., Aono, M.: Quantized conductance atomic switch. Nature. **433**, 47–50 (Jan. 2005)
8. Sakamoto, T., Sunamura, H., Kawaura, H., Hasegawa, T., Nakayama, T., Aono, M.: Nanometer-scale switches using copper sulfide. Appl. Phys. Lett. **82**(18), 3032–3034 (2003)
9. Sakamoto, T., Lister, K., Banno, N., Hasegawa, T., Terabe, K., Aono, M.: Electronic transport in $Ta_2O_5$ resistive switch. Appl. Phys. Lett. **91**(9), 092110–092112 (2007)
10. Tada, M., Sakamoto, T., Miyamura, M., Banno, N., Okamoto, K., Iguchi, N., Hada, H.: Improved OFF-state reliability of nonvolatile resistive switch with low programming voltage. IEEE Trans. Electron Devices. **59**(9), 2357–2362 (2012)
11. Miyamura, M., Nakaya, S., Tada, M., Sakamoto, T., Okamoto, K., Banno, N., Ishida, S., Ito, K., Hada, H., Sakimura, N., Sugibayashi, T., Motomura, M.: Programmable cell array using rewritable solid-electrolyte switch integrated in 90nm CMOS. IEEE ISSCC Dig. Tech. Papers. 228–229 (Feb. 2011)
12. Miyamura, M., Tada, M., Sakamoto, T., Banno, N., Okamoto, K., Iguchi, N., Hada, H.: First demonstration of logic mapping on nonvolatile programmable cell using complementary atom switch. IEEE IEDM Dig. Tech. Papers. 247–250 (Dec. 2012)
13. Miyamura, M., Sakamoto, T., Tada, M., Banno, N., Okamoto, K., Iguchi, N., Hada, H.: Low-power programmable-logic cell arrays using nonvolatile complementary atom switch. In Proc. ISQED. 330–334 (March 2014)
14. Bai, X., Sakamoto, T., Tada, M., Miyamura, M., Tsuji, Y., Morioka, A., Nebashi, R., Banno, N., Okamoto, K., Iguchi, N., Hada, H., Sugibayashi, T.: A low-power Cu atom switch programmable logic fabricated in a 40nm-node CMOS technology. Symp. VLSI Technol. Dig. Tech. Papers. T28–T29 (June 2017)
15. Tada, M., Sakamoto, T., Banno, N., Okamoto, K., Iguchi, N., Hada, H., Miyamura, M.: Improved ON-state reliability of atom switch using alloy electrodes. IEEE Trans. Electron Devices. **60**(10), 3534–3540 (2013)
16. Miyamura, M., Sakamoto, T., Tsuji, Y., Tada, M., Banno, N., Okamoto, K., Iguchi, N., Hada, H.: 0.5-V highly power-efficient programmable logic using nonvolatile configuration switch in BEOL. Proc. ACM/SIGDA Int. Symp. FPGA. 236–239 (Feb. 2015)
17. https://s2.smu.edu/~manikas/Benchmarks/MCNC_Benchmark_Netlists.html
18. Tada, M., Sakamoto, T., Miyamura, M., Banno, N., Okamoto, K., Iguchi, N., Hada, H.: Polymer solid-electrolyte switch embedded on CMOS for nonvolatile crossbar switch. IEEE Trans. Electron Devices. **58**(12), 4398–4406 (2011)
19. http://www.latticesemi.com/Products/FPGAandCPLD/iCE40.aspx

20. Sakamoto, T., Tada, M., Okamoto, K., Hada, H.: Electronic conduction mechanism in atom switch using polymer solid electrolyte. IEEE Trans. Electron Devices. **59**(12), 3574–3577 (2012)
21. Nakaya, S., Miyamura, M., Sakimura, N., Nakamura, Y., Sugibayashi, T.: A non-volatile reconfigurable off-loader for wireless sensor nodes. ACM Sigarch Comput. Architect. News. **40**(5), 87–92 (2012)
22. Tsuji, Y., Bai, X., Miyamura, M., Sakamoto, T., Tada, M., Banno, N., Okamoto, K., Iguchi, N., Sugii, N., Hada, H.: Sub-μW standby power, <18 μW/DMIPS@25MHz MCU with embedded atom-switch programmable logic and ROM. Symp. VLSI Technol. Dig. Tech. Papers. T86–T87 (2014)
23. Sakamoto, T., Tada, M., Tsuji, Y., Makiyama, H., Hasegawa, T., Yamamoto, Y., Okanishi, S., Banno, N., Miyamura, M., Okamoto, K., Iguchi, N., Ogasahara, Y., Oda, H., Kamohara, S., Yamagata, Y., Sugii, N., Hada, H.: Low-power embedded read-only memory using atom switch and silicon-on-thin-buried-oxide transistor. Appl. Phys. Express. **8**(4), 045201 (2015)

# Atomic Switch FPGA: Application for IoT Sensing Systems in Space

**H. Hihara, A. Iwasaki, M. Hashimoto, H. Ochi, Y. Mitsuyama, H. Onodera, H. Kanbara, K. Wakabayashi, T. Sugibayashi, T. Takenaka, H. Hada, M. Tada, M. Miyamura, and T. Sakamoto**

**Abstract** This chapter introduces the applications of atomic switches, which are expected to be used for sensor systems in space. Artificial satellites are regarded as the sensor nodes and communication relay nodes of Internet of Things (IoT). Processor elements using atomic switch field programmable gate array (FPGA) are anticipated as essential element for building embedded system applications for the IoT. We start by looking at the role played by embedded system processors first and point out that leaving from stored program architecture is inevitable and essential to achieve processing speed and low power consumption required for embedded system applications. Space-born sensor application is referred in this chapter as an example of embedded system applications because space systems are somewhat specialized but in many ways typical embedded applications intended to be used for

---

H. Hihara (✉)
NEC Space Technologies, Ltd., Fuchu, Tokyo, Japan

The University of Tokyo, Bunkyo-ku, Tokyo, Japan
e-mail: h-hihara@bc.jp.nec.com

A. Iwasaki
The University of Tokyo, Bunkyo-ku, Tokyo, Japan

M. Hashimoto
Osaka University, Suita, Osaka, Japan

H. Ochi
Ritsumeikan University, Kusatsu, Shiga, Japan

Y. Mitsuyama
Kochi University of Technology, Kami, Kochi, Japan

H. Onodera
Kyoto University, Sakyo-ku, Kyoto, Japan

H. Kanbara
ASTEM RI, Shimogyo-ku, Kyoto, Japan

K. Wakabayashi · T. Takenaka
NEC Corporation, Nakahara-ku, Kawasaki, Japan

T. Sugibayashi · H. Hada · M. Tada · M. Miyamura · T. Sakamoto
NEC Corporation, Tsukuba, Japan

 
M. Aono (ed.), *Atomic Switch*, Advances in Atom and Single Molecule Machines,
https://doi.org/10.1007/978-3-030-34875-5_3

IoT applications which require dependability as an essential feature. We also introduce Generic Processor element (GPE) architecture to exploit atomic switches. GPE architecture is based on the embedded automaton realized with the contribution of high-level behavioral synthesis technology. Practical application implementation on the reconfigurable processor is realized using the matured tool of high-level behavioral synthesis technology.

## 1 Introduction

Sensor nodes are quickly becoming important elements of the Internet of things (IoT) nowadays, that has become social information infrastructure [1]. Such sensor nodes produce huge data. However, due to the capacity limitation of existing communication networks, intelligent processing capabilities like selection, compression and optimization are more and more required before transmitting such vast amount of data to cloud-servers for post-processing. Space system is a typical use-case of IoT sensing node that requires high-level dependability like other high reliability applications such as medical equipment and automotive. Especially the tolerance against so called soft errors such as transient data errors caused by cosmic radiation is a mandatory requirement for sensor nodes used in space. Field programmable gate array (FPGA) using atomic switch is a promising application for space systems because of the radiation hardness of the switches.

A new computational model, which we call "Embedded Automaton" [2], is introduced in this chapter to exploit the merit of atomic switch FPGAs. Conventional microcontroller units (MCUs) used for IoT sensing nodes are designed with the model based on finite automaton. The latent overhead of the implementation based on finite automaton is easily identified by the comparison with Embedded Automaton. An architecture required for processor elements in IoT sensor nodes is discussed by considering the latent overhead. The architecture is different from the so-called stored program architecture. Differences between an MCU and an FPGA as a processor element are also clarified in light of the concept of the Embedded Automaton. Notwithstanding, or rather because of the difference, MCUs and FPGAs are expected to be integrated eventually, because flexible programming capability and efficient implementation are often required in a system at the same time. Such technology trend tells us that an MCU is not the only choice for developing intelligent sensors of Internet of Things (IoT) applications. A dynamically reconfigurable architecture (DRP) [3, 4] is often considered as a suitable candidate for embedded applications, because it possesses both flexible programming capability and implementation efficiency. An advantage of dynamically reconfigurable architecture is ultra-low power consumption. In comparison with a conventional MCU, power consumption of 1/100 is expected for equivalent amount of computational as reported in [5], because most of the chip area can be used for computational resources. Such kind of advantage becomes good combination with atomic switches to realize high processing speed with low power consumption. It is also explained in this chapter that practical implementation of Embedded Automaton is realized by using a matured high-level behavioral synthesis tool called CyberWorkBench (CWB) [6–8], which can handle high level programming language like ANSI-C and SystemC. It is discussed that although MCUs are suitable for many IoT applications, FPGAs and DRPs using atomic switches and high-level behavioral synthesis technology are powerful alternatives for embedded system design.

We discuss a suitable processor element architecture to exploit atomic switches. It is an extension of Flexible Reliability Reconfigurable Array (FRRA) [3, 4], and we call it Generic Processor element (GPE) architecture. The necessity of employing non-stored program architecture is deduced from considerations of the characteristics of processors required for embedded system applications. The requirements for high speed response and low power consumption are essential factors that call for non-stored program architecture—a liberation from the performance bottleneck of stored program architecture. The GPE architecture includes the higher design levels of hierarchy that accommodate functions of higher abstraction like dependability in addition to processor elements (PEs) design in the bottom level of design hierarchy that realizes the same high speed and low power consumption as FPGA does.

## 2 Design Requirements for Edge Computing of IoT Applications in Space

The Internet of Things (IoT) has been envisioned as a fundamental infrastructure that will bring about useful information and knowledge resulting in efficiency and growth in industry and improved comfort and safety in human life. Everything is to be connected through Machine to Machine (M2M) network anytime and anywhere to realize the IoT framework. Wide range of information is collected and accumulated in a system using embedded processors as shown in Fig. 1, which will result in accumulating, sharing and using various kinds of know-how and social knowledge [1].

Sensors, networks, information technology (IT), and robotics are distinguished as key technology elements to make IoT a practical knowledge framework. IoT can be used for supporting so-called lifeline as energy supply, water works, traffic control, logistics, broadcasting, and telecommunication. Namely, IoT is used for constructing social infrastructures such as roads, airports, railways, power plants, factories, etc. Dependability is a mandatory requirement for social infrastructures, and we discuss the dependability requirement through the consideration about space systems applications.

This chapter exemplifies the IoT in space systems and highlights the requirements for embedded processors used in sensor nodes for IoT applications. Space systems, such as artificial satellites, are identified as sensor nodes and communication relay nodes among IoT applications. However, the node size and complexity might be different from those of nodes for environment monitoring, traffic monitoring, home

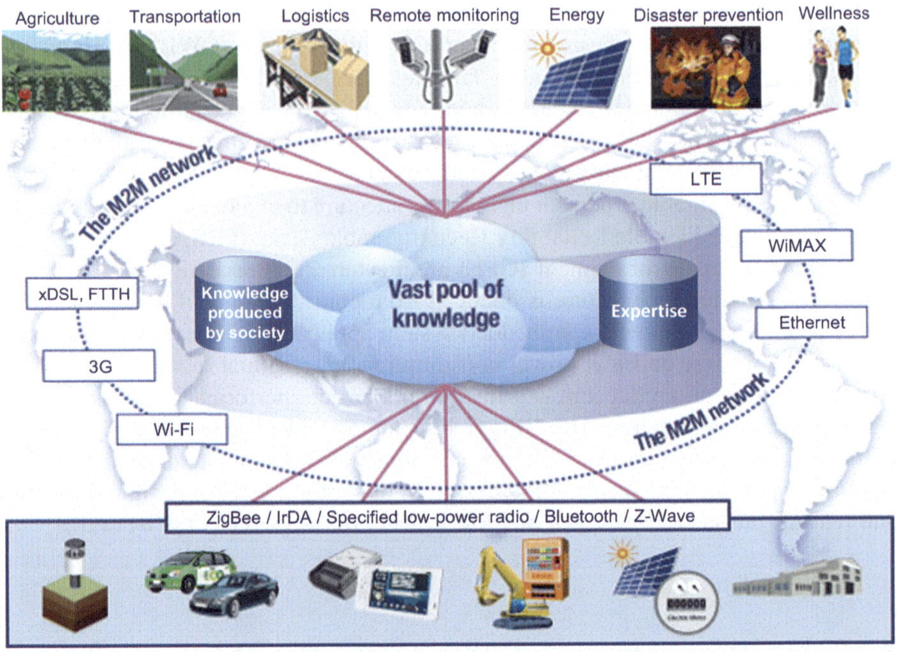


**Fig. 1** IoT using embedded processors. http://www.nec.com/en/global/solutions/nsp/m2m/concept/ [1]

security, etc., and the tolerance against harsh environment in space is one of the most demanding in terms of dependability as IoT applications.

Space systems are integrated with ground network systems that consist of high-performance computers and high-performance network appliances as shown in Fig. 2 [9]. The Internet has already been expanded outside of the Earth's atmosphere. Figure 3 shows the Ethernet Interface Unit (EIU) developed for the International Space Station (ISS) [10]. Such equipment for the Internet is installed inside a structure. On the other hand, sensors used for IoT applications are placed outside of structures. Harsh environment as cosmic radiations and wide temperature range in outer space must be taken into consideration. As for semiconductor devices, soft errors caused by cosmic radiations are major concern, and heat must be dissipated through limited conduction path.

Satellites are placed and utilized in outer space as shown in Fig. 4 [11], and hence the limitation of the capacity of transmission channels between satellites and ground stations must be considered in the system integration. Needless to say, the functionality of the space system must be available at all times even in harsh environment with high level radiation and wide temperature range. Once a successful launch is achieved and the spacecraft is stabilized in the orbit after the initial operational preparation period, it is necessary to consider the external environment such as the temperature difference and radiation environment peculiar to the space. Electronic parts used in spacecraft are exposed to vibration levels ranging from 10 to 20 G and

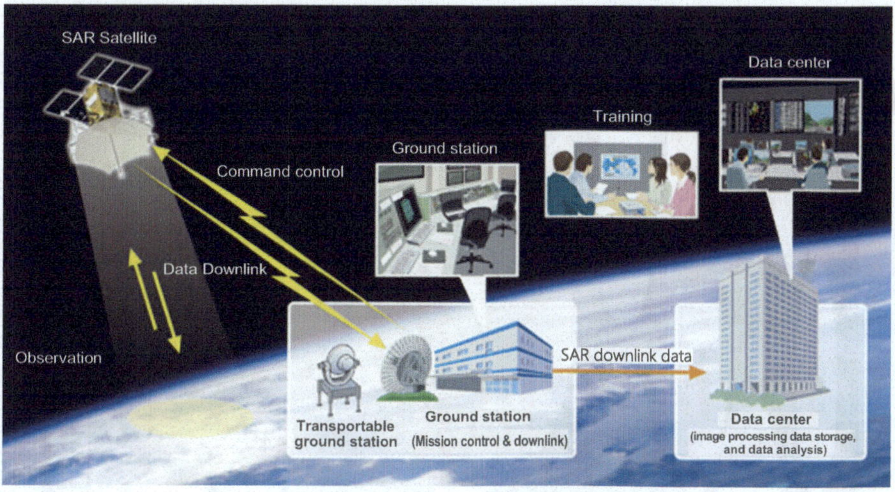


**Fig. 2** IoT implementation using space systems. https://www.nec.com/en/global/solutions/space/remote_sensing/ [9]

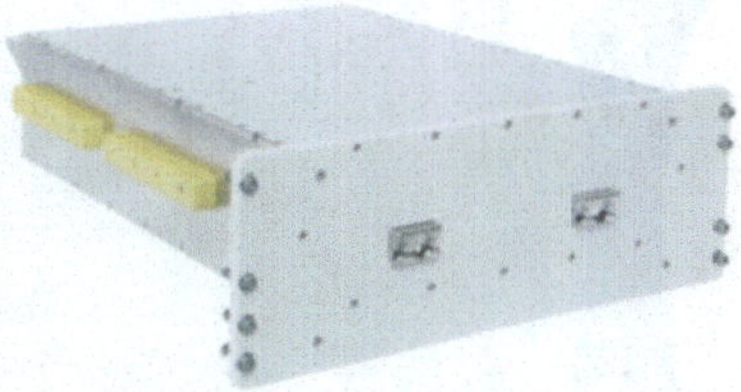

(a)

(b)

**Fig. 3** Ethernet Interface Unit in the International Space Station, (**a**) Ethernet Interface Unit (EIU) (**b**) International Space Station (ISS) © JAXA [10]

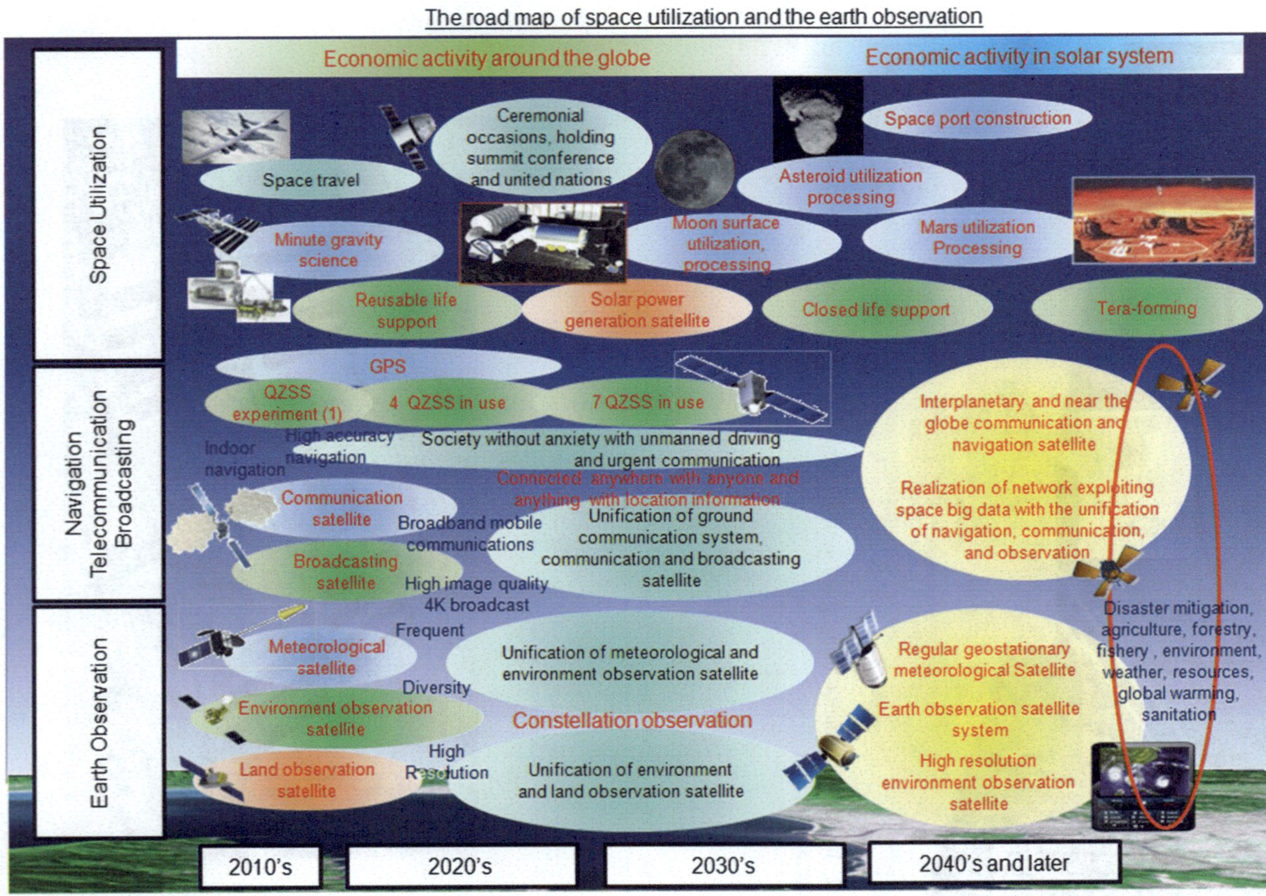


**Fig. 4** The road map of space systems application and the Earth observation system. http://www.scj.go.jp/ja/member/iinkai/kanji/pdf22/siryo201-5-10-7.pdf [11]

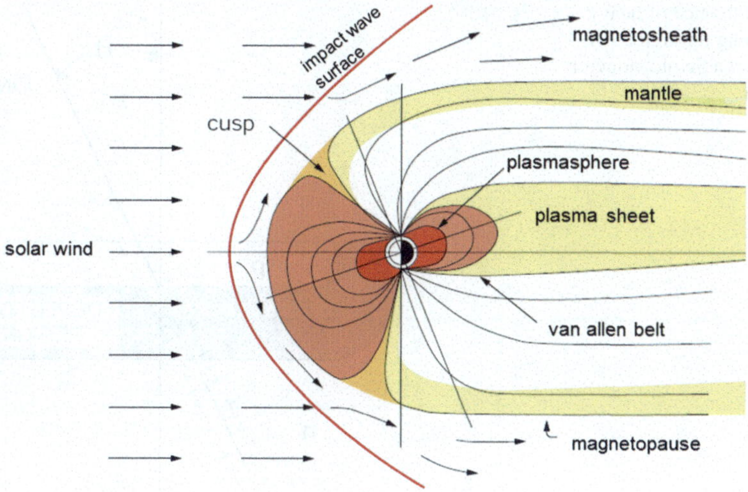


**Fig. 5** Radiations captured by the Earth [12]

temperature changes ranging from –30 to +60 °C or higher when launched with a rocket. Regarding the temperature difference, thanks to the progress of recent spacecraft system development and design technology, onboard equipment can be operated within a controlled environment to keep the temperature inside the satellite. The inside of a satellite can be kept within an appropriate and fixed temperature range. However, considering the entire lifecycle including launch, transportation, etc., the temperature range for the guaranteed operation of onboard equipment still reaches 70–90 °C. As for a radiation environment, once going out of the atmosphere, it is exposed to a high level of radiation peculiar to the outer space. Soft errors often found in semiconductor devices must be taken into account during the operation of the space system.

The influence of cosmic radiation on semiconductor devices is summarized as follows [12]. The source of space radiation environment is high energy particles existing in outer space, such as gamma rays contained in the vicinity of the planet and solar wind, and heavy particles contained in galactic cosmic rays, etc. High level gamma ray mainly leads to degradation of device characteristics such as shift of gate threshold voltage level and increase of leakage current depending on total dose rate. As for heavy particles, it is necessary to consider the influence of protons, neutrons, $\alpha$ (alpha) particles, heavy ion particles, etc. Among them, the radiation carried by solar wind is dominant quantitatively. These radiations are captured by the Earth's magnetosphere as shown in Fig. 5 [12]. The occurrence of electrical charge due to the penetration of charged particles through the device causes transient errors. Even permanent errors are sometimes caused by charged particles, and a parasitic diode inside CMOS device is sometimes activated due to an excessive charge. The activated short circuit leads to latch up and other faults.

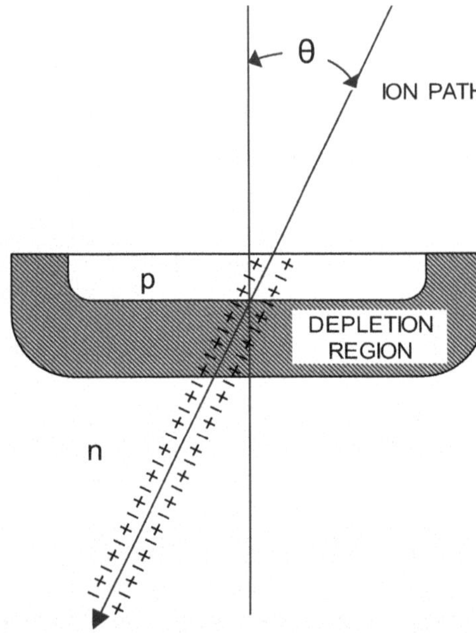


**Fig. 6** Influence of heavy ions passing through a PN Junction of a Semiconductor Device

The influences of high energy particles on semiconductor device are as follows.

1. Total dose effect

Total dose effect is a phenomenon that the performance of a semiconductor device deteriorates depending on the total amount of radiation of the environment in which the device is exposed. Total dose effect is regarded as permanent degradation, and the performance deteriorates due to the threshold voltage shift of transistors. This phenomenon is mainly caused by protons and electrons. Even when a large amount of these particles is received, there sometimes are cases that the effects are relaxed. This phenomenon is called annealing effect. However, it is difficult to suppress the degradation in general. Shielding using metal plates is often employed to reduce this phenomenon.

2. Single Event Upset

High energy particles such as heavy ions or protons from the sun and the center of galaxy hit circuits in Large Scale Integrations (LSIs) such as memory cells, register cells and flip-flops, and the data inside these cells are often inverted. It is a transient error. It is a single event upset (SEU) and is usually called "soft error". Even if correct software or data are stored in the semiconductor memory cells at first, the soft error causes malfunctions on digital LSIs by changing the contents of memory cells. Figure 6 shows how high energy particles collide with and pass through the PN junctions of memory cells of a digital LSI or transistors of a flip-flop. As heavy ions pass through a PN junction, electron-hole pairs are generated by a phenomenon called funneling, and the amount of charge in the memory cell fluctuates. As a result, a transient error (soft error) occurs. As for the case of a memory cell, a bit error occurs unless the charge amount is restored to the predetermined amount which corresponds to the pre-stored value.

3. Single event latch-up

Single event latch-up (SEL) is another phenomenon caused by radiation of high energy particles like SEU described above. Sometimes a semiconductor device such as an LSI often has an unexpectedly formed structure which resemble a thyristor due to a combination of circuit configurations. When high-energy particles hit a thyristor like structure formed on this semiconductor substrate, the circuit of this portion is turned on, and an excessive power supply current flows. It is sometimes referred to as "hard error" because it can lead to permanent failure rather than transient error.

In addition to the above described SEU and SEL, there are other radiation effects such as Single Event Transient (SET) and Single Event Functional Interrupt (SEFI), etc. SET is a transient large current, and the duration of large current is short. However, if a successive flip-flop captures the pulse at the edge of clock, it can lead to SEU. SEFI causes malfunction in the vicinity of high-density memories like flash memories because recent control circuitries inside high density memory devices are complex and are affected by SEEs. Once SEFI occurs, it must be reset or reboot to resume operation. The control circuitries are sometimes harmed permanently by SEFI. These are collectively called Single Event effect (SEE). It is necessary to select radiation hardened devices in which excessive charge can be suppressed. Data refreshing by rewriting the correct data as much as possible is often used to prevent any successive influence from spreading and to eliminate these failures caused by SEE. Various techniques are taken to protect devices from SEEs.

Regarding radiation environments to which space systems are exposed, they were conventionally treated as being unique to outer space. However, due to recent miniaturization of semiconductor processes, soft errors as transient data errors found in semiconductor memory cells caused by natural background radiation on the Earth draw attention in addition to failures caused by radiation emitted from the package of the device [13]. For this reason, opportunities are increasing in which spacecraft onboard equipment design engineers and consumer electronics equipment designers share information about radiation effects. Countermeasures against soft errors that can be employed by the developers of consumer electronics facing with the miniaturization of semiconductor processes are becoming close to the techniques used by spacecraft onboard equipment design engineers. In consequence, not only spacecraft onboard equipment design engineers but also consumer electronics designers have become to be interested in atomic switch. Even though design techniques like various error correction codes are employed, removing the amount of semiconductor memories used as program memories of MCUs and configuration memories of FPGAs is one of the most effective ways to reduce soft errors. Atomic switch is a most promising candidate to reduce such semiconductor memories. Therefore, atomic switch contributes directly to improve reliability of onboard equipment of satellites. It will also be a subject shared with engineers in other industrial domains. In addition, by withdrawing from a stored memory architecture,

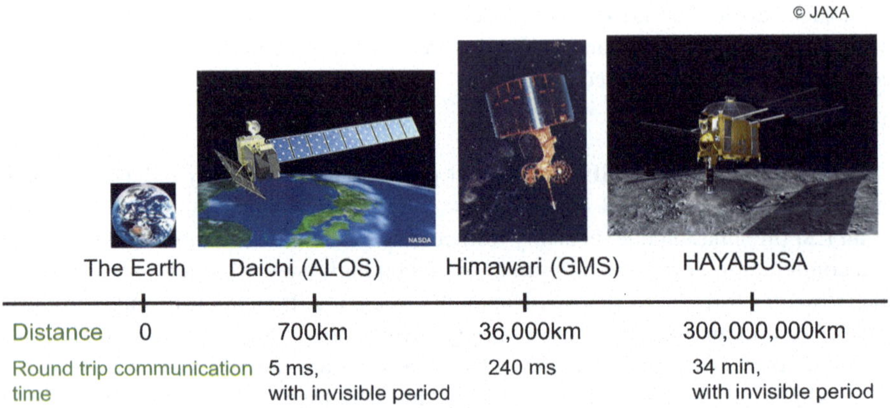


**Fig. 7** Distance and round-trip communication time of typical satellites between the Earth

it is expected that the processing performance can be improved from the processor architecture point of view. These topics are discussed next.

Satellites should be smart even in the harsh environment. They are operated by remote control from the ground stations on the Earth. Distance and round-trip communication time of typical satellites to the Earth are shown in Fig. 7. As the distance becomes longer, the more communication delay time must be considered. As for the example of a deep space probe HAYABUSA, the round-trip communication time was 34 min, and it was difficult to operate the satellite in real time by remote control. Therefore, the satellite must work in itself by using radiation hardened semiconductor devices for microprocessors and memory devices. In addition to that, such satellites have many sensors in them. Transmission capacity of communication channels between ground stations on the Earth and satellites are limited, and onboard data manipulations such as data selection and data compression are required to make the best use of limited transmission capacity.

Examples of satellites with sensors and edge computing functions are shown in Fig. 8. Those satellites equipped with several sensors for various wavelengths, such as visible light, infrared and ultraviolet. The sensor system must have signal preprocessing functions to get high quality data of these spectra through the limited capacity of transmission channels towards ground stations. A high-speed data compression is one of the most needed function. Combinations of FPGAs and MCUs are employed to transmit high quality for those implementations, because radiation hardened MCUs are not high speed enough to compress data using the latest algorithm. AKATSUKI [14], the Venus Climate Orbiter, and HAYABUSA2 [15], the asteroid probe are shown in Fig. 8a, and Fig. 8b respectively. High speed data compression has been realized using the combination of an MCU and FPGAs with distributed memories. Image recognition functions and signal calibration functions are also implemented on HAYABUSA2 by using the combination of an MCU and FPGAs with distributed memories as well.

(a)

(b)

**Fig. 8** Satellites with smart sensors, (**a**) AKATSUKI, (**b**) HAYABUSA2. © JAXA and Akihiro Ikeshita

Dependability and intelligence are demanded for the sensor nodes of IoT applications, because a great number of sensor nodes with embedded preprocessing functions are to be deployed in the field for IoT applications. The same characteristics is required for processor elements embedded in the sensor nodes in satellites to provide intelligent functionalities in the harsh environment in space. On-demand data acquisition is also required in addition to periodical data collection to enhance the value of large amount of data stored in data centers on the Earth. Therefore, sensor units of satellites are required to accommodate interactive capability for data manipulation, data selection and data compression. Even though dependability and intelligent functions are demanded on embedded processor elements used in the sensor units in satellites, small footprint and reduced power consumption are mandatory for space applications. These requirements have been extracted as the reduction of semiconductor memories used as program memories and configuration memories. In consequence, atomic switch has been closed up for our new processor element architecture and control circuitries.

# 3 Adaptation of Atomic Switch FPGAs to IoT Applications in Space

We discuss how to adopt atomic switch FPGA to edge computing for sensing devices in this section. We identified four requirements for a signal processing system architecture of sensor units. The first requirements are low power consumption and low resources. Eliminating large memory cells reduces power consumption, footprint, and weight. The target of elimination are configuration memories of FPGAs and main memory cells of MCUs used to store application program.

Secondly, video rate signal processing throughput is required. An embedded processor element should be capable of processing sensor signals at video rate to eliminate large main memory cells for application programs and large frame memory buffers. One processor should be capable of processing signals from at least one channel. If a processor element can process signals in video rate without using large program memories, we can put one processor element in on one channel, and it is also possible to reduce buffer memory capacity. Thirdly, programming flexibility during the development phase is mandatory for the compensation of detector characteristics. Several kinds of compensation are required for image sensors to enhance dynamic range and/or to normalization. A compensation for a degraded sensing device is also necessary sometimes. Programming capability with high-level programming languages (ex. C language) is anticipated because simulation is a mandatory process for designing onboard sensors. The fourth requirement is reliability, because sensing devices and even edge computing nodes are place in harsh environment in the field.

As for the first requirement, we can exploit the characteristics of embedded application systems to reduce memory related circuitries [16]. Implementing applications as hardware logic, instead of processors with stored memory architecture, does not affect the productivity, while flexibility is required on the development environment.

A dynamically reconfigurable architecture is a proper solution for the issue. It is reported that power consumption of a dynamically reconfigurable processor (DRP) was one hundredth for the same amount of operations compared to a conventional microprocessor based on stored program architecture [5]. High level and behavioral synthesis technology can be used for the application development using DRPs [5, 16]. Operation efficiency is much improved compared to conventional software compilation for fixed ISP hardware of MCUs. Atomic switches exploit the merit of DRPs and high level and behavioral synthesis technologies to reduce power consumption and resources. In consequence, significant improvement of footprint and power efficiency is expected.

Once the first requirement is satisfied, solutions for the requirement contributes to fulfill the second requirement. Customized hardware can be implemented on ASICs augmented by atomic switches. Customization is possible to ASICs without affecting efficiency by using atomic switches. This feature enables processor elements to process signals transmitted through more than or equal to one sensor signal channel.

As for the third requirement, it is necessary for developing embedded processor applications to compensate for the sensing device characteristics, and development efficiency should be maintained. We consider the framework of overall signal processing chain and refer IoT five-layer architecture model [17] as shown in Fig. 9 to realize development efficiency. IoT is not merely regarded as a collection system of sensor data and analysis results. Flexibility, robustness to change, and the ability to maintain the accuracy across the entire system are the key features of the IoT five-layer architecture model. Moreover, the introduction of the layered concept improves the connectivity with other systems. The gradual growth of the system is supported and the means of creating new values can be provided by making it easier

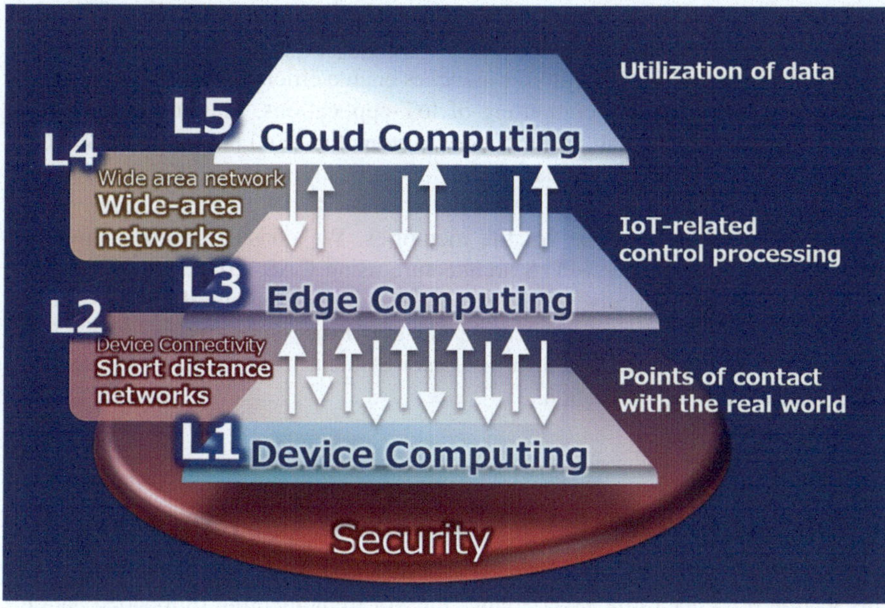


**Fig. 9** IoT five layer architecture model. http://www.nec.com/en/global/solutions/iot/iotplatform/ [17]

to link various IoT systems. Cloud computing for the top-most layer is installed in ground stations on the Earth in the case of space system applications. Existing wide area network extends down link communication channels between satellites and ground stations and provides communication channels for the cloud computing layer. We assign edge computing functions on onboard signal processing units of onboard sensors. The edge computing functions include complex functions such as data selection through feature point extraction, data compression, and signal compensation for optimization. These functions are implemented through the integration of edge computing layer and device computing layer. Conventional MCUs, microprocessors (MPUs), and FPGAs are used for these integrated functions. We identify functions like signal compensation and integral operation of infrared image sensor data are as device computing layer functions. Flexible adoption for each detector characteristics can be realized efficiently in this layer. We place atomic switch augmented ASICs in device computing layer to provide smart signal processing functions in sensing devices.

As for the fourth requirement, which is reliability, device computing functions are installed in sensing devices placed in the field like the orbit in space. Soft errors caused by harsh environment, which is explained in the preceding section, must be considered. Programmability should be prepared without using large memory cells to mitigate soft errors caused by the harsh environment. Even though processor elements were implemented on ASICs, programmability for signal processing is required on hardware circuitry without adding large memory cells. The condition is the same for FPGAs. Large configuration memories should not be used to keep programmability. Programmability can be maintained without large memory cells by using atomic switches. The reconfigurable architecture exploiting high level behavioral synthesis technology and atomic switches enable efficient implementations and high performance for edge computing of IoT applications. The architecture is an embodiment of what we call an Embedded Automaton (EA) [16]. Notwithstanding its low power consumption characteristics, dynamically reconfigurable architecture does not eliminate configuration memories in itself [18, 19]. Physical improvement is required to eliminate those volatile memories. We considered complementary atom switch (CAS) and the FPGA architecture using CAS [20–23] to establish an architecture for dynamically reconfigurable processor element (PE) with atomic switches [24]. We call the architecture Generic Processing Architecture (GPE). GPE is a natural implementation of processor element using atomic switches, and it is explained in the next section.

## 4 Adopting Atomic Switch FPGA to Embedded Automaton

This section presents a four-layer architecture for implementing Embedded Automaton (EA) that is mentioned in the previous section using atomic switches. The design flow considered here aims at designing onboard equipment of space systems. Dependability is a mandatory requirement for this application, and hence the reliability design must be included in the design flow.

### *4.1 Associating Layered Structure Design of a PE with Atomic Switch FPGA*

We proposed a four-layered architecture to design a PE using atomic switch FPGA [2, 16, 25]. It is called "Generic Processor element (GPE) architecture". Each layer corresponds to I/O layer, fine grained layer, coarse grained layer and switch layer, from the bottom to the top, respectively. These layers are shown in Fig. 10 [2]. The bottom layer is I/O interface circuitry. It is implemented with random logic primitives and mixed signal I/O circuitries. The fine grained layer mainly consists of finite state machines and data paths. They are replaceable to follow the context described by high-level languages. This layer includes Look-Up Tables (LUTs). LUTs are used to configure data paths and finite state machines. The coarse grained layer is located on the fine grained layer. This layer includes operation units like Arithmetic and Logic Units (ALUs). The function of an operation unit defined in the coarse grained layer corresponds to an instruction set of a conventional MCU.

I/O blocks, fine grained function blocks, and coarse grained function blocks are connected to each other on the topmost layer "switch layer".

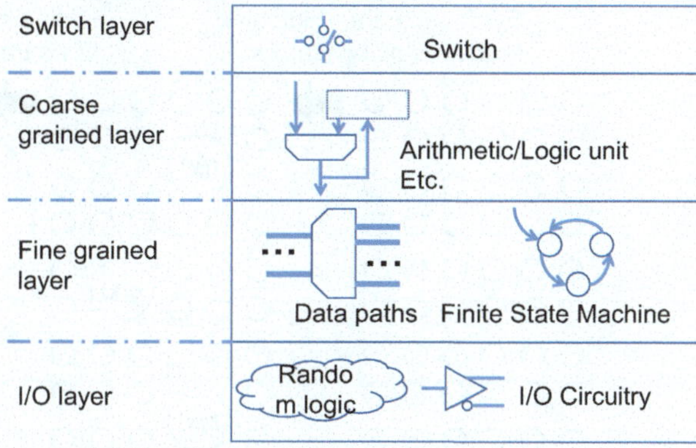


**Fig. 10** Generic Processor element (GPE) architecture [2]

Even though this architecture is different from the stored program architecture, the mathematical model of a finite automaton is maintained and a practical PE can be implemented. Note that PEs without program memories might not correspond to the conventional definition of a MCU or a MPU in terms of programmability. However, any user programs can be implemented on this four-layer architecture using high level and behavioral synthesis tools like CWB. In consequence, practical embedded processors can be implemented using atomic switches.

The above implementation uses LUTs as a conventional FPGA does, and the configuration memories are not necessary because atomic switches are a kind of non-volatile switches [21, 22] to provide programmability without configuration memories. Both program memories and configuration memories are no longer needed by using atomic switches and high level behavioral synthesis tools. This feature is highly suitable for processor elements of device computing layer, because it can eliminate soft errors encountered in harsh environment. For example, it is suitable for space system use, because it can avoid soft errors induced by radiation on orbits in space.

## *4.2 Discussion of a Design Flow to Design Processor Elements Using Atomic Switches*

We proposed a design flow for designing processor elements (PEs) aiming at onboard sensor applications by incorporating GPE architecture [2]. The design flow focuses on sensor unit applications used in harsh environment of outer fields, because the architecture without volatile program memories is suitable for the application.

Figure 11 shows the physical image of the design flow. Primitives such as ASIC block libraries and/or Basic Logic Elements (BLEs) using atomic switches are to be prepared as an initial configuration. Dedicated functions for a specific application are synthesized and aggregated in the design flow by exploiting functional

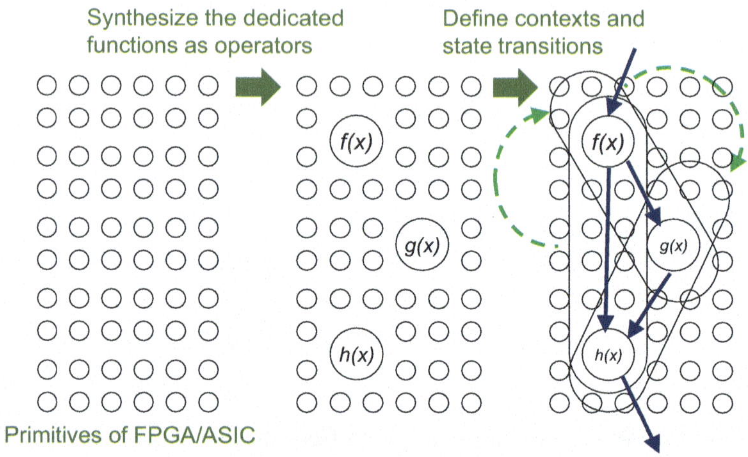


**Fig. 11** Design process

**Table 1** Atomic switch application for each GPE architecture layer

| Layer | Implementation | Remarks |
|---|---|---|
| Switch layer | Routing by open standards, ex. SpaceWire [27] | Routing switches with atomic switches |
| Coarse grained layer | Comparison decision using triple modular redundancy (TMR) [3, 4], CRAFTSYSTEM [28, 29], Software communication [30], etc. | Coarse grained connections with atomic switches |
| Fine grained layer | TMR with a voter, etc. | Atomic switch connections using the framework of robust fabric [3, 4, 16, 31] |

representations of high-level programming languages using a high-level synthesis tool like CWB. Finite state transitions and data paths between function blocks are also defined in the design flow through high-level synthesis in accordance with the context of an application. Inputs and Outputs are defined in HDL in step 4 through behavioral synthesis, and they correspond to the bottom layer of GPE architecture. Dotted lines show state transitions and solid lines show data paths in Fig. 11.

The implementation example has already been demonstrated in our previous work [25]. We implemented signal processing functions of the infrared image sensor of AKATSUKI onto one atomic switch FPGA [26] to evaluate GPE architecture and the design flow. The result showed remarkable power consumption reduction as shown in the next section.

Dependability design is also required in accordance with each layer of GPE architecture. We proposed dependability implementation scheme for GPE architecture [16]. Atomic switches can be used for the implementation design in Switch layer, Coarse grained layer, and Fine grained layer as shown in Table 1.

# 5 An Example Implementation and Evaluation Result

This section describes the implementation example and the evaluation result of GPE architecture using an atomic switch FPGA with CAS configuration [20–23]. We selected an infrared sensor system for the implementation example [26], because infrared sensor system is a major concern for inter-planetary missions, as well as for the Earth observation missions. Infrared sensors are often used to investigate the nature and the formation processes of planets and asteroids. The complementary atomic switches (CAS) enable hardware programming without configuration memories, and almost one-tenth of lower power consumption is expected compared to conventional re-writable FPGAs because of the elimination of configuration memories used in conventional re-writable FPGAs. This enables us to embed a processor element on each infrared signal detector output channel.

## *5.1 Infrared Sensor System Implementation Using an Atomic Switch FPGA*

We selected an infrared sensor system as the motif of evaluation [26]. Infrared sensor systems are high priority payloads for inter-planetary missions like AKATSUKI [14], the Venus Climate Orbiter, and HAYABUSA2 [15], the asteroid probe, to investigate the nature and the formation processes of planets and asteroids. Fast and compact circuitry is required for onboard signal processing equipment to transmit infrared sensor data from a satellite to ground stations due to the following factors. Firstly, output signal calibration of infrared detectors is required to compensate for the intensity of infrared image sensors of satellites and to get high quality data through the limited capacity of transmission channels towards ground stations. Secondly, principal investigators of infrared image sensor data require no degradations on the image data quality, even though the resources of the sensor system are limited. Therefore, lossless image data compression function should be realized without adding remarkable size, mass, and power consumption for onboard electronics of satellites.

The combination of FPGAs and MCUs are employed by AKATSUKI and HAYABUSA2 for signal processing, whereas much smaller size and lower power consumption are demanded for future missions to accommodate larger numbers of sensors. We implemented a processor element with GPE architecture for the requirement, which consists of reconfigurable cluster cores and programmable-logic cells with complementary atomic switches. Extension of operation amount has been achieved by using the reconfigurable function with the support of high level language programming capability. A matured high level and behavioral synthesis tool enabled the architecture design. In addition to that, the programmability of hardware circuitry was achieved without configuration memories by adopting complementary atomic switches. This is a noteworthy advantage for space applications which cannot

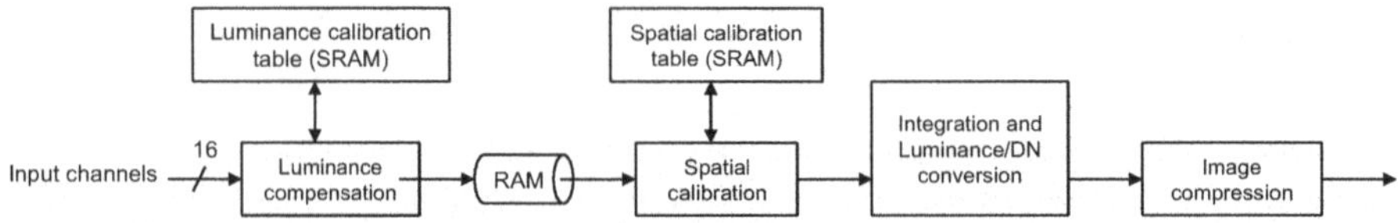


**Fig. 12** The block diagram of a typical onboard infrared image sensor [26]

be found in conventional re-writable FPGAs. Almost 1/10 of lower power consumption is expected than conventional re-writable FPGAs because of the elimination of configuration memories. Thanks to the atomic switches, soft-errors on logic circuit connections are eliminated completely. Signal compensation functions are realized in one chip without accommodating program memories used in conventional MCUs while maintaining compatible operation performance with the combination of conventional FPGAs and MCUs, which enables embedding a processor element on each infrared signal detector peripheral.

## *5.2 Onboard Calibration Functions for Infrared Image Sensors*

A typical processing function block diagram of an onboard infrared image sensor is shown in Fig. 12. Each detector element of the sensor has a 12–16-bit digital data output port. The sensitivity of each pixel is corrected using parameters stored in the table implemented with static random access memories (SRAMs) or flash memories. The corrected data are treated as the luminance of objects represented as 16-bit digital data. Integral data over some frames is calculated to enhance intensity after the above described correction, and detector allocation is relocated as required for post processing. The luminance of each element is compressed as 16-bit digital number (DN) after the onboard calibration. The data are formatted as tiles and compressed by lossless image compression. The compressed data are formatted in accordance with CCSDS recommendation [32].

The function block is usually built into an independent signal processing unit because the size of circuitry is not small enough to build it into the inside of detector units. These onboard signal processing functions are realized with the combination of limited functions on FPGAs and flexible functions on MCUs. The function unit requires a few modules using printed wired boards (PWBs). In consequence, miniaturized processing modules are demanded for future missions.

## *5.3 NanoBridge®*

We used a NanoBridge® FPGA [20–23] for the evaluation. The basic configuration of NanoBridge® FPGA is a complementary atomic switch (CAS) [21, 22]. The CAS

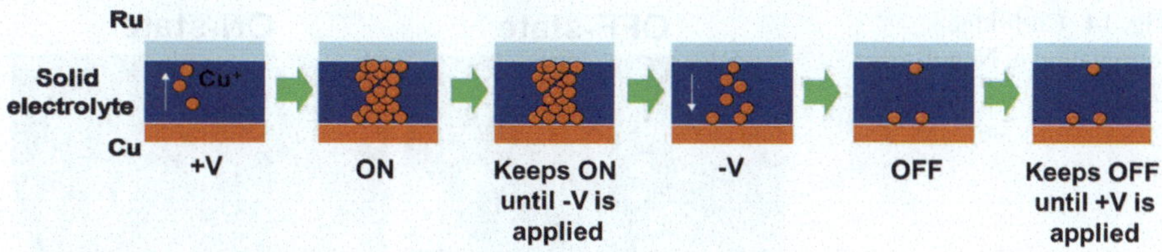


(a)

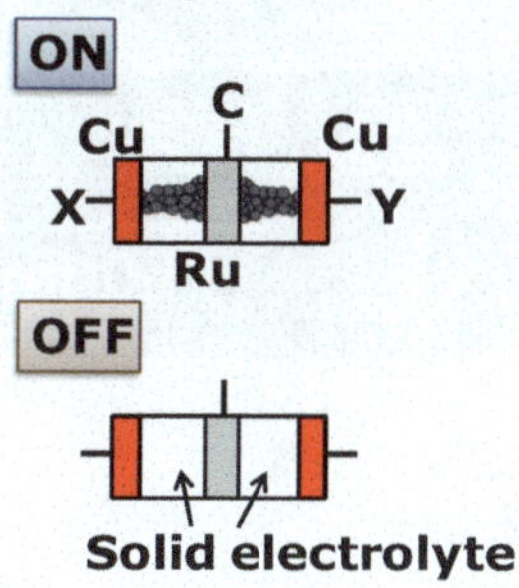


(b)

**Fig. 13** Switching behavior of NanoBridge® [26]

composed of two Cu atomic switches with the polymer solid electrolyte (PSE) sandwiched between two metals (Cu and Ru) as shown in Fig. 13. The switch turns on or off when a nanometer-scale metallic bridge either appears or disappears inside a PSE film by biasing voltages. When a positive voltage is applied to a Cu electrode described in Fig. 13a, the $Cu^{+}$ ions are supplied from the Cu electrode to PSE. The $Cu^{+}$ ions are neutralized and precipitated at the Ru electrode. Subsequently, the precipitated Cu forms conducting bridges between the two electrodes, thus changing the conductance to an ON state. Conversely, by applying a negative voltage to the Cu electrode, the Cu bridge is ionized and disappears, resulting in an OFF state. Each state is nonvolatile and the switching between the two states is repeatable. An LSI using NanoBridge® is fabricated with high reliability configuration. A series of switch of NanoBridge® is implemented in lookup tables and cross bar switches in a FPGA, and high off state impedance is realized. The ON resistance is small enough for operations [20], and the OFF current is below 10 nA. The switching is repeatable up to 1000 times and each state is nonvolatile. The physical formulation is shown in Fig. 14. A programmable FPGA with NanoBridge® is shown in Fig. 15a, and the evaluated device with a plastic package is shown in Fig. 15b. The FPGA does not include a default CPU.

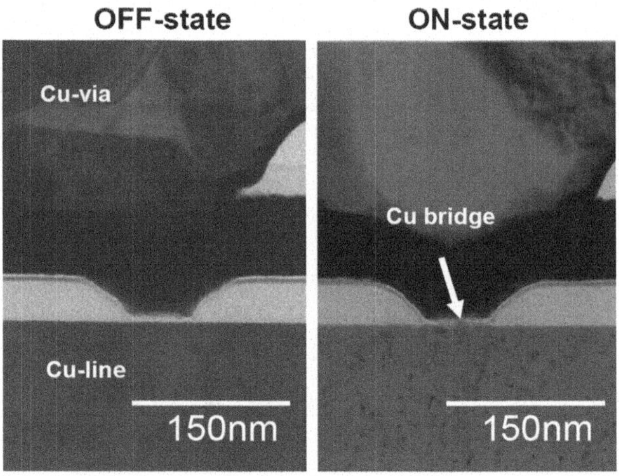


**Fig. 14** Cu bridge formulation in NanoBridge® [26]

## 5.4 Evaluation Result

We implemented signal processing functions of the infrared image sensor of AKATSUKI onto one FPGA chip with NanoBridge® using the architecture described in preceding sections. The original digital electronics is shown in Fig. 16. Its signal processor consists of one central processing unit (CPU) board with a Japan Aerospace Exploration Agency (JAXA) authorized 64bit microprocessor HR5000 and one FPGA board with two Microsemi RTSX72SU FPGAs. Several dedicated operations for an infrared image detector were implemented. They are superposition, mean, median, and normalize functions to improve intensity. These functions were provided for the purpose of the evaluation of functions themselves. These functions were evaluated onboard to see which was the most effective for compensating dispersion of detector pixels.

The functions were implemented in the signal processing unit of AKATSUKI with two modules as described above, and we had a software model of the functions for these modules. The software model corresponded to the structure of one MCU and two FPGAs. We rearrange the structures of functions in accordance with the GPE architecture, which are described in preceding sections and transported the functions on one NanoBridge® FPGA using CWB. These functions have successfully been implemented in one NanoBridge® FPGA. A dedicated arithmetic and logic unit (ALU) was implemented in accordance with step 2 of the design flow of GPE architecture described in the preceding section as shown in Fig. 17. The overview of the test set is shown in Fig. 18.

The maximum power consumption of the original signal processor was estimated as 7.4 W, and it included margin. The nominal power consumption of the original signal processor was estimated as 4.9 W in accordance with the measured value of the whole electronics unit. As for the FPGA implementation, the nominal power consumption of one FPGA used in the original signal processor was estimated as

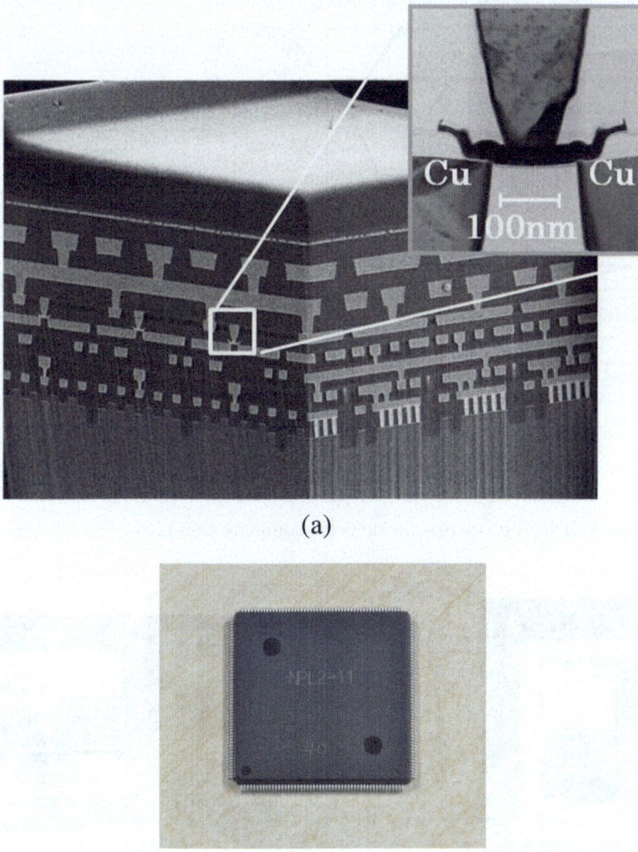


(a)

(b)

**Fig. 15** Programmable FPGA with NanoBridge®, (**a**) cross-sectional FIB-SIM/TEM image, and (**b**) evaluated device with a plastic package [26]

**Fig. 16** The digital electronic of AKATSUKI ©JAXA

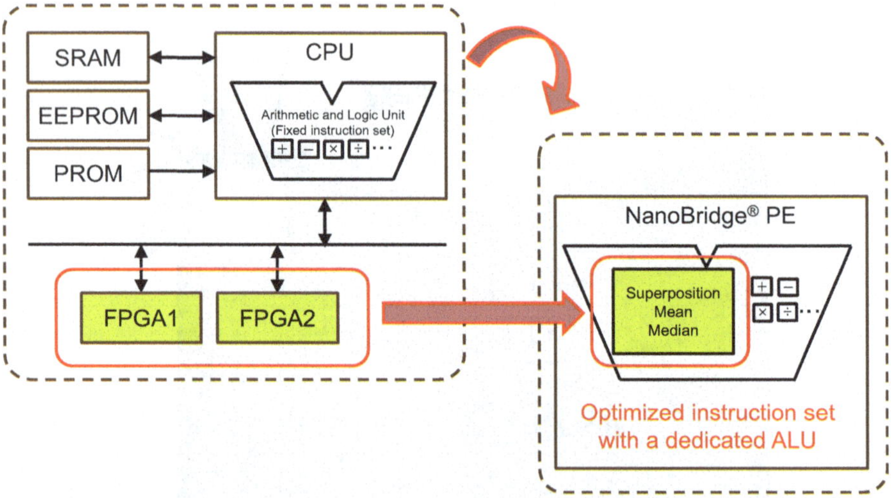


**Fig. 17** One chip dedicated processor for detector compensation [26]

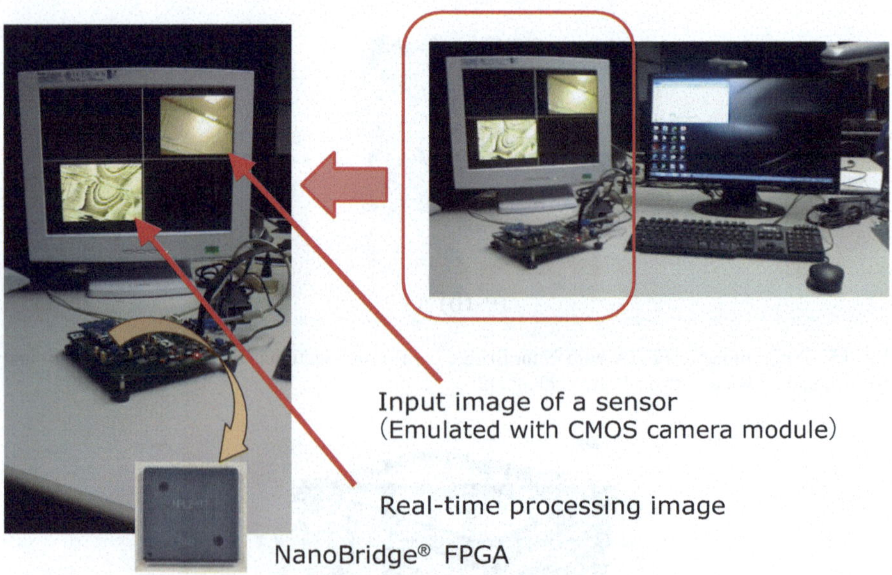


**Fig. 18** Test set overview

500 mW, which was estimated from the power consumption of the whole electronics unit and design schematics.

The measured power consumption of the one-chip processor element implementation on one NanoBridge® FPGA was 18 mW as shown in Table 2 and Fig. 19. It was measured using an LSI tester. Utilization ratio of Look up tables of the NanoBridge® FPGA is 21%. The reduction in circuitry came from the fact that

**Table 2** Size and power consumption reduction derived through the new PE [26]

| Item | Original signal processor | Signal processor with proposed architecture |
|---|---|---|
| Size | Two PWBs, the outside dimension of one PWB is 220 mm × 170 mm | One chip |
| Power Consumption | 4.9 W (nominal value) | 18 mW (measured value) |

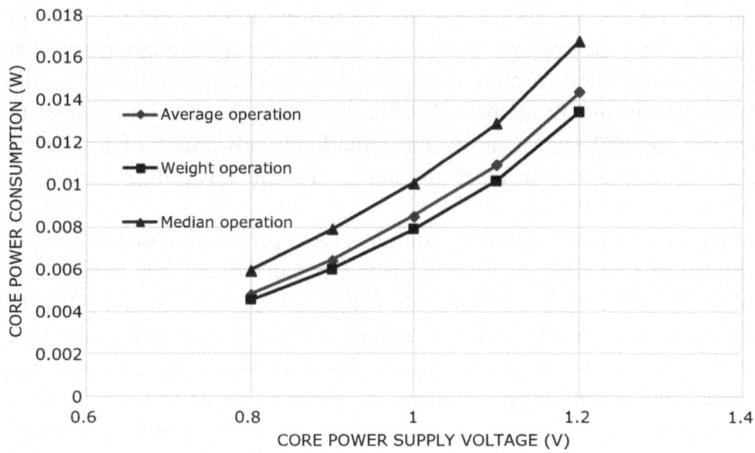


**Fig. 19** Power consumption of NanoBridge® FPGA [26]

whole processor functions are implemented in one chip, and interface circuits over devices are eliminated. This miniaturization was realized by the high level behavioral synthesis tool CyberWorkBench (CWB). The functions of whole signal processor were written in C language. Once an optimized ALU is implemented, the ALU was used in a series of operation process several times, and an optimized circuit was derived through high level behavioral synthesis. As for FPGA design, contribution of NanoBridge® was estimated as 1/56, in which the original power consumption was reduced into 18 mW. The rest of the reduction on power consumption was derived through the new processor architecture, so the reduction ratio based on new architecture was 1/4.9 [26].

## 6 Discussion

An experimental atomic switch FPGA application is introduced in this chapter to show one successful implementation to exploit the merit of atomic switches. The feasibility of a sensing system in space was evaluated experimentally by implementing a real application on a NanoBridge® FPGA. The result shows the practicability of atomic switch FPGAs. Implementation of the processor element architecture using atomic switch FPGA was shown successfully through the evaluation using NanoBridge® FPGA. Power consumption and miniaturization effect were observed using a NanoBridge® FPGA. Significant reduction of power consumption and size was observed through the evaluation of the experiment.

The tape model used in conventional finite automaton were shown to be replaced with Embedded Automaton (EA) through the experimental implementation of GPE architecture using a NanoBridge® FPGA. The implementation process of GPE architecture was verified using NanoBridge® FPGA. The possibility of soft error occurrence is also reduced in space applications by eliminating large memory cells. Some limitation exists for modifying application programs, and dynamic program loading is not easy at present. However, the premises of embedded system are different from those of conventional enterprise system in that unlimited modification is not required, and no fundamental difficulties have been found. Since program memories are reduced significantly, the remarkable advantage of low power consumption, which is found in conventional dynamically reconfigurable processors [5], is expected as well.

As for the system performance of atomic switch processor elements based on the GPE architecture, two kinds of simplification are realized, and real-time performance is improved. Firstly, time is handled as basic data by the system. Therefore, the virtualization of time using software interrupt is no longer necessary. Secondly, concurrency inhered in system design is realized as physically concurrent process, and hence virtualization using serialization of processes or tasks by multi-process operating systems or multi-task operating systems is no longer required. Barrier synchronization of concurrent processes can be performed on fine-grained or coarse-grained implementation. In consequence, simplified synchronization of multiple processes and/or tasks of concurrent real-time systems can be realized.

In the technological road map point of view, system level programmable devices is forecasted to be standardized after the era of System-on-a-chip (SoC) and System in package (SiP) by the extended Makimoto's Wave [33] shown in Fig. 20. That is called as "Highly flexible super integration (HFSI)" in [33]. The right-most surge in the Makimoto's waves shown in [33] is considered as one of the design target of atomic switch FPGAs based on GPE architecture, because the technology trend shown by the Makimoto's waves still holds.

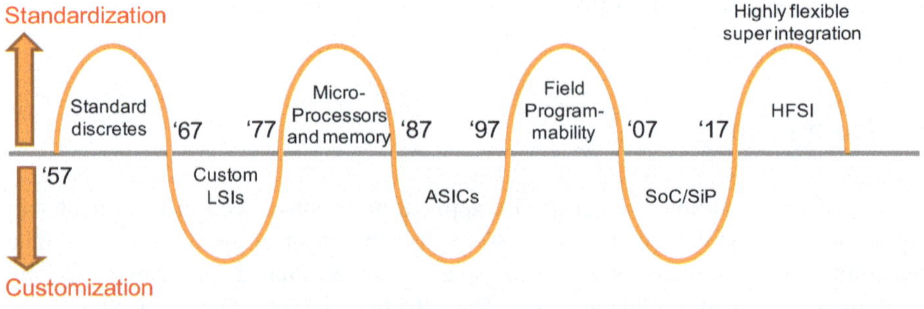


**Fig. 20** Extended Makimoto's Wave [33]

**Acknowledgement** Authors thank Dr. Makimoto for his precious advice. We will proceed the development of atomic switch FPGA based on GPE architecture aiming at to become one of the Highly Flexible Super Integration (HFSI). Authors also thank Dr. Asai, who was Research Supervisor of CREST DVLSI Program, and staff members of JST for their arrangement of the CREST DVLSI (Dependable VLSI) symposium. Some part of the measurement work of power consumption was supported by the Grant-in-Aid for Energy and Environmental New Technology Leading Program of the New Energy and Industrial Technology Development Organization (NEDO) of Japan.

## References

1. http://www.nec.com/en/global/solutions/nsp/m2m/concept/
2. Hihara, H., Iwasaki, A., Hashimoto, M., Ochi, H., Mitsuyama, Y., Onodera, H., Kanbara, H., Wakabayashi, K., Sugibayashi, T., Takenaka, T., Hada, H., Tada, M., Miyamura, M., Sakamoto, T.: Sensor signal processing using high-level synthesis with a layered architecture. IEEE Embed. Syst. Lett. **10**(4), 119–122 (2018)
3. Alnajjar, D., Konoura, H., Mitsuyama, Y., Shimada, H., Kobayashi, K., Kanbara, H., Ochi, H., Imagawa, T., Noda, S., Wakabayashi, K., Hashimoto, M., Onoye, T., Onodera, H.: Reliability-configurable mixed-grained reconfigurable array supporting C-to-array mapping and its radiation testing. In: Proceedings of IEEE Asian Solid-State Circuits Conference (A-SSCC 2013), pp. 313–316 (2013)
4. Konoura, H., Alnajjar, D., Mitsuyama, Y., Shimada, H., Kobayashi, K., Kanbara, H., Ochi, H., Imagawa, T., Wakabayashi, K., Hashimoto, M., Onoye, T., Onodera, H.: Reliability-configurable mixed-grained reconfigurable array supporting C-based design and its irradiation testing. IEICE Trans. Fund. Electron. Commun. Comput. Sci. **E97-A**(12), 2518–2529 (2014)
5. Awashima, T.: Full software implementation of real-time ISDB-T modulator on dynamically reconfigurable SoC using practical co-design environment. In: Invited Talk, CoolChips XIV, Yokohama Industrial Development Corporation, 20th, (2011)
6. Wakabayashi, K., Schafer, B.C.: "All-in-C" SoC synthesis and verification with CyberWorkBench. In: Coussy, P., Morawiec, A. (eds.) High-Level Synthesis from Algorithm to Digital Circuit, vol. XVI, pp. 113–127. Springer, Heidelberg (2008). isbn:978-1-4020-8587-1,. Chapter 7
7. Wakabayashi, K., Okamoto, T.: C-based SoC design flow and EDA tools. IEEE Trans. CAD 1507–1522 (2000)
8. Wakabayashi, K.: CyberWorkBench: Integrated design environment based on C-based behavior synthesis and verification. IEEE VLSI-TSA 173–176 (2005)
9. https://www.nec.com/en/global/solutions/space/remote_sensing/
10. Yokoyama, Y., Kondo, M., Hihara, H., Itoh, T., Kawamata, F.: The development of space-born router for the International Space Station. In: Proceedings of International Conference 2001, pp. 115–110 (2001)
11. Oyama, H., Sakurai, M. (The Japan Society for Aeronautical and Space Sciences), Iwasaki, A. (the University of Tokyo), Takahashi, Y., Toyoshima, M. (National Institute of Information and Communications Technology). The 201st board of governors materials. Science Council of Japan, 19 Sept 2014. http://www.scj.go.jp/ja/member/iinkai/kanji/pdf22/siryo201-5-10-7.pdf
12. Hihara, H.: Thought on ensuring design quality. Design Wave Magazine. CQ Publishing, pp 36–45, February 2006
13. Baumann, R.: The impact of technology scaling on soft error rate performance and limits to the efficacy of error correction. In: Digest IEDM 2002, pp. 329–332 (2002)
14. Fukuhara, T., Taguchi, M., Imamura, T., Nakamura, M., Ueno, M., Suzuki, M., Iwagami, N., Sato, M., Mitsuyama, K., Hashimoto, G.L., Ohshima, R., Kouyama, T., Ando, H., Futaguchi, M.: LIR: Longwave infrared camera onboard the Venus orbiter Akatsuki. Earth Planets Space. **63**, 1009–1010 (2011)

15. Hihara, H., Iwase, K., Sano, J., Otake, H., Okada, T., Funase, R., Kashikawa, R., Higashino, I., Masuda, T.: SpaceWire-based thermal-infrared imager system for asteroid sample return mission HAYABUSA2. J. Appl. Rem. Sens. **8**, 084987-1-13 (2014)
16. Hihara, H., Iwasaki, A., Hashimoto, M., Ochi, H., Mitsuyama, Y., Onodera, H., Kanbara, H., Wakabayashi, K., Sugibayashi, T., Takenaka, T., Hada, H., Tada, M.: Applications of reconfigurable processors as embedded automatons in the IoT sensor networks in space. In: Asai, S. (ed.) VLSI Design and Test for Systems Dependability, 1st edn. Springer, Tokyo (2019)
17. http://www.nec.com/en/global/solutions/iot/iotplatform/
18. Suzuki, M., Hasegawa, Y., Yamada, Y., Kaneko, N., Deguchi, K., Amano, H., Anjo, K., Motomura, M., Wakabayashi, K., Toi, T., Awashima, T.: Stream applications on the dynamically reconfigurable processor. In: Proceedings of 2004 IEEE International Conference on Field-Programmable Technology (IEEE Cat. No.04EX921) (2004)
19. Suzuki, N., Kurotaki, S., Suzuki, M., Kaneko, N., Yamada, Y., Deguchi, K., Hasegawa, Y., Amano, H., Anjo, K., Motomura, M., Wakabayashi, K., Toi, T., Awashima, T.: Implementing and evaluating stream applications on the dynamically reconfigurable processor. In: 12th Annual IEEE Symposium on Field-Programmable Custom Computing Machines (2004)
20. Sakamoto, T., Kaeriyama, S., Mizuno, M., Terabe, K., Hasegawa, T., Aono, M.: NanoBridge technology for reconfigurable LSI. NEC Tech. J. **2**(1), 72–75 (2007)
21. Tada, M., Okamoto, K., Sakamoto, T., Miyamura, M., Banno, N., Hada, H.: Polymer solid-electrolyte (PSE) switch embedded on CMOS for nonvolatile crossbar switch. IEEE Trans. Electron Dev. **58**(12), 4398–4405 (2011)
22. Tada, M., Sakamoto, T., Miyamura, M., Banno, N., Okamoto, K., Iguchi, N., Hada, H.: Improved OFF-state reliability of nonvolatile resistive switch with low programming voltage. IEEE Trans. Electron Dev. **59**(9), 2357–2362 (2012)
23. Miyamura, M., Sakamoto, T., Tada, M., Banno, N., Okamoto, K., Iguchi, N., Hada, H.: Low-power programmable-logic cell arrays using nonvolatile complementary atomic switch. In: 15th International Symposium on Quality Electronic Design, pp. 330–334 (2014)
24. Hihara, H., Tamagawa, N., Imamura, T., Sugaya, H., Sugibayashi, T., Miyamura, M., Sakamoto, T., Tada, M., Hada, H., Wakabayashi, K., Iwasaki, A.: Programmable SpaceWire interface with atom switch: components. In: 2016 International SpaceWire Conference (2016)
25. Hihara, H., Takano, Y., Sano, J., Iwase, K., Kawakami, S., Otake, H., Okada, T., Funase, R., Takada, J., Masuda, T.: Infrared sensor system using robotics technology for inter-planetary mission. In: Proceedings Volume 9608: Infrared Remote Sensing and Instrumentation XXIII (2015)
26. Hihara, H., Iwasaki, A., Tamagawa, N., Kuribayashi, M., Hashimoto, M, Mitsuyama, Y., Ochi, H., Onodera, H., Kanbara, H., Wakabayashi, K., Tada, M.: Novel processor architecture for onboard infrared sensors. In: Proceedings of SPIE, Infrared Remote Sensing and Instrumentation XXIV, vol. 9973, pp. 99730S-1-8 (2016)
27. ECSS-E-ST-50-12C. SpaceWire—links, nodes, routers and networks. ECSS Secretariat, ESAESTEC, Requirements & Standards Division (2008)
28. Hihara, H., Ohtsuka, M., Mizushima, Y., Morisato, T., Ohshima, T., Miyoshi, H., Baba, K.: Space-born fault tolerant computers. Inform. Process. **35**(6), 497–503 (1994)
29. Hihara, H., Yamada, K., Adachi, M., Mitani, K., Akiyama, M., Hama, K.: Autonomous fault tolerant computer for SERVIS-2 satellite. Technical Report of IEICE, DC2002-75, vol. 102, no. 492, pp. 19–24 (2002)
30. Kanekawa, N., Ihara, K.: Space computer systems. Onboard space computers using state-of-art-LSIs. J. IEICE. **73**(11), 1209–1214 (1990)
31. Onodera, H.: Overview of device variations. In: Asai, S. (ed.) VLSI Design and Test for Systems Dependability, 1st edn. Springer, Tokyo (2019)
32. CCSDS.org. The official web site of the Consultative Committee for Space Data Systems. http://public.ccsds.org/default.aspx
33. Makimoto, T.: Implications of Makimoto's wave. IEEE Comput. **46**(12), 32–27 (2013). http://doi.ieeecomputersociety.org/10.1109/MC.2013.294

# An Evaluation of Single Event Effects by Heavy Ion Irradiation on Atom Switch ROM/FPGA

**K. Takeuchi, M. Tada, T. Sakamoto, and S. Kuboyama**

**Abstract** Atom switches embedded in read-only memories (ROMs) and field programmable gate arrays (FPGAs) were investigated in terms of single event effects (SEEs). In this study, their radiation tolerance against single event upset (SEU) was demonstrated with an LET of up to 68.9 MeV/(mg/cm$^2$), irrespective of the voltage conditions or logic states of the cells.

## 1 Introduction

### *1.1 Overview of the Space Radiation Environment*

In the space environment, there are three types of radiation sources that affect electronics in satellite systems: Van Allen radiation belt, solar energetic particles (SEPs) and galactic cosmic rays (GCRs). Figure 1 shows an illustration of the space radiation environment. Semiconductor devices have to survive in this harsh radiation environment.

The Van Allen radiation belts were discovered by the Explorer I satellite in 1958 and analyzed by Van Allen et al. [1] and McIlwain [2] revealing that high energy radiation particles are trapped by Earth's magnetic field. The belts extended in a donut-like shape along the equator, consisting of an inner (electrons and protons) and outer (electrons) double-layered structure. Figure 2 shows contour plots of the trapped proton (left) and electron (right) fluxes with energies above 1 MeV and 10 MeV, respectively, as predicted by SPENVIS code [3]. The actual shapes of the radiation belts are distorted by solar wind. These belts would be 6.6–10 times as large as the Earth's radius (about 6400 km) on the sun side, while being extended up to 1000 times like a comet's tail on the other side [4].

K. Takeuchi (✉) · S. Kuboyama
Research and Development Directorate, Japan Aerospace Exploration Agency (JAXA), Tsukuba, Ibaraki, Japan
e-mail: takeuchi.kozo@jaxa.jp

M. Tada · T. Sakamoto
System Platform Research Laboratories, NEC Corp., Tsukuba, Ibaraki, Japan



M. Aono (ed.), *Atomic Switch*, Advances in Atom and Single Molecule Machines,
https://doi.org/10.1007/978-3-030-34875-5_4

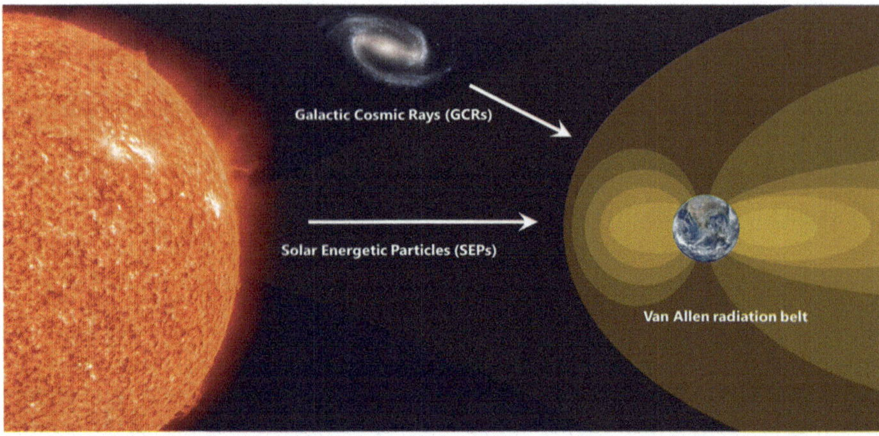


**Fig. 1** Illustration of the space radiation environment

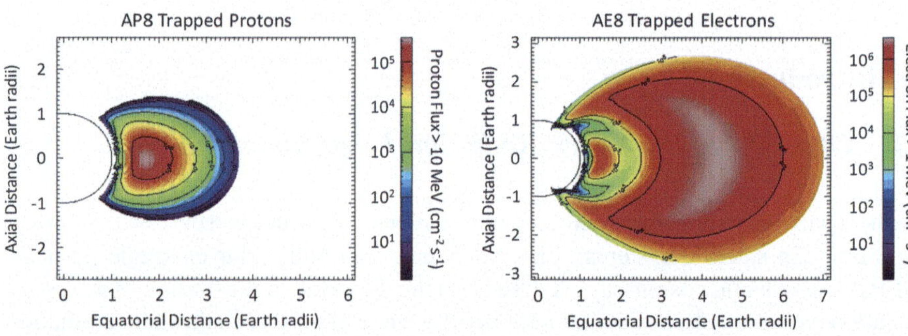


**Fig. 2** Trapped proton (left) flux population with energy >10 MeV and the trapped electron (right) flux population with energy >1 MeV as predicted by the AP-8 model for solar maximum conditions [5]

There are two types of solar events—coronal mass ejections (CMEs) and solar flares. Solar energetic particles (SEPs) are high energy protons, electrons and other ions released from the sun's surface as these events. Solar activity is characterized by the number of black spots (also known as the Wolf number) and changes in the approximately 11-year solar cycle, as illustrated in Fig. 3. It is known that possibility of Solar flares occurrence would probably increase at "solar maximum" (i.e., period during which black spots on the sun increase). The energy spectrum of each solar flares is uniquely different, and thus difficult to predict due to its stochastic nature [4].

Galactic cosmic rays (GCRs) are the nucleus of energetic heavy ions that come from outside of our solar system (e.g., other galaxies, supernovas). As illustrated in Fig. 4, these ions are composed of 83% protons, 13% alpha particles, 3% electrons and 1% heavy ions, which include all of the naturally occurring elements in the periodic table. Ion flux represents a decay-like exponential function as ion energy increases. The energy range is between $10^7$ and $10^{20}$ eV. Though GCRs always

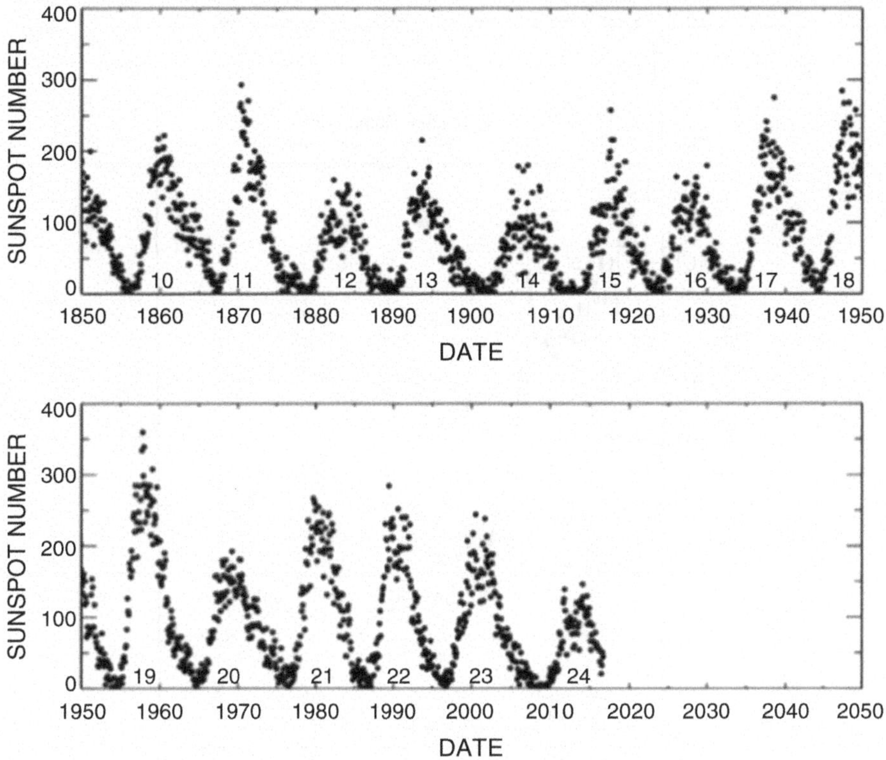


**Fig. 3** Monthly averages of the sunspot numbers [6]

present in interplanetary space, they are modulated by solar activities with an approximately 11-year cycle near the Earth. Ion flux less than 10 GeV will be screened and reduced by magnetic shielding and particles from the sun during the "solar maximum" period. On the other hand, the flux of GCRs will increase during the "solar minimum" period [4, 5].

## *1.2 Radiation Effects on Semiconductor Devices*

There are two types of interactions involving X-rays, gamma rays, electrons, protons, neutrons and heavy ions with the materials—ionizing energy loss and non-ionizing energy loss (NIEL). As shown in Table 1, by their characteristics, the radiation effects on semiconductor devices are classified—total ionizing dose (TID), displacement damage (DD) and single event effects (SEEs).

This section discusses the SEEs in detail. SEEs are the result of interactions involving materials with radiation via the ionizing energy loss process. SEEs are divided into two categories—the so-called "soft" and "hard" errors. Although both classifications are ambiguous, soft errors are generally non-destructive errors, while

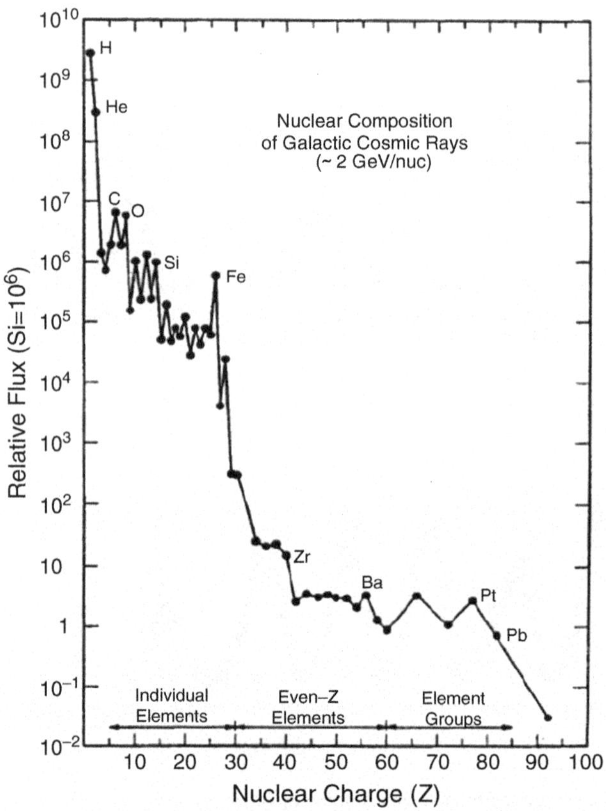


**Fig. 4** Relative flux of GCRs by nuclear charge (Z) [7]

**Table 1** Classification of radiation effects

| | Short term (caused by single ion hit) | Long term (cumulative effect) |
|---|---|---|
| Ionizing energy loss | Single event effects (SEEs) | Total ionizing dose (TID) |
| Non-ionizing energy loss (NIEL) | – | Displacement damage (DD) |

hard errors are destructive. Soft errors include single event upset (SEU), transient (SET) and functional interrupt (SEFI). Because the soft errors are non-destructive, local and temporary, the systems where such phenomena occur can be recovered by system rebooting. In contrast, hard errors include single event latch-up (SEL), gate burnout (SEB) and gate rupture (SEGR), which result in permanent failure. Though a system in which SEL occurred may be recovered by rebooting, SEL is generally classified as a hard error. Figure 5 shows the classification of SEEs.

The primary cause of these SEEs is the charge generated by energetic particles striking in a semiconductor. Figure 6 shows an illustration of reverse biased p-n

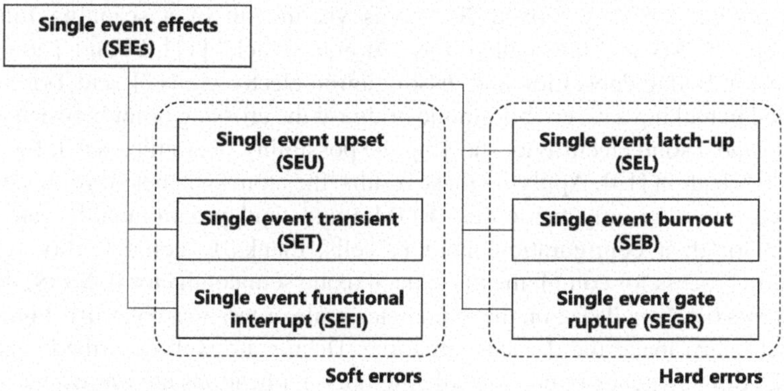


**Fig. 5** Classification of SEEs

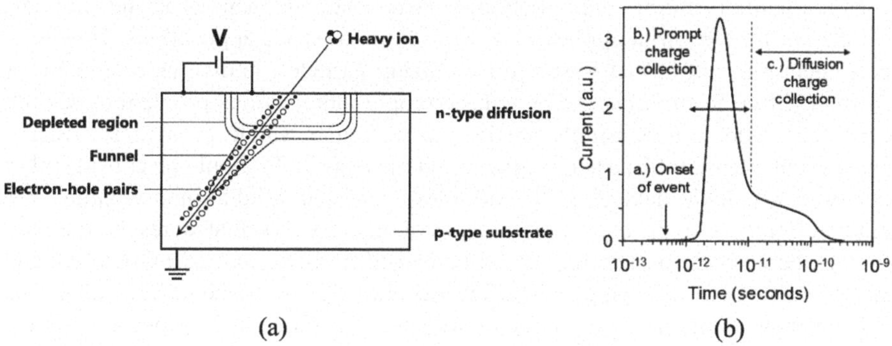


**Fig. 6** (**a**) Illustration of reverse-biased p-n junction. (**b**) Transient current collected as a function of time [8]

junction that collects a deposited charge. At the very early phase of this event (lasting less than 0.1 picoseconds), a cylindrical track of an electron-hole pair with a sub-micrometer radius is generated along the particle trajectory. If this pair traverses the depleted region of the junction, the carriers are rapidly collected by the local electric field within a few tens of picoseconds. This drift collection is followed by the diffusion-dominants collection phase until the excess carriers have been recombined or diffused out (at hundreds to thousands of picoseconds) [8]. An important feature of this phenomena is the distortion of the local electric field into a funnel shape (i.e., the so-called funneling or field-funneling effect) [9]. It is known that this funneling enhances the drift collection of the carriers [10].

## *1.3 Introduction to Atom Switches*

Atom switches are a programmable electrically conductive bridge that forms between two metals grown by electrochemical phenomenon. Historically, Eigler et al. reported that they moved Xe atoms via the tip of a scanning tunneling microscope (STM), and thus called it as "Atomic switch" [11]. Li and Tao demonstrated metal bridge deposition and dissolution in electrolyte [12], and Terabe et al. succeeded in making a reversible atomic bridge at the crossing point between a metal electrode and a solid electrolyte, showing the possibility of atomic switch for "nano-device" fabrication [13]. Applying these results, the atom switches were successfully integrated into read-only memories (ROMs) and field programmable gate arrays (FPGAs) for their configuration memory cells, thanks to compatibility with the fabrication process for complementary metal oxide semiconductor (CMOS) devices. The features (i.e., small size on the nanometer-scale order, non-volatility, high on/off resistance ratio) make the devices attractive. Details are well described separately [14, 15]. Atom switches in non-volatile memory applications are known as conductive bridge random access memory (CBRAM), electrochemical metallization cell (EMC) or programmable metallization cell (PMC) memory [16].

Non-volatile memories are widely used as data storage memory in the commercial market for terrestrial applications, as well as for space applications. However, these memories are in fact sensitive to ionizing particles, known as Single Event Effects (SEEs). Such vulnerability is becoming a more serious problem as scaling techniques become increasingly smaller. In particular, single event upset (SEU), single event latch-up (SEL) and single event transient (SET) should be considered in memories and logic devices [17]. Atom switches that exhibits non-volatility are totally different from the conventional types of memory cells that stores the memory state as the amount of electric charges contained in them. Instead of such electric charge, the resistive changing switches that employing electrochemical mechanisms and ion transportation are used for these cells. Therefore, it is expected to have non-vulnerability to the ionizing particles.

Previous studies have reported gamma-ray tolerance [18] for Ag/GeS2/W-based CBRAM [19]. And other studies have reported on an Ag-based test chip that exhibited SEUs due to upsets of the access transistors [20]. The Ag-based stand-alone memory chip was vulnerable to bit upsets during dynamic write/read tests [21].

This work investigated a Cu-based conductive bridge or atom switches in terms of SEEs. Radiation test results were presented as a preliminary evaluation for applying these technologies for space applications.

## 2 Atom Switch

### *2.1 Overview*

The atom switch is a nano-scale switch that electrochemically controls the connection/disconnection of an electrically conductive Cu-bridge. Figure 7a shows a cross-sectional illustration (left) and transmission electron microscopy (TEM) image (right) of an atom switch cell that is fabricated through the Cu Back-end-of-the-line (BEOL) process in the CMOS fabrication process. Figure 7b shows schematic

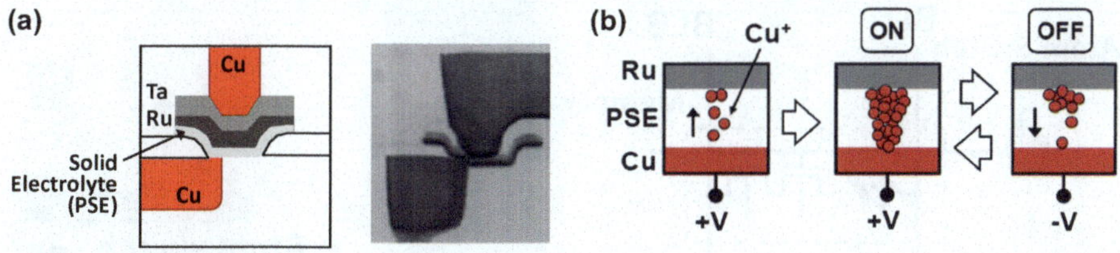


**Fig. 7** (**a**) Cross-sectional illustration (left) and TEM image (right) of atom switch cell. (**b**) Schematic diagrams of switching mechanism [22]

diagrams of the resistive-change mechanism. The atom switch has an Ru/Polymer solid electrolyte (PSE)/Cu sandwich structure. The Cu line edge is used for the Cu electrode. This electrode structure induces the field enhancement effect, resulting in an accurate positioning of the conductive bridge. The formation/annihilation of the Cu bridge causes low/high resistance states that correspond to each digital state.

## 2.2 Atom Switch Memory

Since the high on/off ratio of the atom switch enables the low-voltage, memory operation without a sense amplifier, the minimum operating power of 18.26 μW/MHz at 0.39 V was demonstrated with the embedded atom switch ROM [22].

Another embedded atom switch ROM was fabricated in a five-stage pipelined 32-bit RISC CPU by using 65 nm-node Silicon-On-Thin-Buried-oxide (SOTB) CMOS technology [23]. In the fabrication process, conventional transistors not on SOTB are also integrated for peripheral circuits and programming drivers for ROM that requiring high-voltage (HV) operation shown in Fig. 8.

## 2.3 Atom Switch FPGA

An atom switch was also applied for the configuration switch in the FPGA as complementary atom switch (CAS), where the atom switches were connected in series in the opposite direction [24, 25]. Figure 9 shows (a) schematic images of CAS that show ON and OFF states, (b) illustration of the CAS-based memory cell, (c) illustration of the routing selector, and (d) programmable logic tile in the FPGA utilizing the CAS-based elements. Compared to the SRAM-based FPGA, this CAS-based FPGA demonstrated 60% lower power consumption and three times faster operation [24].

The FPGA is composed of a 64 by 64 programmable logic tile array. Each logic tile is composed of two pairs of D-Flip-Flop (D-FF) and a CAS-based four-input look-up table (LUT), and a 51 by 20 crossbar CASs matrix for inter-logic tile connection. The logic tiles are interconnected with each other through the crossbar matrix. In each

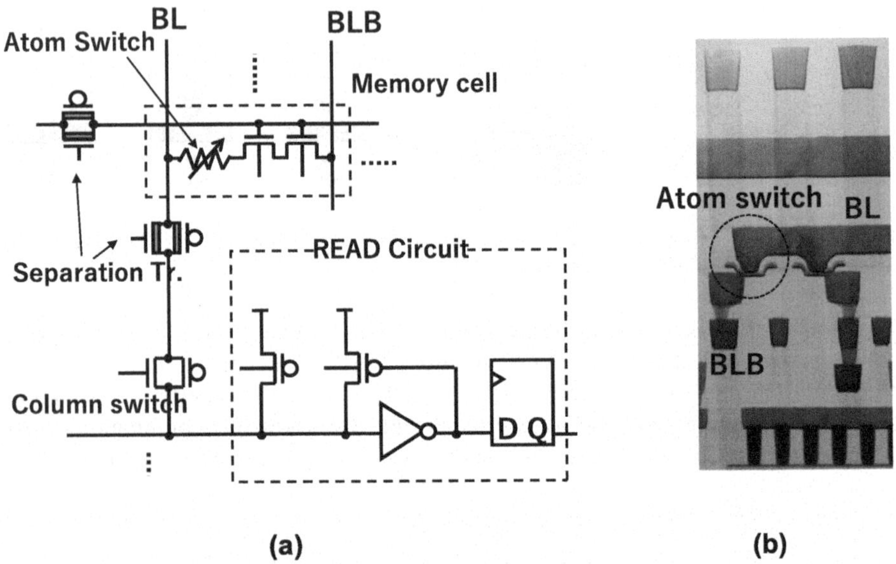


**Fig. 8** (**a**) Block diagram of ROM. (**b**) Atom switch integration on Metal-4 layer [23]

programmable logic tile, 100 CASs for the LUTs and 1020 CASs for the crossbar matrix are embedded. Thus, the FPGA chip contains about nine million atom switches.

## 3 Experimental Setup

Both ROM and the FPGA were evaluated for radiation tolerance by using the Takasaki Ion Accelerators for Advanced Radiation Application (TIARA) at the National Institutes for Quantum and Radiological Science and Technology (QST). A cocktail heavy ion beam with energy of 3.52–3.83 MeV/amu was used for the evaluation.

The atom switches in the ROM and the FPGA were programmed for either the ON or OFF state and validated before and after the irradiation by using the LSI tester. Each ROM and FPGA was de-capped by chemical processes to expose the chip surface directly to the incident ions, and mounted on a custom-made test board with an IC socket. Figure 10 shows the chip images of the ROM (left) and the FPGA (right). The test board was placed into a vacuum chamber at the radiation facility. Control signal and power cables were connected via feed through terminals. The power supply voltage and current were recorded by the HP 6629A Precision System Power Supply located outside the vacuum chamber. “0xA” was written to ROM (i.e., states “0” and “1” evenly written to ROM). ROM was controlled by a lap-top PC via signal cables, and read back at one-second intervals during an irradiation run (dynamic mode) or read back just after irradiation (static mode). The FPGA was

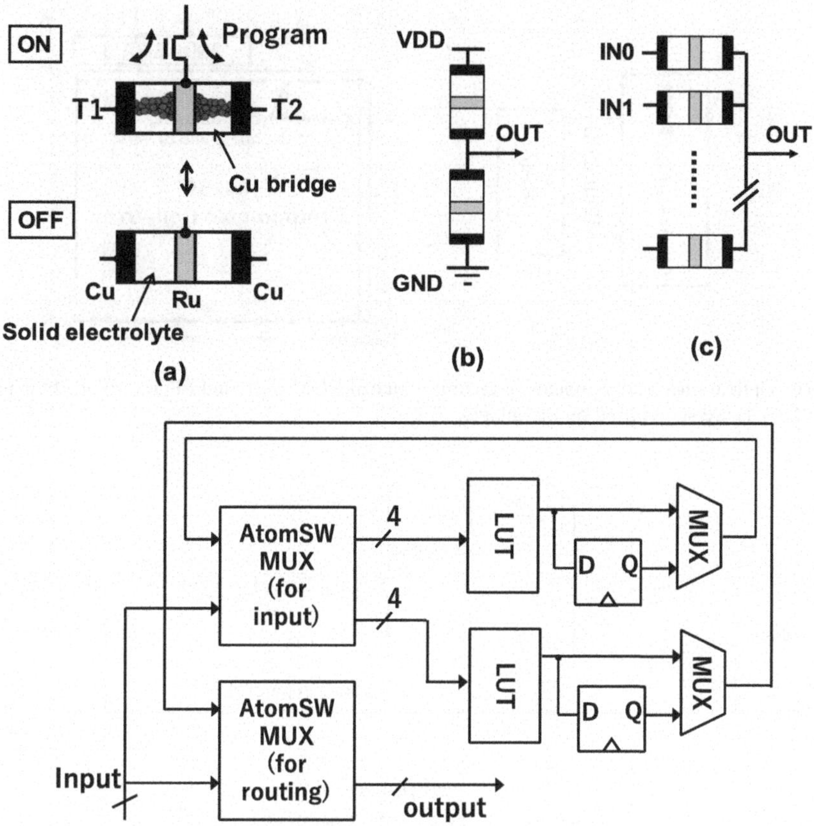


**Fig. 9** (**a**) Schematic images of CAS that show ON and OFF states, (**b**) illustration of CAS-based memory cell which is used in LUT (**c**) illustration of CAS-based routing MUX [24, 25], (**d**) schematic diagram of programmable logic tile, consisting of two pairs of FF and four-input LUTs [26]

configured to map the buffer chains before irradiation, and 1 MHz clock and 100 kHz input signals were applied by the AFG 3101C Function Generator during irradiation. Output signals were monitored using the Tektronix DPO7254C Digital Phosphor Oscilloscope. Figure 11 shows the schematic diagrams of (a) the buffer chain circuit configured in the FPGA, which is composed of 1022 stages of unit buffer blocks in serial connection for a chain, with eight chains connected in parallel, and (b) a unit buffer block in the buffer chains. The input/output (I/O) logic shown in Fig. 11 is composed of D-FF and a level shifter, which translates I/O signal voltage and internal logic voltage (e.g., 3.3–1.8 V in this case).

The devices were evaluated by using Xe and Kr ions. Linear energy transfers (LETs) of Xe and Kr was calculated by using SRIM [27] and found to be 68.9 and 40.3 MeV/(mg/cm$^2$) at the Si surface, respectively. Table 2 lists the irradiation conditions; Table 3 lists the test conditions.

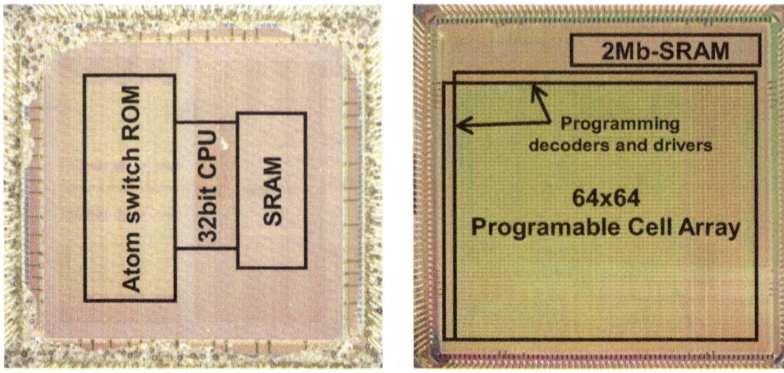


**Fig. 10** Optical microscope images of chemical etched ROM (left) and FPGA (right). Functional block layouts are superposed on the images

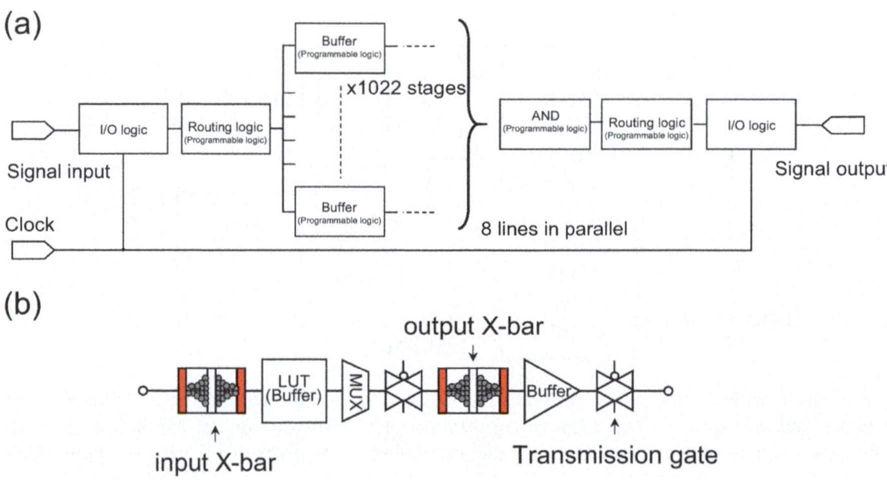


**Fig. 11** Schematic diagram of (**a**) buffer chain circuit configured in FPGA and (**b**) a unit buffer block in the buffer chains

**Table 2** Irradiation conditions

| Ion | $^{129}Xe^{25+}$ | $^{84}Kr^{17+}$ |
|---|---|---|
| Net energy [MeV] | 398 | 289 |
| LET at Si surface [MeV/(mg/cm$^2$)] | 68.9 | 40.3 |
| Range in Si [μm] | 35.0 | 37.3 |
| Total fluence | $10^6$–$10^7$ [p/cm$^2$] | |

**Table 3** Test conditions

| Item | Atom switch | |
|---|---|---|
| | ROM | FPGA |
| # of atom switch | 131 k | 9.175 M (4.587M CAS) |
| Voltage applied to each cell during irradiation | 0 V except read-out timing | Continuously applied |
| Data | "0" & "1" in half | Buffer chain circuit |
| Expected cell area [cm$^2$] | $1.5 \times 10^{-11}$ | |
| Expected # of ion hits for a fluence of $10^7$ [p/cm$^2$] | 19 | 1376 |

# 4 Results and Discussions

## 4.1 ROM

During irradiation, some read-out errors apparently attributable to the CMOS circuit were observed in dynamic mode. However, no memory state change was observed up to 68.9 MeV/(mg/cm$^2$) in both static mode and dynamic mode, and the state was checked on the PC just after irradiation and validated by the LSI tester. Even though in dynamic mode, no voltage is applied to each cell except for read-out timing during most of the time, with results similar to those obtained in static mode. Another study about SEE susceptibility of CBRAM reported SEU immunity under a certain condition (i.e., "static" and "read-only" test mode) [20]. These results including this work suggest that these non-volatile technologies have intrinsic SEU tolerance. Moreover, there was a slight and negligible change in the reading window of ROMs. Figure 12 shows a comparison of shmoo plot between before (left) and after (right) irradiation. The black arrow indicates the difference between before and after irradiation, which could be negligible. It should be noted that only 19 particles were expected to hit somewhere in the atom switches on ROM, since the expected cell area (conductive bridge area) of the atom switch is $1.5 \times 10^{-11}$ cm$^2$.

Single event latch-up (SEL or a sudden increase in the current), on the other hand, was frequently observed. Once SEL was observed, we interrupted the irradiation and reset the power, and then restarted irradiation.

The estimated upper limit of the SEU cross section ($\sigma_{SEU}$) for the atom switch was calculated by following equation,

$$\sigma_{SEU} \leq \frac{1}{fluence \times bit} \left[cm^2/bit\right] \tag{1}$$

where fluence is the total fluence in p/cm$^2$ and bit are the total bits bedded in ROM, respectively. Figure 13 shows the estimated $\sigma_{SEU}$ with blue down triangle for Xe and Kr ions. Each plotted point indicates the value calculated for each irradiation run,

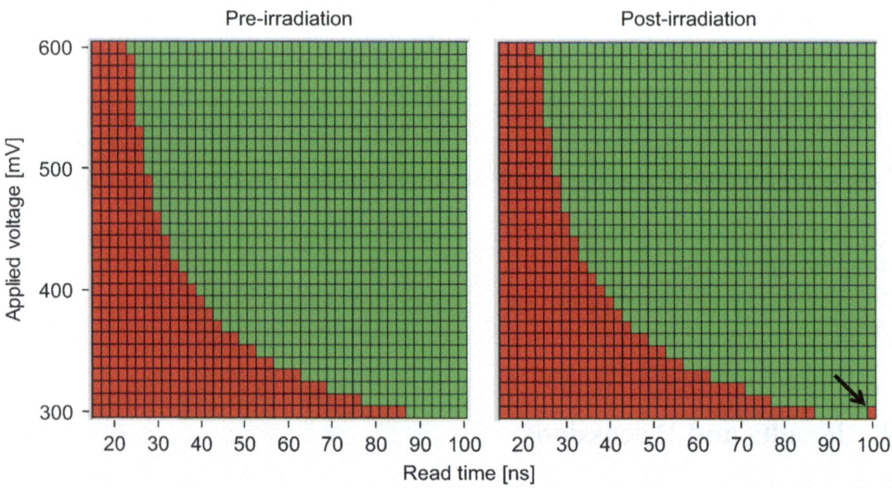


**Fig. 12** Shmoo plot of ROM between applied voltage and read time (green: pass, red: fail). The black arrow indicates the difference between pre-irradiation(left) and post-irradiation(right)

and dose not represents statistical fluctuations. Though actual ion hits to the cell are relatively limited, estimated SEU cross sections were much smaller than the actual cell area regardless of the cell state, in the cases where voltage was not applied to each cell in irradiations to ROM.

## *4.2 FPGA*

No SEU was also observed up to 68.9 MeV/(mg/cm$^2$) on the FPGAs as validated by the LSI tester after irradiation to ensure that the configured circuit had not changed. About 1376 particles were expected to hit somewhere in the atom switches on the FPGA, assuming that the cell area (conductive bridge area) of the atom switch was atom switch is $1.5 \times 10^{-11}$ cm$^2$.

Apparent SEL signatures were also observed, however. Once SEL was observed, the reset-check-restart sequence was done similar to that for ROMs. In Fig. 13, the estimated upper limit of SEU cross sections was also much smaller than actual cell area (showed in the red plot), even if the power supply voltage was applied to each CAS during irradiation to the FPGA. Conversely, single event transient (SET) pulses caused by ions hitting the CMOS circuit (which is not hardened against radiation) were observed. The SET cross-section was about $10^5$ times larger than the atom switch itself, and was dependent on LET (data not shown). Even though the atom switch would be fine, SET could cause the logic status to fail. Radiation hardening by design (RHBD) technique is necessary for CMOS logics, so that the CAS FPGA will work in space applications.

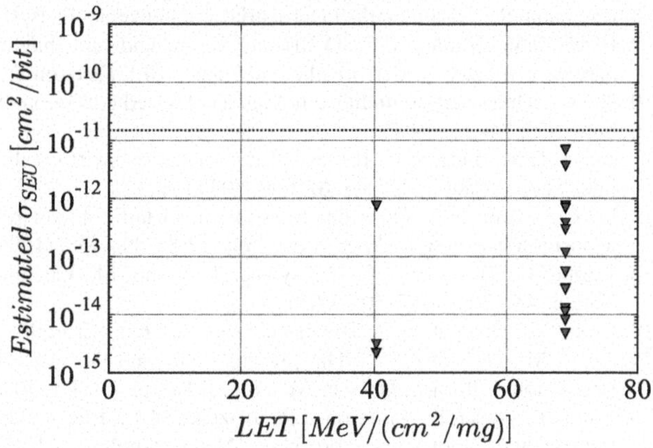


**Fig. 13** Actual cell area (dashed line) and estimated SEU cross section (blue and red down triangles for ROM and red for FPGA, respectively). Down triangles represent each estimated value that is expected to be less than the point in the figure

# 5 Conclusion

It was demonstrated that the estimated upper limit of the SEU cross sections of the atom switch for heavy ions was much smaller than the expected area of the conductive bridge in the atom switch, regardless of the voltage applied to the switch. So far there is no evidence of SEU vulnerability nor physical damage in the atom switch, and thus we could conclude that the atom switch cell itself is an inherently radiation-tolerant structure for SEU caused by heavy ions.

Other evaluations, such as for total ionizing dose(TID) and displacement damage (DD) evaluations would be required to obtain further information. A technique for hardening against radiation is thus necessary for conventional CMOS circuits, in order to use the atom switch ROM and the CAS FPGA in space.

**Acknowledgement** The authors wish to gratefully acknowledge valuable assistance provided by the members of the TIARA accelerator operation group at QST. The authors would also like to thank Ryoei Technica Corp. for its technical support.

# References

1. Van Allen, J.A., McIlwain, C.E., Ludwig, G.H.: Radiation observations with satellite 1958 ε. J. Geophys. Res. **64**(3), 271–286 (1959)
2. McIlwain, C.E.: Coordinates for mapping the distribution of magnetically trapped particles. J. Geophys. Res. **66**(11), 3681–3691 (1961)
3. SPENVIS—Space environment, effects, and education system. https://www.spenvis.oma.be/
4. Goka, T. et al.: Space Environment Risk Dictionary (2006)
5. Pellish, J.: Single-event and total dose testing for advanced electronics. IEEE NSREC short course (2012)
6. Marshall Space Flight Center, NASA. https://solarscience.msfc.nasa.gov/SunspotCycle.shtml
7. Mewaldt, R.A.: Galactic cosmic ray composition and energy spectra. Adv. Space Res. **14**(10), 737–747 (1994)

8. Baumann, R.C.: Landmarks in terrestrial single-event effects. IEEE NSREC short course (2013)
9. Hsieh, C.M., Murley, P.C., O'Brien, R.R.: A field-funneling effect on the collection of alpha-particle-generated carriers in silicon devices. IEEE Electron Device Lett. **2**(4), 103–105 (1981)
10. Hu, C.: Alpha-particle-induced field and enhanced collection of carriers. IEEE Electron Device Lett. **3**(2), 31–34 (1982)
11. Eigler, D.M., Lutz, C.P., Rudge, W.E.: An atomic switch realized with the scanning tunnelling microscope. Nature. **352**(6336), 600–603 (1991)
12. Li, C.Z., Tao, N.J.: Quantum transport in metallic nanowires fabricated by electrochemical deposition/dissolution. Appl. Phys. Lett. **72**(8), 894–896 (1998)
13. Terabe, K., Hasegawa, T., Nakayama, T., Aono, M.: Quantized conductance atomic switch. Nature. **433**(7021), 47–50 (2005)
14. Aono, M., Hasegawa, T.: The atomic switch. Proc. IEEE. **98**(12), 2228–2236 (2010)
15. Aono, M., Hasegawa, T.: The present status and the future of the atomic switch. In: 2011 International Electron Devices Meeting. IEEE, pp. 30.1.1–30.1.4 (2011)
16. Valov, I., Waser, R., Jameson, J.R., Kozicki, M.N.: Electrochemical metallization memories—fundamentals, applications, prospects. Nanotechnology. **22**(25), 254003 (2011)
17. Cressler, J.D.: Extreme Environment Electronics. CRC, Boca Raton, FL (2012)
18. Gonzalez-Velo, Y., Barnaby, H.J., Kozicki, M.N., Gopalan, C., Holbert, K.: Total ionizing dose retention capability of conductive bridging random access memory. IEEE Electron Device Lett. **35**(2), 205–207 (2014)
19. Jameson, J.R., Blanchard, P., Cheng, C., Dinh, J., Gallo, A., Gopalakrishnan, V., Gopalan, C., Guichet, B., Hsu, S., Kamalanathan, D., Kim, D., Koushan, F., Kwan, M., Law, K., Lewis, D., Ma, Y., McCaffrey, V., Park, S., Puthenthermadam, S., Runnion, E., Sanchez, J., Shields, J., Tsai, K., Tysdal, A., Wang, D., Williams, R., Kozicki, M. N., Wang, J., Gopinath, V., Hollmer, S., Van Buskirk, M.: Conductive-bridge memory (CBRAM) with excellent high-temperature retention. In: 2013 IEEE International Electron Devices Meeting. IEEE, pp. 30.1.1–30.1.4 (2013)
20. Mahalanabis, D., Barnaby, H.J., Kozicki, M.N., Bharadwaj, V., Rajabi, S.: Investigation of single event induced soft errors in programmable metallization cell memory. IEEE Trans. Nucl. Sci. **61**(6), 3557–3563 (2014)
21. Chen, D., Wilcox, E., Berg, M., Kim, H., Phan, A., Figueiredo, M., Seidleck, C., LaBel, K.: Single-event effect performance of a conductive-bridge memory EEPROM. IEEE Trans. Nucl. Sci. **62**(6), 2703–2708 (2015)
22. Okamoto, K., Tada, M., Sakamoto, T., Miyamura, M., Banno, N., Iguchi, N., Hada, H.: Conducting mechanism of atom switch with polymer solid-electrolyte. Tech. Digest. **1**, 279–282 (2011)
23. Sakamoto, T., Tsuji, Y., Tada, M., Makiyama, H., Hasegawa, T., Yamamoto, Y., Okanishi, S., Maekawa, K., Banno, N., Miyamura, M., Okamoto, K., Iguchi, N., Oda, H., Kamohara, S., Yamagata, Y., Sugii, N., Hada, H., Ogasahara, Y.: A silicon-on-thin-buried-oxide CMOS microcontroller with embedded atom-switch ROM. IEEE Micro. **35**(6), 13–23 (2015)
24. Miyamura, M., Sakamoto, T., Tada, M., Banno, N., Okamoto, K., Iguchi, N., Hada, H.: Low-power programmable-logic cell arrays using nonvolatile complementary atom switch. In: 15th International Symposium on Quality Electronic Design. IEEE, pp. 330–334 (2014)
25. Miyamura, M., Sakamoto, T., Tsuji, Y., Tada, M., Banno, N., Okamoto, K., Iguchi, N., Hada, H.: 0.5-V highly power-efficient programmable logic using nonvoltile configuration switch in BEOL. In: 23rd ACM/SIGDA International Symposium on Field-Programmable Gate Arrays (2015)
26. Miyamura, M., Tada, M., Sakamoto, T., Banno, N., Okamoto, K., Iguchi, N., Hada, H.: First demonstration of logic mapping on nonvolatile programmable cell using complementary atom switch. In: 2012 International Electron Devices Meeting. IEEE, pp. 10.6.1–10.6.4 (2012)
27. The stopping and range of ions in matter. http://www.srim.org/

# Nanoscale Electrochemical Studies: How Can We Use the Atomic Switch

**Ilia Valov, Tsuyoshi Hasegawa, Tohru Tsuruoka, Rainer Waser, and Masakazu Aono**

**Abstract** Scanning tunnelling microscopy (STM) and in particular the Atomic Switch concept provides a unique opportunity for site invariant, atomically resolved studies on the electrochemical reaction kinetics, combined with close monitoring of the surface morphology. This approach can be applied on particular active or defect sites, but also to larger surface areas, where electrochemically induced changes on the surface states and nanocluster dynamics are monitored by current-voltage (*I*-*V*) spectroscopy and/or current-displacement (*I*-*z*) spectroscopy.

The redox reaction kinetics of areas or sites with different electrocatalytic activity can be distinguished and independently characterized. By modification of the Buttler-Volmer equation, we use switching time as a kinetic parameter instead of current density, where this approach is independent on the electronic partial conductivity.

Thus, the Atomic Switch concept allows for redox kinetics studies with highest lateral, mass and charge resolution.

## 1 Introduction

Since its introduction and invention by Binnig and Rohrer [1, 2], scanning tunnelling microscope (STM) has been used for characterization and manipulation of electronically conductive surfaces providing unique opportunity for achieving atomic lateral resolution and precision [3–5]. STM is now an inevitable tool for high resolution in situ and ex situ studies in the fields such as surface science and (electro)catalysis. More recent developments have transferred this typically UHV technique to liquids, establishing the electrochemical STM [6, 7]. In most of its applications STM is used as a non-invasive tool for observation and/or characterization e.g. to image

I. Valov (✉) · R. Waser
Research Centre Jülich, Electronic Materials (PGI-7), Jülich, Germany
e-mail: i.valov@fz-juelich.de

T. Hasegawa · T. Tsuruoka · M. Aono
International Center for Materials Nanoarchitectonics (MANA), National Institute for Materials Science (NIMS), Tsukuba, Ibaraki, Japan

 
M. Aono (ed.), *Atomic Switch*, Advances in Atom and Single Molecule Machines,
https://doi.org/10.1007/978-3-030-34875-5_5

adsorption/desorption processes, structural and morphological evolutions, reconstruction of surfaces, changes in the electronic structure, etc. In two particular situations STM can also play the active role—atomic/cluster manipulations and the Atomic Switch. In the first case due to physical interactions and local heating generated between tip and substrate single (or group of) atoms can be removed/added and/or dragged on surfaces [5]. This technique has been further developed for nanostructuring i.e. STM lithography [8–10].

The Atomic Switch is the second and maybe the latest development in applications of STM. It has been invented by Terabe et al. and uses STM for inducing electrochemical transformations at surfaces, related with mass and charge transfer [11–13]. This technique is applied in the information technology for storing information and performing neuromorphic operations [14, 15].

A limiting factor for the applications of STM is the requirement for the samples to be electronically conductive. Thus, samples with insufficient electronic conductivity e.g. insulating oxides and purely ionic solid electrolytes are in principle not accessible by this technique. To avoid the restriction, several different approaches have been used. For example, insulating oxides can be reduced prior to STM experiments. Reduction can be achieved either by variation of the deposition conditions, or by annealing in reducing atmosphere. Alternatively, oxides can also be deposited very thin (few monolayers) or can be doped with foreign elements in order to increase their electronic conductivity to a sufficient level, however, without changing their structure. Many examples can be found using various oxides such as NiO, $SrTiO_3$, $TiO_2$, BaO, ZnO, Nb-STO, $CeO_2$ etc. but also for super ionic conductors [16–22].

In this chapter the use of the STM in electrochemical studies in the role of an active electrode will be demonstrated and discussed. Based on the Atomic Switch approach and also using it in combination with the other STM modes this technique provides the opportunity for site-invariant electrochemical studies with highest lateral, mass and charge resolution, using time as critical kinetic parameter. Theoretical discussion and particular examples will highlight this application.

## 2 (Sub-)Nanoscale Electrochemical Studies

Electrochemical studies at the nano- and sub-nanoscale are essential for mechanistic understandings of processes occurring in variety of applications within the fields of energy conversion and storage, sensors, (electro)catalysis, smart devices, nanoelectronics and information technology [23]. The small dimensions of electrodes and solid electrolytes and related small amounts of exchanged mass and charge challenge the sensitivity of the modern equipment. In addition, electrochemical experiments in nanoscale cells are accompanied by extreme conditions: Applying voltages even in the lower range of up to 1 V, corresponds to e.g. electric fields in the range of $10^7$–$10^8$ V m$^{-1}$ and current densities varying between $j \sim (10^4$–$10^9)$ A cm$^{-2}$. Moreover, non-trivial effects can appear, deviating from expectations and theories formulated for macroscopic samples. Thus, even insulating thin films of materials such as $Ta_2O_5$, $HfO_2$, $SiO_2$ etc. turn at nanoscale from high-k materials with insulating macroscopic properties into ionic or mixed ionic-electronic electrolytes at the nanoscale, blurring borders of macroscopic definitions for insulators, semiconductors and electrolytes [24].

The development of the field of nanoelectrochemistry has been especially accelerated by re-discovering the resistive switching memories and their applications not only as non-volatile memory but as building units for neuromorphic applications, alternative logic operations and beyond von Neumann computing [25, 26]. One of the reasons for this acceleration is the focus of both academia and industrial interest on the nanoscale processes. In other areas where electrochemistry plays an inevitable role such as energy conversion and storage or catalysis, industrial research and development targets typically middle and large scale applications, whereas academia aims microscopic understandings. In contrast, in the case of resistive switching memoires (and the Atomic Switch as a part of it) the interest and efforts of both academia and industry are concentrated on research and development of devices, cells and structures with lowest possible size, fastest kinetics, and highest scalability. This joint interest is especially productive and pushes forward the equipment development, fundamental research and technology/applications.

In that sense the Atomic Switch is used not only for information processing and storage, but can be a powerful tool for fundamental research of nanoscale electrochemical processes with highest precision.

## 3 Using the Atomic Switch

As any STM based technique, the Atomic Switch in generally can work only with materials providing a sufficient electronic conductivity. In a macroscopic sense solid materials (excluding metals) can be categorised as insulators, semiconductors/mixed ionic-electronic conductors and ionic conductors. As discussed above after appropriate treatment such as doping, annealing etc., most solid thin films can be used for STM studies. An example illustrating some of these materials related to the use of Atomic Switch is shown in Fig. 1.

Metals and highly doped semiconductors are usual materials choice for STM studies. However, within the Atomic Switch concept also ionic conductors and even materials considered macroscopic insulators can be modified in a way to become suitable for this technique without significantly changing their ionic transport characteristics and preserving their crystallographic structure.

For example $RbAg_4I_5$ is a superionic solid conductor with the highest reported room temperature ionic ($Ag^+$) conductivity, being measured in the order of 0.1 $\Omega^{-1}$ $cm^{-1}$ [27, 28]. It provides a very suitable model system for studies on kinetics of electrode processes in solids. However, the low electronic conductivity of this material, suppressing the quantum mechanical tunnelling, is in principal not supporting STM studies. A way to avoid this restriction is using a doping approach, selectively increasing the electronic conductivity, without influencing the $Ag^+$ partial conductivity. It has been shown that small amount of Fe (0.1 at.%) introduced in $RbAg_4I_5$ are able to

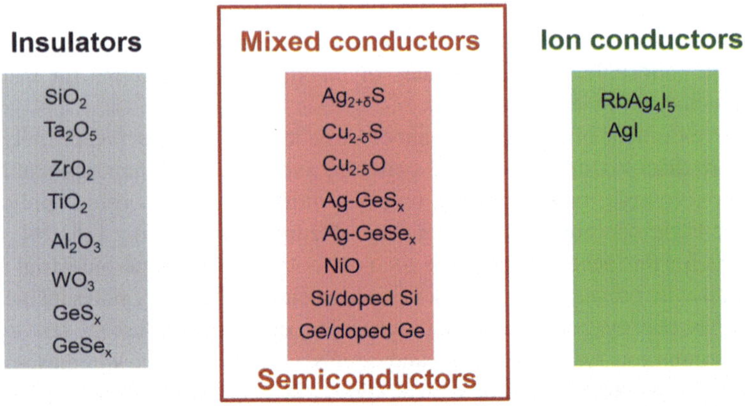


**Fig. 1** Macroscopic classification of materials/compounds in respect their transport properties. The column in the middle shows materials that can be commonly used for STM studies. The other classes of materials require particular modifications to become suitable for Atomic Switch experiments

increase the electronic conductivity by orders of magnitude. In the same time the material remains superionic conductor with an ionic transference number of $t_{\text{ion}} > 0.99$. Thus, the crystallographic structure and the superionic properties are preserved, but the electronic conductivity in the order of $\sigma_e = (10^{-2} - 10^{-3})\sigma_{\text{ion}}$ was sufficient to enable STM imaging and experiments [21]. This approach has allowed to use the full functionalities of the STM on $RbAg_4I_5$ thin films, including imaging, *I-z* spectroscopy and Atomic Switch experiments.

Different approach has been used with oxides such as $Ta_2O_5$, $HfO_2$ and $TiO_2$. In a macroscopic sense they are high-k materials at room temperature and per definition not suitable for STM. In order to use STM on these materials the thickness of the studied layers should be reduced down to 3–5 nm and they should be annealed in vacuum in order to ensure some level of reduction, without reaching the level of decomposition. After such treatment the samples were able to be studied by scanning tunnelling microscope using all its analytical modes [29].

Thus, even "non-classical" materials such as insulators and purely ionic conductors can be modified in a way that enables Atomic Switch and in general STM studies.

## 4 The Atomic Switch as a Fundamental Approach for Electrochemical Studies

The atomic switch configuration is providing a unique opportunity to perform electrochemical studies with highest precision, lateral, mass and charge resolution. It combines both the technical advantages of the STM (imaging, morphology and electronic changes, *I-z* spectroscopy, STS spectroscopy) and the possibility to modify parameters not accessible by any other technique (e.g. by variation of the tunnelling distance). The STM tip can be also effectively used as an active electrode inducing electrochemical redox reactions at the surface of the solid electrolyte. Based on this new approach the reaction kinetics of several materials such as $RbAg_4I_5$, $Ag_2S$, $Cu_2S$, $TaO_2$, $HfO_2$ and $TiO_2$ has been studied [21, 29–32].

## 4.1 Theoretical Considerations

### 4.1.1 Imaging and Electrochemical Reactions

On a first place STM can be used to image the surface of the solid film and precisely (with atomic resolution) select the position for applying voltage and studying particular process(es).

For studies on macroscopic samples, the evaluated properties of surfaces are averaged and the main efforts are focused on the preparation of an ideal, defect free surface. Even mono-crystalline surfaces are showing microscopic defects, possible dislocations, missing atoms, monoatomic steps, etc., being energetically not equivalent, thus predetermining different local reaction kinetics. In that sense, STM is providing the unique opportunity to find and select defect-free position or particular defect site and perform locally electrochemical studies on it, without influencing and/or gathering information from neighbouring domains.

The sensitivity of the STM towards surface morphology inhomogeneity allows to follow the formation and dissolution of small number of atoms (even single atoms), providing an important advantage to observe finest changes induced at the surface. This sensitivity is determined by the exponential dependence of the tunnelling current on the tip-sample distance. The tunnelling current $I_{\text{tunnel}}$ is given by the equation:

$$I_{\text{tunnel}} \sim e^{-2z\sqrt{\frac{2m_e}{(h/2\pi)^2}\Phi}} \tag{1}$$

where $z$ is the width of the energy barrier, $m_e$ is the effective mass of the electron, $h$ is the Planck constant and $\Phi$ is the work function (representing the height of the tunnelling barrier). In the case of STM, the distance between tip and sample surface is corresponding to the width of the energy barrier $z$. Therefore, even very small changes in the distance $z$ e.g. formation of a new atom beneath the tip can be easily detected by the increased tunnelling current.

In case of purely electronic conducting sample surfaces e.g. metals, changes in the surface morphology can appear (or can be induced) only at sufficiently high voltages due to the physical in nature tip-sample interactions. However, if we have an ionic or mixed ionic-electronic conductor also electrochemical redox reactions can be induced, that leads to formation of a new phase. In that respect the main

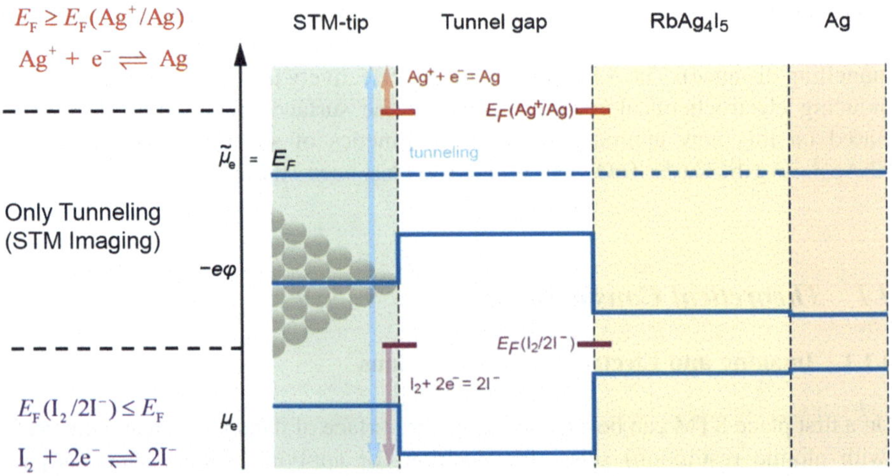


**Fig. 2** Atomic switch (STM) configuration and energy level positions of the half-cell redox reactions at equilibrium state. The electrochemical potential of electrons $\widetilde{\mu}_e$ is given by $\widetilde{\mu}_e = \mu_e - ze\varphi$ ($\mu_e$ is their chemical potential and $\varphi$ is the electric potential) and corresponds to their Fermi energy

question that appears is at which energetic conditions e.g. applied voltage, such reactions will be induced. Figure 2 is visualizing these conditions assuming constant temperature.

At equilibrium no voltage is applied and no driving force for electrochemical reactions is present. Approaching a tunnelling distance, the Fermi level of electrons of the STM tip and the sample, or in terms of thermodynamics their electrochemical potentials $\widetilde{\mu}_e$(tip), $\widetilde{\mu}_e$(Ag), equilibrate and the condition $E_F$(tip) = $E_F$(Ag) or, respectively, $\widetilde{\mu}_e$(tip) = $\widetilde{\mu}_e$(Ag) is fulfilled presuming no net current flow (at zero voltage). The main potential drop is concentrated between the STM tip and $RbAg_4I_5$ surface.

An applied voltage shifts the Fermi level of the STM tip relative to the sample ($\Delta E_F$(tip) = $-e\Delta\varphi$) and an effective tunnelling current is induced. Electrons coming at the $RbAg_4I_5$ surface with an energy lower than the Fermi energy of the redox reaction:

$$Ag^+ + e^- \rightleftharpoons Ag \tag{2}$$

i.e., $E_F < E_F(Ag^+/Ag)$ can only tunnel but cannot contribute to the electrochemical reaction.

After exceeding the energy level of $E_F > E_F(Ag^+/Ag)$ the tunnel electrons can overcome the energy barrier for the $Ag^+/Ag$ redox reaction and Ag atoms will be formed at the electrolyte surface.

At the same moment in order to keep the electroneutrality within the solid electrolyte at the $RbAg_4I_5$/Ag interface, an equal amount of Ag (from bottom electrode) will be oxidized to $Ag^+$ entering the $RbAg_4I_5$.

Thus, at $E_F < E_F(Ag^+/Ag)$ one can perform STM imaging and at $E_F > E_F(Ag^+/Ag)$—electrochemical measurements of the reaction kinetics.

Same discussion applies to the other half-cell reaction i.e.:

$$I_2 + 2e^- \rightleftharpoons 2I^- \tag{3}$$

Applying a sufficiently positive voltage to the tip will result in oxidation of $I^-$ ions with formation of iodine.

Thus, the general condition for STM imaging without inducing electrochemical reactions is given by:

$$E_F(I_2/2I^-) < E_F < E_F(Ag^+/Ag) \tag{4}$$

For higher, respectively lower Fermi levels electrochemical reactions of reduction of $Ag^+$ or oxidation of $I^-$ will be induced at the bare surface.

The Atomic Switch approach combined with scanning tunnelling spectroscopy (STS) and current-displacement (*I-z*) spectroscopy and imaging provides a unique opportunity for studying site-invariant electrochemical processes with highest possible resolution. It also allows to observe in situ or ex situ the stability and the dynamics of the formed atomic clusters and relate them to the electrochemical characteristics.

### 4.1.2 Using Time as a Kinetic Parameter

The methods used for studying kinetics of electrochemical electrode processes can be formally divided into stationary (or steady-state), dynamical and quasi-stationary (e.g. impedance spectroscopy) methods. To evaluate the reaction kinetics a precise knowledge on the partial ionic current is essential. Determining the partial ionic current (or the ionic transference number) is often a main challenge in electrochemistry. This especially applies to nanoscale systems. At small dimensions, structure and properties deviate from the expected macroscopically defined quantities and effects such as field assisted migration, leakage electronic currents and possibility for electron tunnelling through the electrolyte should be taken into account. In fact, the main contribution to the total current is the electronic partial current. Additionally, in case of field assisted migration (exponential dependence of the ion velocity from electric field and particle charge) the transference number can vary depending on the magnitude of the applied voltage. Thus, in these systems there is a principle difficulty to determine the ionic transference number and therefore to evaluate the Faraday reaction kinetics. One possible solution of this problem is using the Atomic Switch approach and introducing time (instead of current) as critical kinetic parameter. As an example can be considered the Butler-Volmer equation, describing the current density ($j$) – voltage ($\Delta\varphi$) relation for charge transfer limited electrode processes. Its classical form is given by:

$$j = j_0 \left[ \exp\left( \frac{(1-\alpha) z_i e}{k_B T} \Delta\varphi \right) - \exp\left( -\frac{\alpha z_i e}{k_B T} \Delta\varphi \right) \right] \tag{5}$$

with $j_0$ being exchange current density, $k_B$—the Boltzmann constant, $T$—the temperature, $e$—the electron charge, $z_i$ the number of exchanged electrons and $\alpha$ denoting the transfer coefficient for the cathodic process.

In this form Eq. (5) cannot be applied to STM studies or to studies in nanoscale systems due to the high uncertainty in determining the ionic transference number. However, the Atomic Switch approach allows to modify the equation replacing the current density by the parameter switching time ($t_s$). To illustrate the approach, first the principle way the Atomic Switch operates should be illustrated, as depicted in Fig. 3.

During step 1 the surface of the solid electrolyte is scanned/imaged at positive tip voltages (fulfilling condition 4) in order to select the position (or site) for the experiments. After this step is completed, a negative voltage is applied ensuring $E_F > E_F(Ag^+/Ag)$. This is inducing an electrochemical reaction (reaction (2)) and a metallic nucleus is formed at the electrolyte surface. This nucleus continues to grow into direction of the STM tip and short circuits the tunnel gap. The moment that the nucleus (or filament) short circuits the tunnel gap is related to a rapid increase of the conductivity. The time for completing this process can be precisely detected by measuring the conductivity (current) reaching the value of $G_0 = 2e/h = 78\ \mu S$ (or resistance of 12.9 kΩ). Technically, the time resolution of the modern equipment is fully capable in detecting times within or even below picosecond range.

Thus, the critical parameter in the Atomic Switch approach is the time for formation of a quantum point contact between the tip and the metallic nucleus (step 4 in Fig. 3).

The current density $j$ from Bulter-Volmer equation can be expressed and related to the switching time according to:

$$j = \frac{I}{A} = \frac{Q}{A \cdot t_s} \tag{6}$$

where $I$ is the absolute current value, $A$ is the electrode surface area (being for the time of experiment constant), $Q$ is the total charge related to reaction (2) and $t_s$ is the time for establishing the quantum point contact (the switching time). As we measure the switching time, the only unknown parameter is the charge $Q$ i.e. the number of electrons that are used in reaction 2. Figure 4 shows how $Q$ can be estimated in a good approximation.

The tip-to-sample distance is fixed prior to applying the voltage and in the typical case for $RbAg_4I_5$ was set to 1 nm. To fill this gap (distance) a row of 4 Ag atoms are required. This means that $4e^-$ is the minimum number of charges. However, ordering atoms in a chain and keeping this configuration is physically unrealistic,

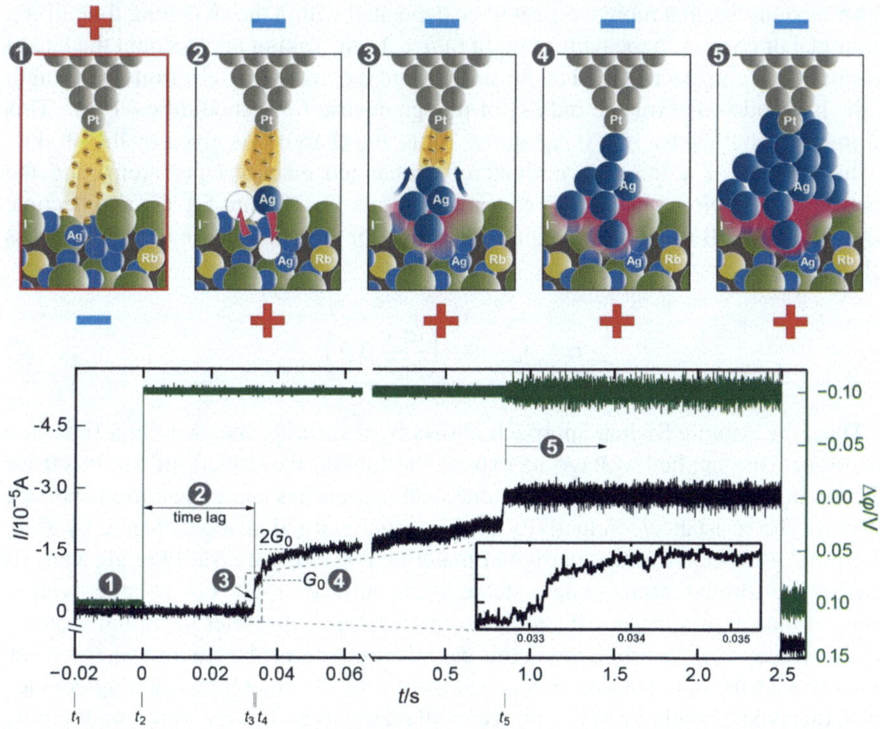


**Fig. 3** Schematic presentation of physicochemical processes during Atomic Switch operation and the related current-time characteristics. Current-time dependence at applied voltage of –100 mV. The number (1–5) notation relates the five regions defined in the current-time characteristics to the microscopic model (upper part of the graph) for the sequence of individual steps during the switching. The inset shows the first quantum step at $G_0$. The sharpness of the current increase is determined by the rate of the filament growth. The figure is reproduced from [21]

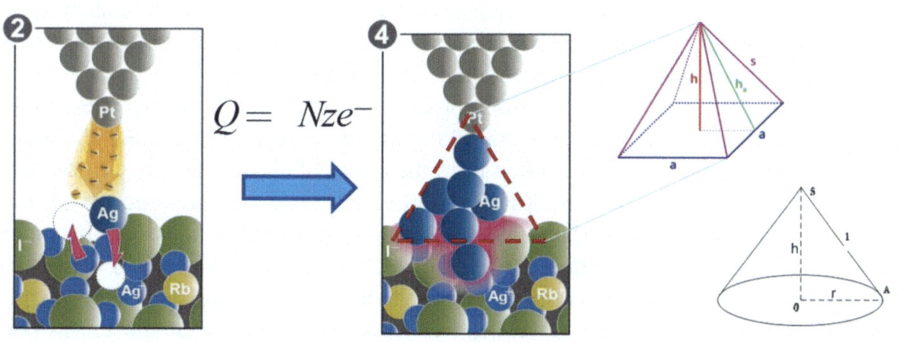


**Fig. 4** Estimation of the charge used within the time for formation of quantum point contact

so we account for, that more atoms will be deposited within the switching time filling pyramidal or conical shape with a height of $h = 1$ nm. Taking into account the lateral distribution of the electric filed at the surface and the tunnelling electrons, the length of the base side-edge (or the radius) of this geometric form should be ~1 nm. This volume can be filled by 15–20 Ag atoms. Thus, the charge $Q$ is given by the product of the number of atoms, the number of exchanged electrons per atom, and the electron charge. Now knowing $Q$ we can substitute Eq. (6) into Eq. (5) and become the time-derived Butler-Volmer equation, which for the cathodic (reduction) process will be:

$$t_s = t_0 \cdot \exp\left(\frac{\alpha z_i e}{k_B T}\Delta\varphi\right) \tag{7}$$

Thus, the Atomic Switch approach allows by measuring the switching time as a function of the applied voltage to extract the kinetic parameters of the electrode reaction in a manner as reliable as for classical approaches using the current-voltage relation. The resistance/conductivity of the tunnel gap will always change by short circuiting, irrespective on the particular material. It is also not disturbing the analysis whether the formed atoms remain stable at the surface or not. For volatile components one can expect (especially at higher temperatures) that after the initial moment of contact they will desorb. Even in this case the time-derived technique/analysis can be applied, as the only parameter we need is the time of achieving switching/contact. Thus, this type of studies can be applied to all materials with some ionic conductivity and being able to support electrochemical reactions.

### 4.1.3 Using STM Tip to Modify the Transfer Coefficient α

Using STM is allowing for another degree of freedom for tuning the experimental conditions during electrochemical experiments, inaccessible by any other technique. This is namely the possibility for variation of the transfer coefficient $\alpha$. Figure 5 visualizes this by showing the potential energy—distance plot.

The charge transfer during electrode processes occurs within the electric double layer. This processes and the effects of the structure of the dense and diffusive double layer were intensively studied in in the classical electrochemistry in the last century e.g. [33, 34]. In the simplest case (provided no specific adsorption) the ion approaches the electrode surface and the distance has the thickness of the outer Helmholtz plane (OHP). At this position the electron transfer takes place, where the energy barrier also depends on the value of this distance. For classical electrochemical cell the thickness of the OHP depends only on the nature of the electrolyte (e.g. solvent, ions, etc.) and electrode material. In one of the earliest definitions the transfer coefficient was introduced as a symmetry factor [35] representing the geometrical position of the energy barrier maximum (for a charge transfer) within the OHP. In that sense the transfer coefficient $\alpha$ is fixed at constant other parameters.

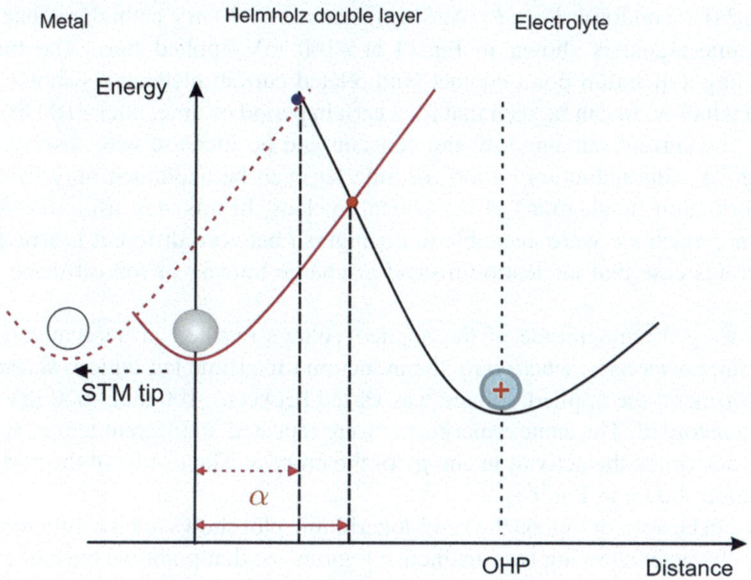


**Fig. 5** Energy-distance plot for a charge transfer reaction. In the case of tip-to-sample distance can be varied (adjusted), using tunnelling current fulfilling the condition of Eq. (4)

In the case of Atomic Switch (STM) one can easily change the distance between the STM tip and the electrode surface and in this way to change the thickness of the OHP, and therefore the position of the energy barrier maximum i.e. the transfer coefficient. This provides the unique opportunity of studying the electrode processes tuning the transfer coefficient $\alpha$, but keeping the same value of the applied potential and electrolyte composition.

## 4.2 *Examples for Studying Electrochemical Processes Using the Atomic Switch*

The Atomic Switch approach as described and discussed above, combined with other STM based techniques has been successfully applied for studies on electrochemical processes and reaction kinetics at solid electrolytes and oxides surfaces. These studies have demonstrated the power of this method and have provided information on the kinetics of different redox reactions.

### 4.2.1 $Ag^+$ Reduction at $RbAg_4I_5$ Surface

The kinetics of the cathodic reduction of $Ag^+$ (reaction 2) at the surface of $RbAg_4I_5$ was studied in details by the Atomic Switch technique. It has been found that the reaction 2 (i.e. condition $E_F > E_F(Ag^+/Ag)$) starts at ~ –70 mV cathodic voltage. The current-time signal is shown in Fig. 3 at –100 mV applied bias. The time for establishing a quantum point contact (and related current increase) is shown in the zoomed window. In can be seen that for a certain period of time, after –100 mV were applied, the current remains low and constant and no increase was observed (step 2 in Fig. 3). This induction period (or time lag) can be explained only by a slow phase formation (nucleation) of the critical nucleus. In this way using the Atomic Switch approach we were also able to distinguish between different kinetic limitations, in this case that nucleation instead of charge transfer or ion diffusion is rate limiting.

Increasing the magnitude of the applied voltage resulted in qualitatively same current time responses, where only the induction time (time lag region) shortens. In this experiment, the applied voltage was varied between –100 and –600 mV and $t_s$ has been recorded. The same experiments were repeated at different temperatures in order to determine the activation energy of the process. The results of the performed analysis are shown in Fig. 6.

As it can be seen in Fig. 6a the semi-logarithmic plot shows not a continuous but a discrete character showing two distinctive regions. At that point we have to account that approaching nano- and/or sub-nano dimensions the number of ions/atoms we exchange with the solid electrolyte is reduced below ~20. Thus, instead of the classical theory of nucleation we have applied the atomistic theory [36]. The nucleation equation was modified by substation with Eq. (6) as follows [21]:

$$t_s = t_0 \exp\left(\frac{(N_c + \alpha)e\Delta\varphi}{kT}\right) \tag{8}$$

where $N_c$ is the number of the atoms constituting the critical nucleus.

Depending on the voltage range $N_c$ has been determined to be 1 and 0, respectively, the transfer coefficient $\alpha = 0.2$ and the activation energy of the process was $\Delta G_a = \sim 1$ eV.

Except determining the kinetic parameters of the reaction, we were able also to image Ag clusters and also to observe their life-time. Figure 7a–c shows STM images of formed clusters. The smallest stable cluster we were able to image is of a height of roughly 3 nm. We used also *I-z* to monitor smaller clusters and cluster dynamics. It has been observed that small nuclei are spontaneously dissolved after the applied voltage was removed (Fig. 7d), corresponding to unstable nucleus configuration. Applying higher voltages (currents) results in a stabilization of the clusters with a height of approximately 1 nm (Fig. 7e). Higher voltages (over 200 mV) and currents lead to stable nuclei of larger size (Fig. 7f).

Thus, we were able to determine the kinetic parameters and rate limiting step for $Ag^+/Ag$ redox reaction and also the conditions for stability of the deposited clusters.

Atomic Switch experiments discussing the switching time and mechanism and reaction kinetics have been also performed for mixed ionic-electronic solid electrolytes using $Ag_2S$ and $Cu_2S$ as model systems [30, 31]. In these experiment no

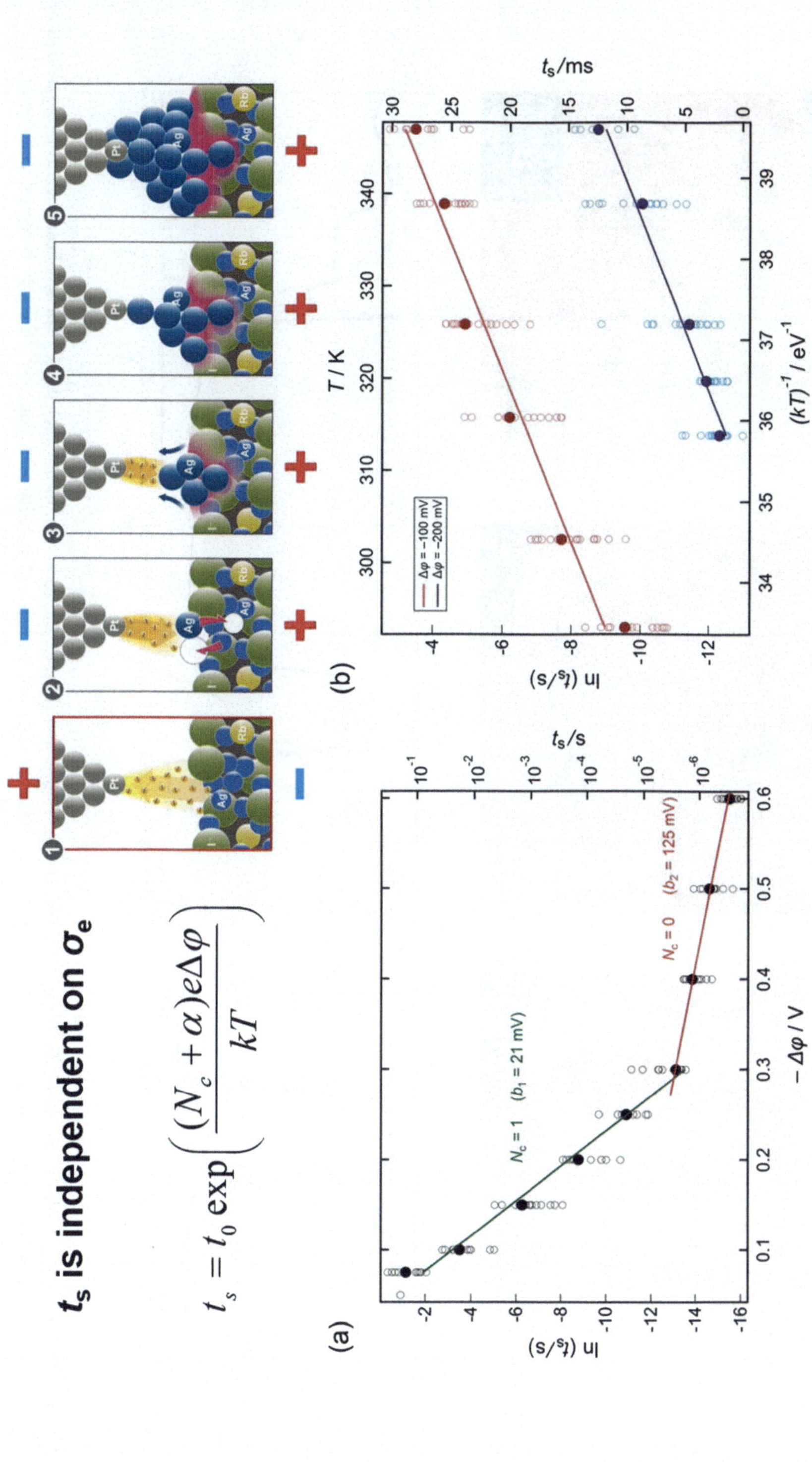


**Fig. 6** Analysis of electrode kinetics of $Ag^+$ reduction at the $RbAg_4I_5$ surface. Plots of applied voltage vs. time (**a**) and Arrhenius type plot (**b**) are shown. The figure is modified from [21]

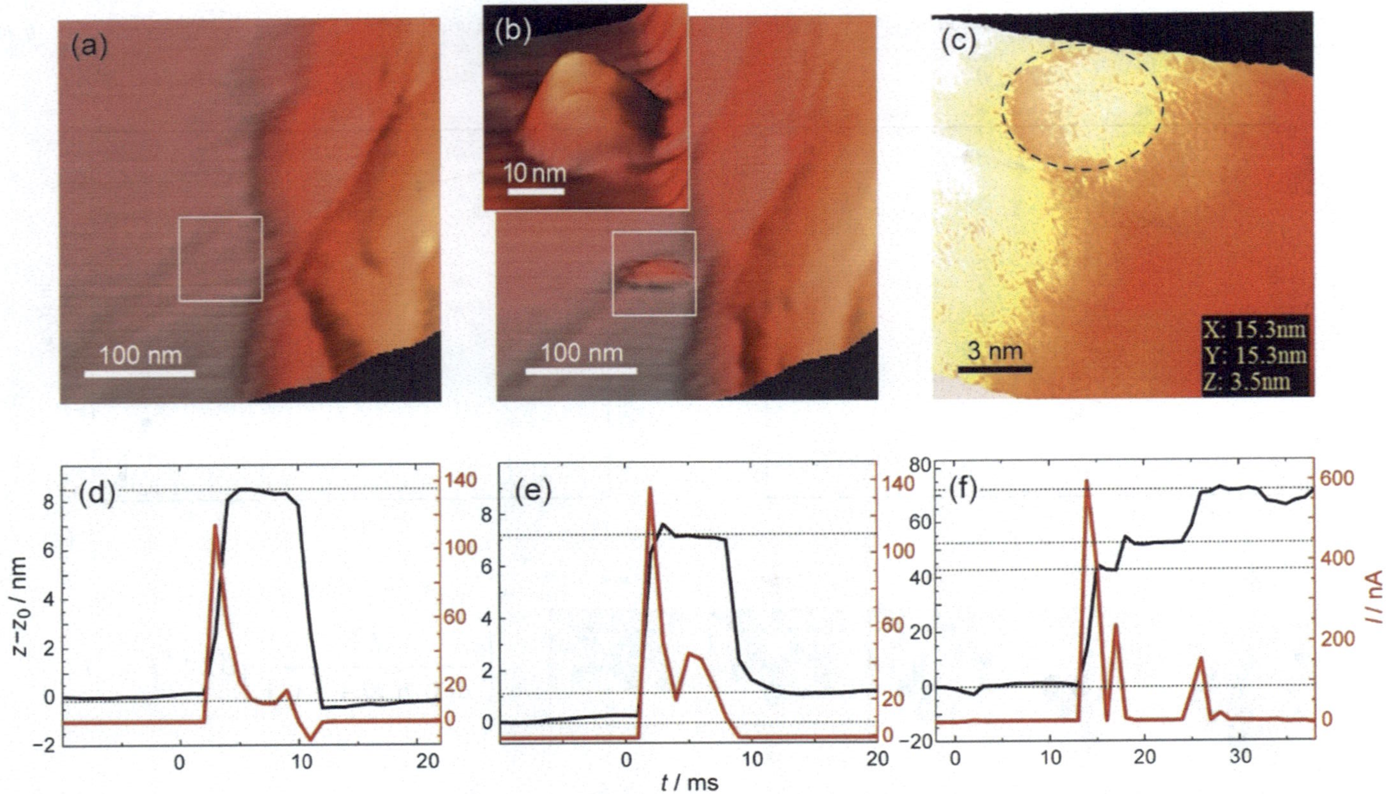


**Fig. 7** STM images of electrochemically formed Ag clusters. (**a**) Before and (**b**) after the formation of the nucleus. (**c**) The smallest nucleus being able to be imaged within its stability time. (**d–f**) *I-z* spectroscopy showing different stages of growth and stability of Ag nuclei. The figure is reproduced from [21]

modification/doping of the thin film electrolytes was necessary, because the intrinsic electronic conductivity was sufficient for electron tunnelling.

These results have demonstrated the potential of the Atomic Switch (and STM in general) as powerful tool for studies of electrochemical processes with an ultimate lateral, mass and charge resolution.

### 4.2.2 Redox Processes on Oxides Studied by Atomic Switch

The concept of the Atomic Switch has been further developed and applied on oxides such as $Ta_2O_5$, $HfO_2$ and $TiO_2$. These samples have been adapted for Atomic Switch studies by annealing in vacuum in order to slightly increase their electronic conductivity. As discussed in the previous section the electronic conductivity is not influencing the kinetic measurements of the ionic redox reactions.

Applying different voltage polarities to the STM tip, it has been possible to distinguish between two different partial redox reactions, indicating that different ionic species are involved in the electrode reaction process. The system used for these experiments was $Ta_2O_{5-x}$ thin film deposited on 300 nm metallic Ta. As in the case of $RbAg_4I_5$ we firstly applied negative voltage to the tip and have observed formation of metallic nuclei at the oxide surface. We were also able to detect the formation of quantum point contacts for each of the experiments. Figure 8 shows STM images recorded after three consecutive switching events.

Using negative tip voltage we formed clusters of metallic Ta as also confirmed from the STS sweeps (sweep 3) performed immediately after the cluster formation. It shows a linear relation between voltage and current, thus evidencing for a metallic phase. The formed nucleus is however prone to re-oxidation and if one waits for ~15 min the metallic nucleus is covered by oxide scale (sweep 1 black curve) showing restored band gap. This surface oxide can be easily reduced back to metallic state (sweep 1 red curve).

Applying a positive bias to the tip results in a completely different electrical responses and images at the $Ta_2O_{5-x}$ surface. No formation of quantum point contact was observed at these conditions. The form and morphology of the modified areas were much broader and semiconducting instead of metallic properties were detected as evident from the STS sweeps, indicating that no metallic phase has been formed.

Based on these analysis it has been concluded that during the cathodic process (negative tip bias) the main reacting species are metal cations that are reduced to form metallic nuclei, whereas during the anodic process (positive tip bias) oxygen ions have been oxidized and oxygen has been released. Thus, depending on the applied voltage polarity different ionic species are undergoing electrochemical reaction at the interface.

Similar experiments were performed with other oxides such as $HfO_{2-x}$ and $TiO_{2-x}$ [32].

These conclusions were further supported and additional details on these processes were provided by using STM on larger surface areas on the oxide surfaces. Applying a negative voltage to the tip caused a reduction of the selected area. The

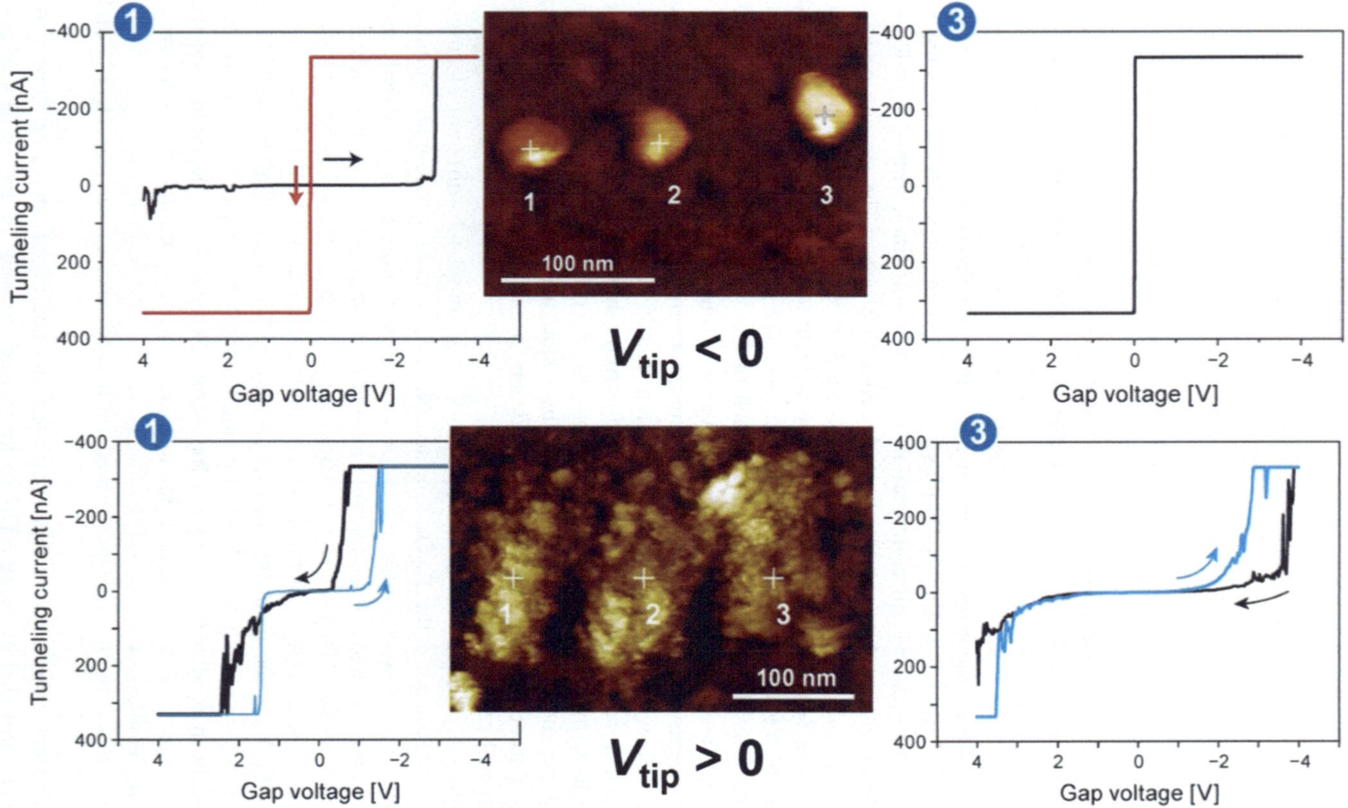


**Fig. 8** Atomic switch experiments on $Ta_2O_{5-x}$ surface using negative (upper image) and positive voltage polarity on the tip. Scanning tunnelling spectroscopy has been used to characterize the electronic structure of the modified surfaces. The figure is modified from [32]

reduced surface could be reversibly re-oxidized by applying a positive tip voltage. In Fig. 9 the recorded images for these experiment are presented.

The modified areas have been studied by ST spectroscopy to identify the changes in the electronic structure and conclude on the nature of induced modifications. As it can be seen from the *I-V* plot in Fig. 9 the reduced state shows purely metallic behaviour with liner relation between tunnelling current and applied voltage. After re-oxidation the metallic atoms are oxidized and the bandgap of the oxide was restored. Thus, we have confirmed that Ta-ions can be reduced at the surface of TaOx and reversible re-oxidized without inducing irreversible changes on the surface morphology.

Similar experiments have been performed by initially applying positive voltage to the STM tip. In this case the surface has been again modified but instead expected oxidation a clearly pronounced reduction of the selected area was detected. These reduction could be reversibly removed by applying a negative tip voltage. The STM images can be seen in Fig. 10.

The observed changes were explained by removing oxygen ($2O^{2-} - 4e^{-} = O_2$) from the $Ta_2O_{5-x}$, thus causing an effective reduction of the surface. An additional peak was observed close to the unoccupied states, indicating defect states in the band gap. Calculations on the hexagonal $Ta_2O_5$ structure revealed that occupied states mainly correspond to O 2p levels and unoccupied states to Ta 5d and 6s levels. Creating oxygen vacancies in the structure leads to an extra peak in LDOS close to the unoccupied states due to electron localization on Ta 5d and 6s states. The extra peak found in our experimental d*I*/d*V* data is consistent with this computational LDOS analysis on $Ta_2O_5$. The characteristic peak close to the unoccupied states and smaller band gap on LRS confirm resistive switching by an oxygen vacancy mechanism [29].

In these large scale experiments we were able also to distinguish that the predominance of one of the ionic species i.e. $Ta^{x+}$-cations or $O^{2-}$-anions in the redox reactions is strongly influenced by the stoichiometry of the oxide. Thus, more strongly reduced $Ta_2O_{5-x}$ shows cation-type switching (as in Fig. 9), whereas less reduced $Ta_2O_{5-y}$ ($y < x$) allows only anion-type switching. Therefore, it can be concluded that the level of non-stoichiometry also influences significantly the ionic transference numbers. These conclusions were supported by theoretical calculations showing the same trend.

Redox reactions, effects of polarity of the applied voltage and the influence of oxygen molecules have also been studied with purely electronic oxides e.g. $SrRuO_3$ that however, allow STM-tip induced ionic redox reactions. The electrical and structural properties of $SrRuO_3$ are very sensitive to the oxygen non-stoichiometry in the material and distinguishing between these effects is very challenging. The complex processes that were observed during anodic or respectively cathodic polarization were able to be resolved only by a combination of STM imaging, spectroscopy as well supported by PEEM and AFM analysis.

Under ultra-high vacuum conditions applying of positive tip bias resulted in irreversible modifications (reduction) of the surface. Two processes, different in nature, were able to explain these initial observations—(1) purely structural changes

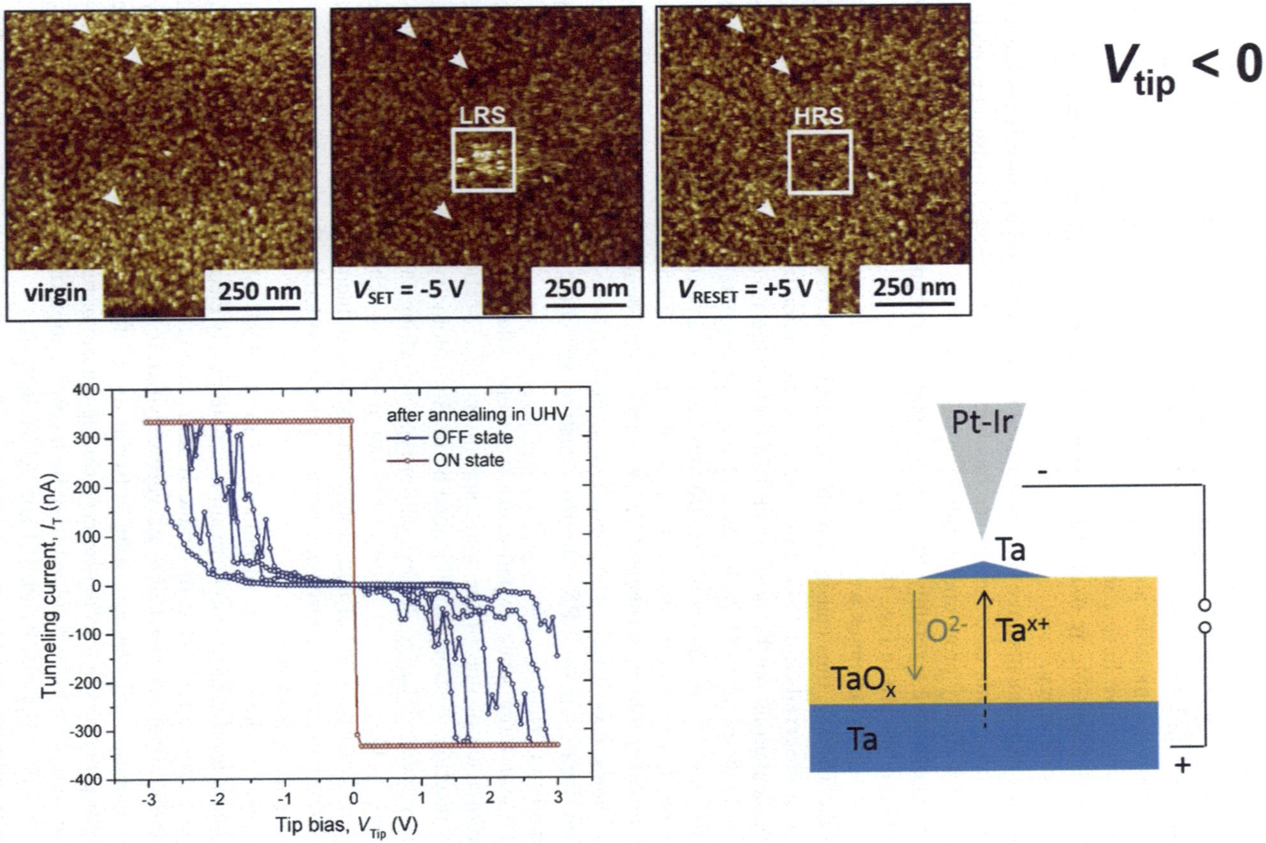


**Fig. 9** STM experiment with electrochemical polarization causing reduction of the initial surface (left) at $V_{set} = -5$ V (middle) and re-oxidation at positive tip bias of $V_{reset} = 5$ V. The $Ta_2O_5$ film (deposited by RF sputtering on 100 nm metallic Ta) has been reduced in the vacuum chamber by annealing in vacuum at 300°C for 5 min in order to enable STM experiments. The figure is reproduced from [29]

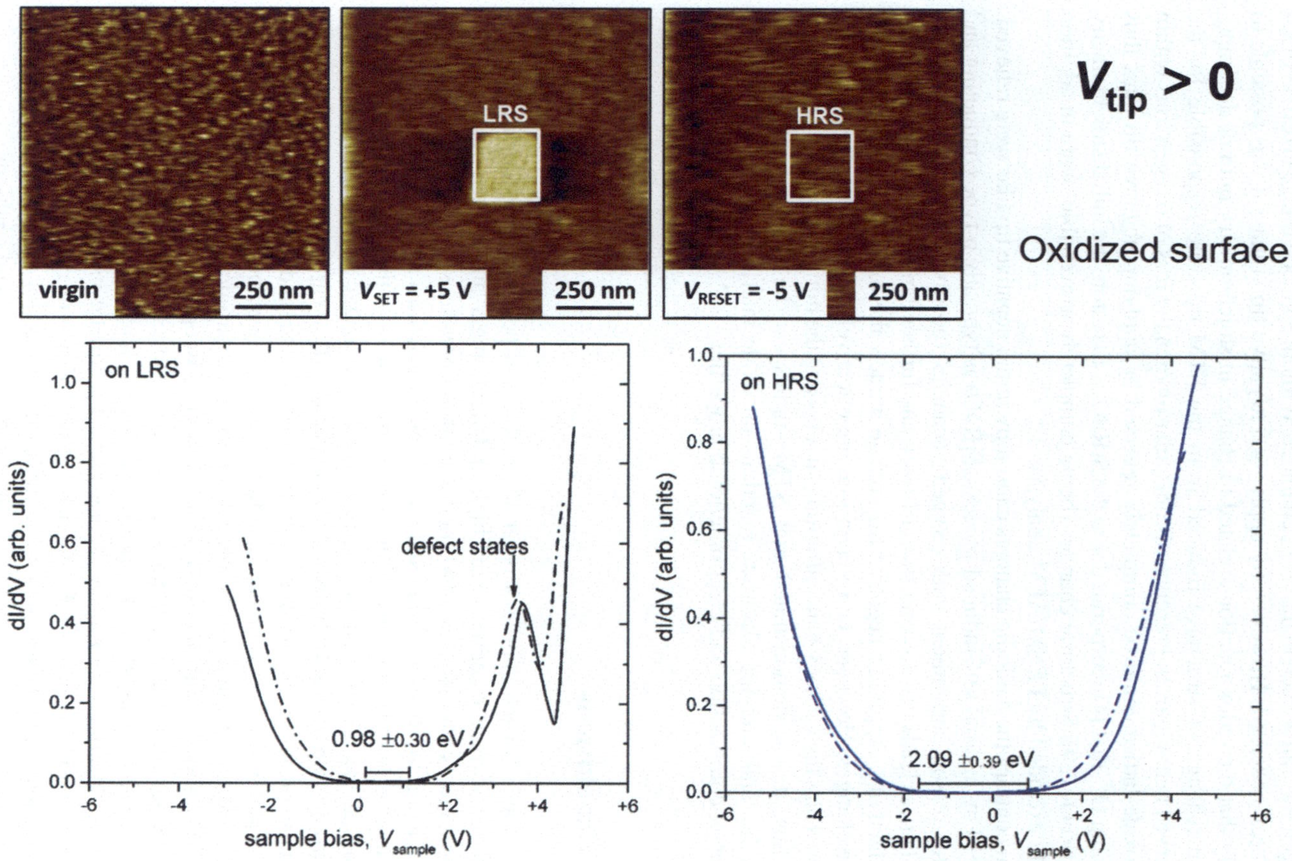


**Fig. 10** STM experiment with electrochemical polarization causing reduction of the initial surface (left) at $V_{set} = 5$ V (middle) and re-oxidation at negative tip bias of $V_{reset} = -5$ V. The $Ta_2O_5$ film (deposited by RF sputtering on 100 nm metallic Ta) has been reduced in the vacuum chamber by annealing in vacuum at 300°C for 5 min in order to enable STM experiments. The figure is reproduced from [29]

within the $SrRuO_3$ or (2) redox reactions, related to change of stoichiometry. Both possibilities are related to loss of oxygen ions. To resolve this issue small amount of oxygen was introduced in the STM chamber. Without applied voltage no effect of $O_2$ has been observed. However, by applying a negative tip voltage it was possible to verify that the redox reactions related to change in stoichiometry but not decomposition (change of structure) induced from the STM tip were responsible for the observed behaviour. In presence of molecular oxygen the loss of oxygen ions within the oxide lattice could be completely reversed by applying negative tip bias, thus forming $O^{2-}$ that is incorporated into the $SrRuO_{3\text{-}x}$ to restore the initial stoichiometry of $SrRuO_3$ [29]. Structural changes (being completely irreversible) were in addition safely excluded by PEEM/AFM analysis.

In contrary the cathodic changes caused by initial negative tip bias were relieved, trigged by internal self-induced process. $SrRuO_3$ is typically p-type conducting and of this reason enrichment with oxygen vacancies (positive relative charge in the sub-lattice) leads first to conditions of $p \sim n$ and further can lead to change to n-type conductivity. Applying negative tip voltage attracts the oxygen vacancies causing these changes. However, no oxygen ions are effectively lost and after removing the applied bias the vacancy concentration can relax and the initial profile is restored [29].

Thus, it was demonstrated that STM studies can be very powerful tool for inducing and studying electrode reactions on oxide interfaces, using materials of different composition and electronic/ionic properties.

## 5 Conclusions

The Atomic Switch approach is an alternative and powerful way to study electrochemical surface reactions with highest lateral, mass and charge resolution. In combination with classical STM modes it allows precise, site-invariant selection of the reaction location, avoiding statistical signal deviations averaged over properties of larger electrode surfaces. Instead of current, it uses the switching time (short circuiting the tunnel gap) as a critical kinetic parameter and is insensitive to the electronic partial conductivity of the materials. The variation of the tip-sample distance provides the unique opportunity to define and/or change the transfer coefficient of the charge transfer reactions, by keeping all other parameters constant.

The Atomic Switch can be used not only on electronically conducive materials but also on ionic solid electrolytes and even macroscopically insulating oxides after appropriate doping and/or treatment. The presented examples on difference classes of materials such as $RbAg_4I_5$, $Ta_2O_5$ and $SrRuO_3$ have demonstrated the ability of this method not only to allow to determine the kinetic parameters of a particular redox reaction but also to distinguish between different reacting species and rate limiting steps. This technique will be further developed for even atomically resolved experiments and expand the variety of studied materials and reactions.

## References


1. Binnig, G., Rohrer, H.: Surf. Sci. **126**, 236 (1983)
2. Binnig, G., Rohrer, H.: Helv. Phys. A. **55**, 726 (1982)
3. Güntherodt, H.-J., Wiesendanger, R.: Scanning Tunneling Microscopy I: General Principles and Applications to Clean and Adsorbate-Cover Surfaces. Springer, Berlin (1994)
4. Wiesendanger, R.: Scanning Probe Microscopy and Spectroscopy: Methods and Applications. Cambridge University Press, Cambridge (1994)
5. Voigtländer, B.: Scanning Probe Microscopy. Springer, Berlin (2015)
6. Kolb, D.: Surf. Sci. **500**, 722 (2002)
7. Gentz, K., Wandelt, K.: Chimia. **66**, 44 (2012)
8. Garfunkel, E., Rudd, G., Novak, D., Wang, S., Ebert, G., Greenblatt, M., Gustafsson, T., Garofalini, S.: Science. **246**, 99 (1989)
9. Iwasaki, H., Yoshinobu, T., Sudoh, K.: Nanotechnology. **14**, R55 (2003)
10. Liu, Y., Zhang, J.: Physica B. **405**, 1890 (2010)
11. Terabe, K., Hasegaw, T., Nakayama, T., Aono, M.: Riken Rev. **37**, 7 (2001)
12. Terabe, K., Hasegawa, T., Nakayama, T., Aono, M.: Nature. **433**, 47 (2005)
13. Terabe, K., Nakayama, T., Hasegawa, T., Aono, M.: J. Appl. Phys. **91**, 10110 (2002)
14. Hasegawa, T., Terabe, K., Tsuruoka, T., Aono, M.: Adv. Mater. **24**, 252 (2012)
15. Ohno, T., Hasegawa, T., Tsuruoka, T., Terabe, K., Gimzewski, J.K., Aono, M.: Nat. Mater. **10**, 591 (2011)
16. Esch, F., Fabris, S., Zhou, L., Montini, T., Africh, C., Fornasiero, P., Comelli, G., Rosei, R.: Science. **309**, 752 (2005)
17. Diebold, U.: Appl. Phys. A Mater. Sci. Process. **76**, 681 (2003)
18. Setvin, M., Aschauer, U., Hulva, J., Simschitz, T., Daniel, B., Schmid, M., Selloni, A., Diebold, U.: J. Am. Chem. Soc. **138**, 9565 (2016)
19. Castell, M.R., Wincott, P.L., Condon, N.G., Muggelberg, C., Thornton, G., Dudarev, S.L., Sutton, A.P., Briggs, G.A.D.: Phys. Rev. B. **55**, 7859 (1997)
20. Castell, M.: Surf. Sci. **516**, 33 (2002)
21. Valov, I., Sapezanskaia, I., Nayak, A., Tsuruoka, T., Bredow, T., Hasegawa, T., Staikov, G., Aono, M., Waser, R.: Nat. Mater. **11**, 530 (2012)
22. Sterrer, M., Freund, H.: Surf. Interface Sci. **3**, 229–278 (2014). Volume 3: Properties
23. Valov, I., Lu, W.D.: Nanoscale. **8**, 13825–13827 (2016)
24. Valov, I., Lu, W.: Nanoscale. **8**, 13828 (2016)
25. Waser, R., Aono, M.: Nat. Mater. **6**, 833 (2007)
26. Valov, I.: ChemElectroChem. **1**, 26 (2014)
27. Bredikhin, S.I., Bondarev, V.N., Boris, A.V., Pikhitsa, P.V., Weppner, W.: Solid State Ionics. **81**, 19 (1995)
28. Scrosati, B.: J. Appl. Chem. Biotechnol. **21**, 223 (1971)
29. Moors, M., Adepalli, K.K., Lu, Q., Wedig, A., Bäumer, C., Skaja, K., Arndt, B., Tuller, H.L., Dittmann, R., Waser, R., Yildiz, B., Valov, I.: ACS Nano. **10**, 1481 (2016)
30. Nayak, A., Tamura, T., Tsuruoka, T., Terabe, K., Hosaka, S., Hasegawa, T., Aono, M.: J. Phys. Chem. Lett. **1**, 604 (2010)
31. Nayak, A., Tsuruoka, T., Terabe, K., Hasegawa, T., Aono, M.: Nanotechnology. **22**, 235201/1 (2011)
32. Wedig, A., Luebben, M., Cho, D.-Y., Moors, M., Skaja, K., Rana, V., Hasegawa, T., Adepalli, K., Yildiz, B., Waser, R., Valov, I.: Nat. Nanotechnol. **11**, 67 (2016)
33. Frumkin, A.N.: Z. Phys. Chem. **164A**, 121 (1933)
34. Delahay, P.: Double Layer and Electrode Kinetics. Wiley, New York (1965)
35. Erdey-Gruz, T., Volmer, M.: Zeitschrift für Physikalische Chemie A (Leipzig). **150A**, 203 (1930)
36. Milchev, A.: Nanoscale. **8**, 13867 (2016)

# Atomistic Simulations for Understanding Microscopic Mechanism of Resistive Switches

**S. Watanabe and B. Xiao**

**Abstract** In this chapter, we describe the results of our first-principles simulations to investigate the switching mechanism of amorphous $TaO_x$ (a-$TaO_x$) based resistive switching devices. For the Cu/a-$Ta_2O_5$/Pt atomic switch, we first discuss the atomic structure of the conductive filaments, focusing on the exploration of possible thinnest filament structure. Then we discuss the structures of interfaces between metal electrodes and a-$Ta_2O_5$, which are important in understanding Cu ion supply for the switching. For the Pt/a-$TaO_x$/Pt resistive switch, we discuss the nature of the conductive filaments and diffusion behaviors of active ions. Here we point out the importance of Ta-Ta bonding and the non-negligible contribution of Ta diffusion under certain conditions.

## 1 Introduction

The resistive switches based on metal oxide have got wide attention because of their excellent retention, long endurance and low power consumption. Such devices are composed of an insulator layer ($HfO_x$, $TiO_x$, $TaO_x$, *etc.*) between two electrodes [1–3]. On the basis of different switching behaviors, the devices could be divided into the following two categories: (1) unipolar switches (such as Pt/$TaO_x$/Pt), where the insulator material is sandwiched between two inert electrodes [4], and the switching is realized by changing the magnitude of the applied electric field with keeping its polarity; (2) bipolar switches (including the atomic switches such as Cu/$Ta_2O_5$/Pt), which have asymmetric structures with inert and oxidizable electrodes [5], and change their states by applying the electric fields with different polarities.

S. Watanabe (✉)
Department of Materials Engineering, The University of Tokyo, Tokyo, Japan
e-mail: watanabe@cello.t.u-tokyo.ac.jp

B. Xiao
School of Chemistry and Chemical Engineering, Yantai University, Yantai, China

 
M. Aono (ed.), *Atomic Switch*, Advances in Atom and Single Molecule Machines,
https://doi.org/10.1007/978-3-030-34875-5_6

Among the oxide-based resistive switching devices, those including amorphous-$TaO_x$ (a-$TaO_x$) were widely studied, which exhibit excellent performance [3–5]. In general, the switching of these devices could be attributed to the forming/rupture of conduction filaments (CFs). However, the atomistic details of switching mechanisms of a-$TaO_x$-based resistive switches has been unclear until recently. More specifically, (1) in a-$TaO_x$-based bipolar switches such as Cu/a-$Ta_2O_5$/Pt, the component of CF had been experimentally identified to be Cu [5], while the atomic structure of CF was unclear: the ionization of interface Cu had been found to be the rate-limiting step during the switching process of the device, while the detail atomic structure near the interface region was still unknown, too [5]. (2) In the Pt/a-$TaO_x$/Pt resistive switch, the atomic component of CF, that is, whether it consists of O vacancies or Ta clusters, was still unknown [4]; on the other hand, how the active ions (such as O and Ta) diffuse in a-$TaO_x$ was also unclear.

We tackled with the above issues via first-principles simulations based on the density functional theory (DFT) [6–11]. In this chapter, we discuss the atomic structures of the CFs, the interface structures, and the diffusion behaviors of active ions in a-$TaO_x$-based resistive switching devices on the basis of our simulation results.

# 2 Computation Methods and Models

## *2.1 Methods*

The calculation of structure relaxation, electronic properties and molecular dynamics (MD) simulations were carried out using the Vienna ab initio simulation package (VASP) [12, 13]. The projector augmented-wave (PAW) [14] method and the generalized gradient approximation (PW91) [15] were adopted to describe the atomic core electrons and electron-electron interactions, respectively.

The calculations of electronic transport properties were carried out via the Atomistix Tool-kit (ATK) program [16]. A numerical atomic basis set, a single-$\zeta$ basis with polarization, was used to solve the Kohn–Sham equations. The Perdew-Burke-Ernzerhof (PBE) functional form within the generalized gradient approximation (GGA) was adopted for electron-electron interaction [11].

## *2.2 Structure of Amorphous $Ta_2O_5$*

In real tantalum-oxide-based resistive switching devices, $TaO_x$ is always in the amorphous phase. Therefore, we performed most of our simulations assuming the amorphous $TaO_x$ (a-$TaO_x$) phase. We first built an a-$Ta_2O_5$ (i.e. a-$TaO_{2.5}$) structure by melt-quenching method using VASP. More specifically, the initial crystalline $\delta$-$Ta_2O_5$ structure was melted by MD simulation at 6000 K for 9 ps, and

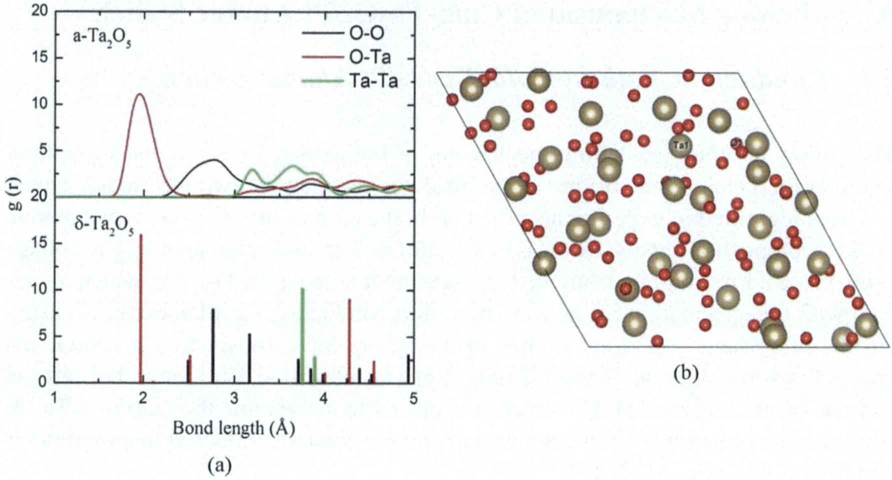


**Fig. 1** (**a**) Partial pair correlation functions of amorphous and crystal $Ta_2O_5$ (reproduced from Ref. [6]) and (**b**) the top view of the a-$Ta_2O_5$ structure

subsequently quenched down to room temperature (300 K) with the rate of 4 K/3 ps. After that, the as-generated amorphous structure was further equilibrated at room temperature for 9 ps.

The amorphous model generated in this way has the following structural features [6]. As seen in the calculated partial pair correlation functions of amorphous and crystalline $Ta_2O_5$ (Fig. 1a), the first peak in a-$Ta_2O_5$ is located at 1.93 Å, which corresponds to the O-Ta bond length and is similar to that in δ-$Ta_2O_5$. On the other hand, the bond length of Ta-Ta (3.20 Å) in a-$Ta_2O_5$ is shorter than that in δ-$Ta_2O_5$ (3.67 Å), which is attributed to the existence of considerable $Ta_2O_2$ quadrangle units in a-$Ta_2O_5$.

Next, we pay attention to the coordination numbers (CNs) of Ta and O atoms in a-$Ta_2O_5$. Here, a pair of Ta and O is regarded as bonded when their interatomic distance is less than the sum of the covalent radii plus a tolerance factor of 0.10 Å, with the covalent radii of Ta = 1.70 Å and O = 0.73 Å. As for Ta atoms, their first shell is coordinated with O atoms by the formation of octahedral $TaO_6$ and pyramidal $TaO_5$ units with their ratio of 20:9. In the case of O atoms, both $OTa_3$ and $OTa_2$ structures could be found, with their ratio of 25:55. It should be emphasized that these structural parameters, bond lengths and CNs, agree with the previous theoretical and experimental studies [12–14], indicating the reliability of the a-$Ta_2O_5$ structure obtained in our studies. In addition, we have also examined the effects of quenching speed on the structure of a-$Ta_2O_5$. It was found that the structural parameters of a-$Ta_2O_5$ only slightly change with the quenching rate among 4/3 K/fs, 2/3 K/fs, and 1/3 K/fs [6].

Using this a-$Ta_2O_5$ model, we have explored the atomic structures of CFs in Cu/a-$Ta_2O_5$/Pt and Pt/a-$TaO_x$/Pt resistive switches [6, 7], the interface structures of Cu/a-$Ta_2O_5$ and Pt/a-$Ta_2O_5$ [8], and the diffusion behaviors of metal and O ions in a-$TaO_x$ [9–11].

# 3 Switching Mechanism of Cu/a-$Ta_2O_5$/Pt Atomic Switch

## 3.1 *Conduction Path in Cu/a-$Ta_2O_5$/Pt Atomic Switch*

The widely accepted switching mechanism of the atomic switch is the formation/rupture of a metal atomic bridge in the insulator between the two electrodes, which has been identified in experiments [17–19]. In the case of the a-$Ta_2O_5$-based atomic switch, composition analyses of a-$Ta_2O_5$ film before and after applying a voltage show that Cu ions migrate from the Cu electrode into the a-$Ta_2O_5$ film, which agrees well with the speculation of the formation of a conducting Cu filament in a-$Ta_2O_5$. On the other hand, previous studies in our group have shown that a conduction channel is formed in the crystal $Ta_2O_5$ film via interstitial Cu atoms, but not via oxygen vacancies [20, 21]. However, it is desirable to examine the case of a-$Ta_2O_5$ since, as mentioned before, the amorphous phase is usually adopted in experiments and prototype devices [12, 13].

### 3.1.1 Single Cu Atomic Chains in a-$Ta_2O_5$

In our study to explore the possible conductive paths in the a-$Ta_2O_5$ structure [6], both atomic positions and lattice constants of Cu doped a-$Ta_2O_5$ structures were fully relaxed. Keeping the previous studies using the Cu-doped crystalline $Ta_2O_5$ [20, 21] in mind, we examined both alternate Ta-Cu atomic chain and continued Cu atomic chain as the candidates of conductive path in a-$Ta_2O_5$ via interstitial Cu atoms or substituting O by Cu atoms. After structural optimization, both the alternate Ta-Cu and continued Cu atomic chains were found in a-$Ta_2O_5$ via substituting O with Cu atoms as shown in Fig. 2a,b. The analysis of the density of states (DOS) indicates the formation of conductive, delocalized defect states near the Fermi level, which can serve as conductive paths. However, both structures were easily destroyed after MD simulations at 500 K. Cu atoms prefer bonding together to single atomic chain arrangements, and tend to form cluster or nanowire structures. This gives us a hint for the formation of Cu filaments in a-$Ta_2O_5$.

### 3.1.2 Cu Nanowires in a-$Ta_2O_5$

Cu structures with bigger diameters such as Cu nanowires have been extensively studied both theoretically and experimentally in environments other than the a-$Ta_2O_5$ matrix [22, 23]. In the present studies [6], two relatively thin Cu nanowires with the interlaced trigonal and tetragonal packing were chosen to be inserted into the a-$Ta_2O_5$ along c-axis (perpendicular to the a-b plane in the unit cell). Although the structures of these two Cu nanowires changed in some degree after structural relaxation, the continued Cu-Cu bonding structures were still observed, and the subsequent MD simulations further confirm their stability. However, we should note

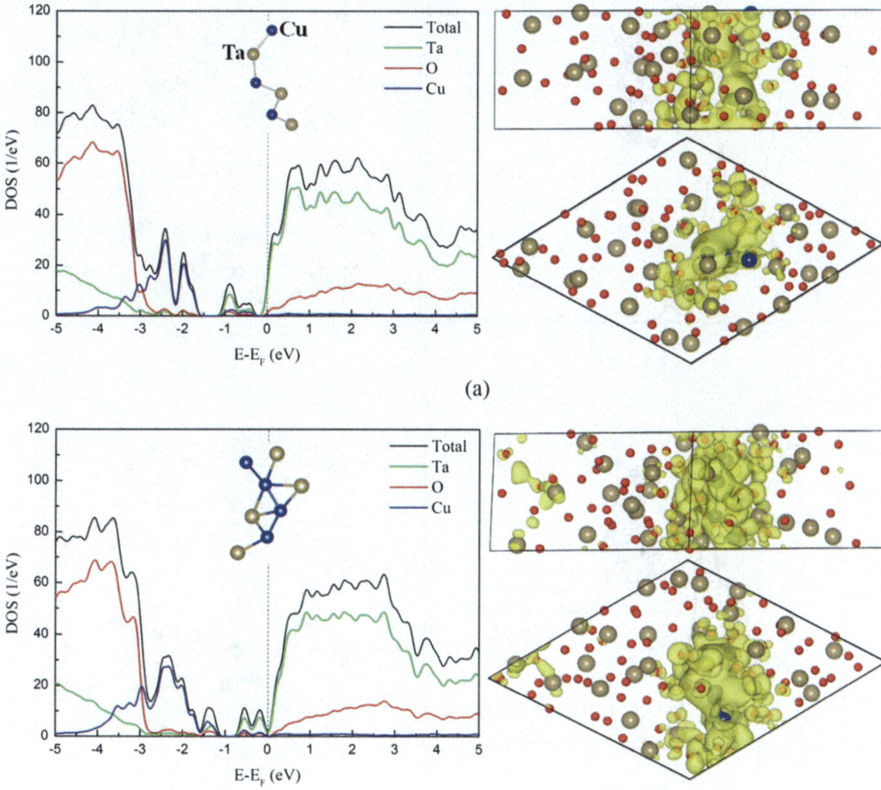


**Fig. 2** Density of states (adapted from Ref. 6) and local density of states near the Fermi level of a-$Ta_2O_5$ with (**a**) alternate Ta-Cu and (**b**) continued Cu-Cu atomic chains

that MD simulations may be insufficient to confirm their stability due to the short time scale compared with that in the practical use. Thus, we also examined their thermodynamic stability, by evaluating the insertion (formation) energies of both structures [6]. The calculated values of insertion (formation) energies are −0.45 (0.48) and −0.29 (0.54) eV, respectively, which suggests that (1) such thin Cu filaments could be stable in a-$Ta_2O_5$, while (2) the Cu atoms in thin filaments are less stable than that in the bulk system. It should be noted that the Cu nanowire structures have been widely observed in experiments [23], even though they are less stable than the Cu bulk system. Thus, we can expect that both of the structures (Fig. 3a,b) can exist in a-$Ta_2O_5$ for sufficiently long periods.

Next, we examine the electronic properties of the Cu nanowires inserted in a-$Ta_2O_5$. In the local density of states near the Fermi level shown in Fig. 3a,b, the conductive paths could be clearly observed. Our projected DOS analyses reveal that the defect states are mainly ascribed to the Cu atoms or Cu-Cu bonding structures.

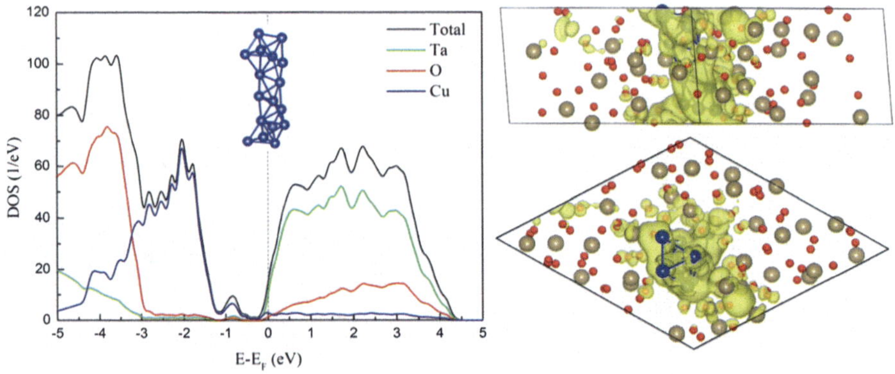


(a)

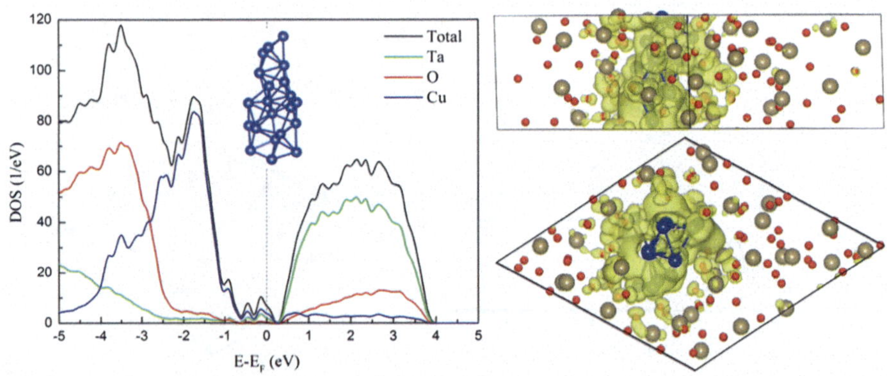


(b)

**Fig. 3** Density of states (adapted from Ref. 6) and local density of states near the Fermi level of a-$Ta_2O_5$ with interlaced (**a**) trigonal and (**b**) tetragonal packed nanowires

Thus, we can say that the conduction in the a-$Ta_2O_5$-based atomic switch is ascribed to the formation of continued Cu-Cu bonding.

### 3.1.3 The Thinnest Cu Filament in a-$Ta_2O_5$

Nowadays, downscaling of nanodevices becomes more and more challenging due to physical limitations and increasing processing complexity. So we have explored the thinnest Cu filament that could exist in the a-$Ta_2O_5$, which would be helpful in downscaling of the Cu/$Ta_2O_5$/Pt atomic switch. For this purpose, the following scheme was adopted: starting from the a-$Ta_2O_5$ structure containing a thick Cu nanowire (Fig. 3a), some of Cu atoms were removed from the system each time, and the MD simulation at room temperature and structure relaxation were carried out after each removal step. The removal process was continued until the Cu filament becomes disconnected in a-$Ta_2O_5$. As can be seen in Fig. 4, the thinnest Cu filament generated in this way is composed of the three-membered ring structures with linear

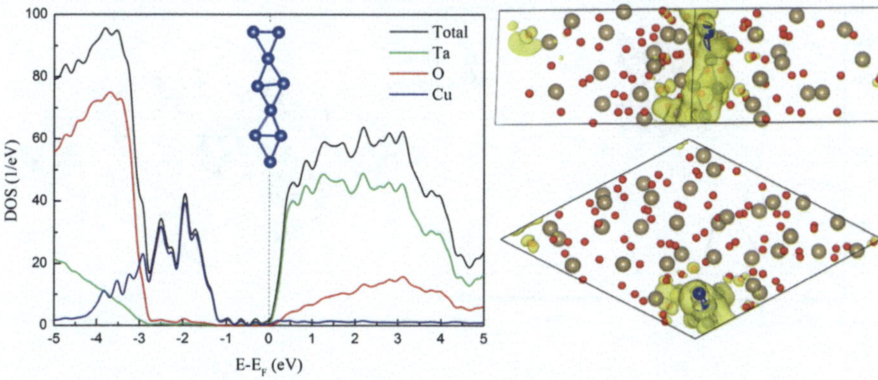


**Fig. 4** Density of states (reproduced from Ref. [6]) and local density of states near the Fermi level of a-$Ta_2O_5$ with the thinnest Cu nanowire in a-$Ta_2O_5$

arrangement. The formation energy of this Cu filament in a-$Ta_2O_5$ is 0.38 eV per supercell. The DOS and partial charge density analyses (Fig. 4) reveal that the thinnest Cu filament is conductive, which agrees well with the "Cu-Cu bonding" conductive mechanism mentioned above.

### 3.1.4 Cu Filament in a-$Ta_2O_5$ with Nanopore

Experimental studies suggest that a-$Ta_2O_5$ films sputtered at room temperature generally have nanoporous structures consisting of piled or small grains, and Cu ions are most likely to migrate along the grain boundaries than inside the grains due to the relatively low migration barriers at the former [5]. Thus the Cu filament is most likely to be formed inside the nanopores. Meanwhile, we would like to emphasize that the atomic density of Cu in our previous calculations is rather high, and it is necessary to investigate the a-$Ta_2O_5$ structure with low Cu concentration. Accordingly, a big stoichiometric a-$Ta_2O_5$ with a nanoporous structure (a-$Ta_{124}O_{310}$) was constructed based on the previously constructed a-$Ta_2O_5$ model (i.e. the model used in the studies of the previous subsections). In this case, most of Ta atoms near the sidewall of nanopore are unsaturated with the O coordination numbers of only 4 or 5. Subsequently, a thicker Cu nanowire with interlaced centered-hexagon packing was inserted into the nanopore. After structural optimization, the conductive path was clearly observed on the Cu filament as shown in Fig. 5. The defect states near the Fermi level have the contributions from both Cu and Ta atoms, indicating the existence of certain Ta-Cu bonds on the sidewall of Cu filaments. It should be noted that in this case (Fig. 5), DOS around the Fermi level are more delocalized as compared with all the others in the present studies due to the increased thickness of Cu filaments. In fact, the increase of number of conduction channels with the increase in the thickness has been reported for isolated Cu nanowires and Cu nanowires encapsulated in a boron nitride nanotube [24]. So we predict that the thick Cu filament is more conductive than the others.

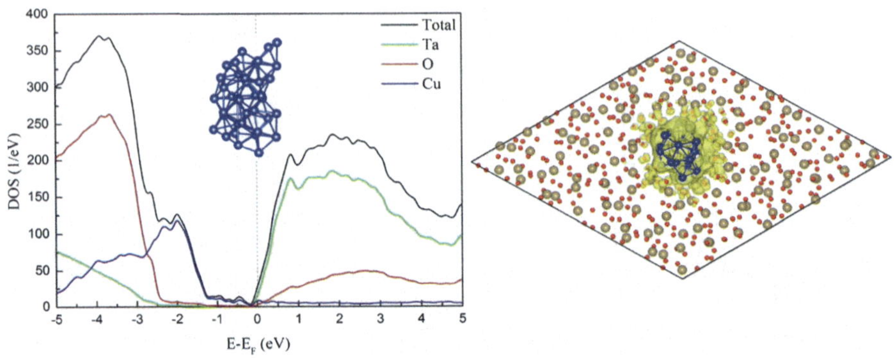


**Fig. 5** Density of states (DOS) and local density of states near the Fermi level of a-$Ta_2O_5$ with the interlaced centered-hexagon packed Cu nanowire in a-$Ta_2O_5$

It is also worth mentioning that the insertion (formation) energy of Cu filament in a-$Ta_2O_5$ (Fig. 5) is −0.94 (−0.46) eV. Accordingly, we can say that the thick Cu nanowire structure in a-$Ta_2O_5$ is thermodynamically stable.

### 3.1.5 Transport Properties of Cu/a-$Ta_2O_5$/Pt with and Without Cu Filament

Next, we examine the electronic and transport properties of Cu nanowires in the Cu/a-$Ta_2O_5$/Pt heterostructure shown in Fig. 6 [6]. We can see that the electron conduction in Cu/a-$Ta_2O_5$/Pt heterostructure is poor due to the insulating nature of a-$Ta_2O_5$. On the other hand, the electron conduction is enhanced by inserting the Cu filament in the a-$Ta_2O_5$ between two Cu and Pt electrodes as seen in Fig. 6c. In addition, when a thicker Cu nanowire with the interlaced centered-hexagon packing is used as the conductive filament, the transmission coefficient further increases (Fig. 6d), which agrees with the thickness dependence of the conduction in Cu nanowires [24]. It is also worth noting that the Cu/a-$Ta_2O_5$/Pt with a discontinued Cu filament (Fig. 6b) has smaller conduction than those with the Cu filaments bridging two electrodes.

In summary of this subsection, we have examined the structure of conduction filaments (CFs) in the Cu/a-$Ta_2O_5$/Pt resistive switch from first principles. Our results reveal that Cu nanowires with various diameters are stable in the a-$Ta_2O_5$ and can serve as CFs. In this case, the Cu-Cu bonding mainly contributes to the conductive, delocalized defect states.

## *3.2 Interface Structures of Cu/a-$Ta_2O_5$/Pt*

In the widely accepted switching mechanism of the Cu/a-$Ta_2O_5$/Pt resistive switch [17–19], there are three main processes: (1) the ionization of Cu at the Cu/a-$Ta_2O_5$ interface, (2) the migration of Cu ions through the a-$Ta_2O_5$ layer, and (3) the

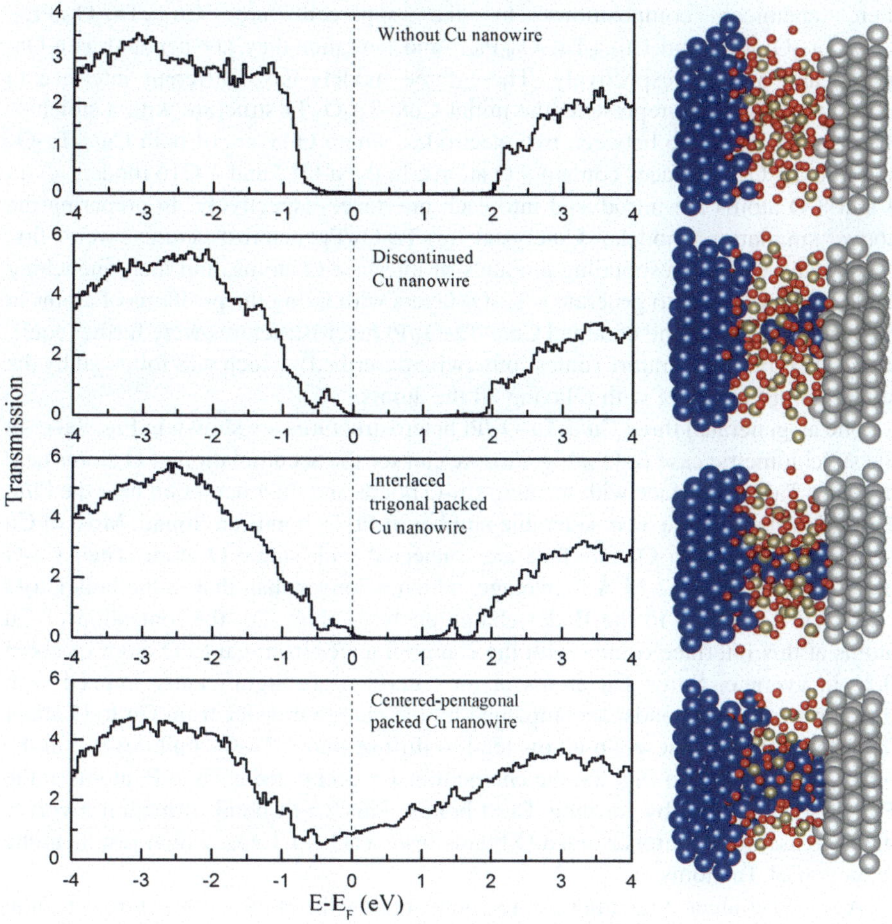


**Fig. 6** Transmission spectra and the corresponding structures for the Cu/a-$Ta_2O_5$/Pt with and without the Cu filament between two electrodes (adopted from Ref. [6])

precipitation of Cu on Pt electrode [5]. The initial ionization of Cu atoms in the Cu/a-$Ta_2O_5$ interface is important in the performance of the switch, in the sense that the ionized Cu atoms are the main source of Cu ions used in the subsequent processes.

Considering the above, in this subsection, we pay attention to the interface structures and electronic properties of Cu/a-$Ta_2O_5$/Pt heterostructure, and their dependence on the interface O concentration and temperature [8].

### 3.2.1 Interface Structures and Electronic Properties of Cu/a-$Ta_2O_5$/Pt Structure

To deepen the understanding on the structural features near the interface between Cu (or Pt) and a-$Ta_2O_5$, three types of Cu/a-$Ta_2O_5$/Pt heterostructures were constructed:

their chemical compositions in the supercell are $Cu_{108}Ta_{32}O_{80}Pt_{84}$, $Cu_{108}Ta_{32}O_{88}Pt_{84}$ and $Cu_{108}Ta_{32}O_{96}Pt_{84}$, and hereafter they are denoted as a-O8, a-O12 and a-O16, respectively. These three models have different interface O concentrations: a-O8 represents the initial Cu/c-$Ta_2O_5$/Pt structure with a stoichiometric c-$Ta_{32}O_{80}$ slab between two electrodes, where O layers of both Cu/c-$Ta_2O_5$ and Pt/c-$Ta_2O_5$ interfaces contain 8 O atoms; In the a-O12 and a-O16 models, extra 4 and 8 O atoms are introduced into each interface, respectively. In preparing the above simulation models, Cu/crystal (c)-$Ta_2O_5$/Pt heterostructures were first constructed with corresponding amounts of interface O atoms, and melt-quenching cycle was performed to generate a-$Ta_2O_5$ layers with fixing the positions of atoms in the electrodes. Then the obtained Cu/a-$Ta_2O_5$/Pt heterostructures were further equilibrated at room temperature (unless otherwise described), which was followed by the structural optimization with relaxing all the atoms.

The as-generated three Cu/a-$Ta_2O_5$/Pt heterostructures are shown in Fig. 7a–c. In the stoichiometric case (a-O8, Fig. 7a), we can see the accumulation of O atoms near the Cu/a-$Ta_2O_5$ interface with forming Cu-O bonds and their depletion near the Pt/a-$Ta_2O_5$ interface where a considerable number of Pt-Ta bonds are found. Most of Cu atoms at the Cu/a-$Ta_2O_5$ interface are connected with single O atom. Their Cu-O bond length is about 2.15 Å in average, which is longer than that in the bulk $Cu_2O$ (1.85 Å). According to the Bader charge analyses (Fig. 8a), the ionization of Cu atoms at this interface occurs with the charge transfer from Cu to O atom of about 0.2e in average. Some Cu atoms at the interface are significantly ionized with forming two Cu-O bonds, accompanied by the charge transfer from Cu to O about 0.4e. Interestingly, these Cu atoms tend to diffuse into a-$Ta_2O_5$ bulk layer. On the other hand, as seen in Fig. 8a, the charge transfer occurs from Ta to Pt atoms at the Pt/a-$Ta_2O_5$ interface by forming Ta-Pt bonds. This Ta-Pt bond formation involves the decrease in the number of Ta-O bonds from $TaO_5$ to $TaO_4$, which results in the reduction of Ta atoms.

As the O content at the interface region increases, the interface structures of Cu/a-$Ta_2O_5$ show larger distortion as seen in Fig. 7b (a-O12) and 7c (a-O16). Interestingly, Ta-O-O bonding structures are seen at the interface region due to the high interface O content, and the O ions in Ta-O-O bond hold less amount of electrons than the others, as can be seen in the Bader charges in a-O16 model (Fig. 8b). It is worth noting that a similar Ta-O-O bonding structure has been identified in Cu/a-$Ta_2O_5$/Pt atomic switch in experiment [25]. It is also noted that, in the a-O12 and a-O16 structures, considerable amount of Cu atoms at the Cu/a-$Ta_2O_5$ interface have diffused into the a-$Ta_2O_5$ bulk region with forming two (predominant) or three Cu-O bonds. The corresponding Cu-O bond length is about 1.90 Å in average, which is a little longer than that in bulk $Cu_2O$ (1.85 Å). For example, in the a-O16 structure, the Cu atoms at the interface region are oxidized with losing about 0.55 electrons in average (Fig. 8b). Since the Bader charge values in the Cu/a-$Ta_2O_5$/Pt systems are about half of the corresponding formal values (for example, the Bader charges of Ta and O atom are about +2.5 and $-1$e, respectively, while the formal charges are Ta (+5e) and O ($-2$e) ions in a-$Ta_2O_5$), the Cu atoms at the interface region with 0.55e could be regarded as $Cu^{1.1+}$.

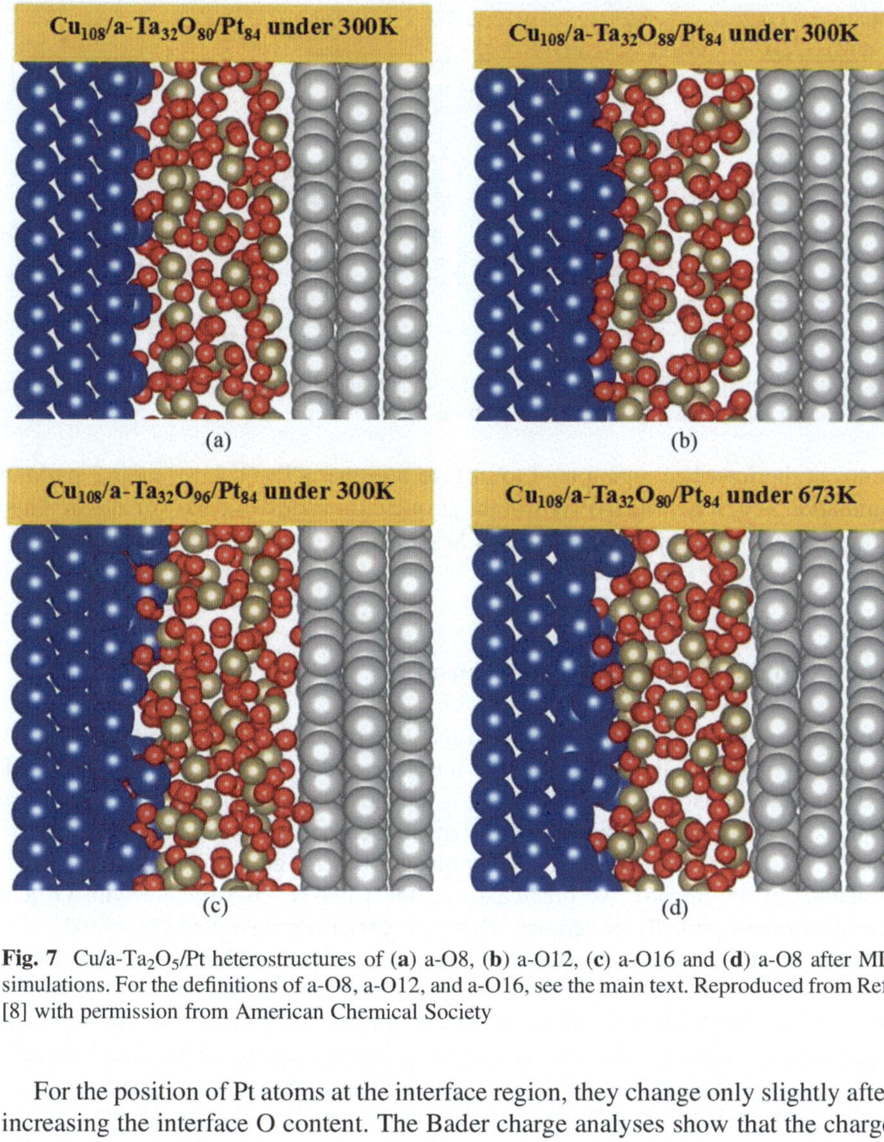


**Fig. 7** Cu/a-$Ta_2O_5$/Pt heterostructures of (**a**) a-O8, (**b**) a-O12, (**c**) a-O16 and (**d**) a-O8 after MD simulations. For the definitions of a-O8, a-O12, and a-O16, see the main text. Reproduced from Ref. [8] with permission from American Chemical Society

For the position of Pt atoms at the interface region, they change only slightly after increasing the interface O content. The Bader charge analyses show that the charge transfers from Pt to O and Ta atoms are about 0.20 and 0.25e, respectively.

Next, we discuss the effect of temperature, because the temperature is considered to be one of the most important factors to affect the switching process of Cu/a-$Ta_2O_5$/Pt device. For example, with the increase of operation temperature, the absolute values of SET voltage of resistive switches gradually decrease [5], which could be partly attributed to the faster diffusion of Cu ions in the devices at high temperature. Meanwhile, the ionization of Cu at the interface region may also be enhanced due to the faster Cu diffusion into the a-$Ta_2O_5$ layer at high temperature. We have examined this point using the a-O8 model. As can be seen in Fig. 7d, the Cu

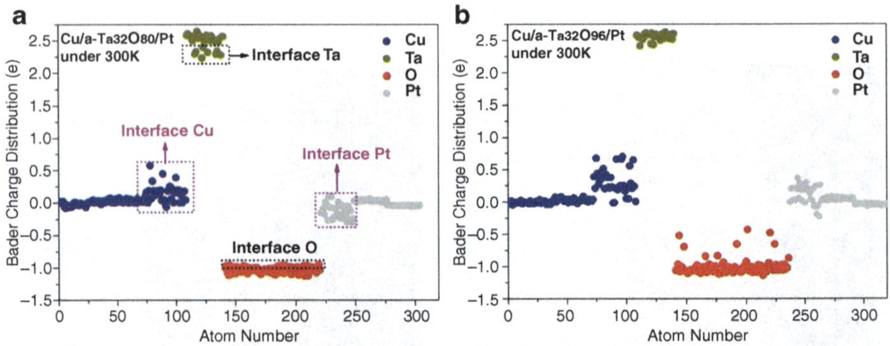


**Fig. 8** Bader charge distributions of (**a**) Cu/a-$Ta_{32}O_{80}$/Pt and (**b**) Cu/a-$Ta_{32}O_{96}$/Pt. Adapted from Ref. [8] with permission from American Chemical Society

atoms at the interface region tend to diffuse into the a-$Ta_2O_5$ layer after 2 ps MD simulation at 673 K, which results in the strong oxidization of these Cu atoms. This confirms that the temperature is one of key factors to control the ionization of Cu atoms at the Cu/a-$Ta_2O_5$ interface.

### 3.2.2 Stability of Cu/a-$Ta_2O_5$/Pt Structure

We next examine the stability of Cu/a-$Ta_2O_5$/Pt structures with various interface O concentrations [8]. Figure 9 shows the interface energies of the a-O8, a-O12 and a-O16 models. It is found that the a-O12 is the most stable among the three within a wide range of O chemical potential. As mentioned before, the content of O at the Cu/a-$Ta_2O_5$ interface is obviously higher than the other regions in the Cu/a-$Ta_2O_5$/Pt structure. In this regard, we predicted that the Cu/a-$Ta_2O_5$ structure with O-rich interface is energetically preferable, where the $Cu_2O$ structure could be found.

Next, we discuss the electronic properties of Cu/a-$Ta_2O_5$/Pt heterostructure. Here, we consider only the a-O12 model for simplicity. As seen in Fig. 10, the DOS of the interface regions is finite around the Fermi level. This means that the interface regions show metallic behavior, and the gap states responsible for this behavior mainly come from the interfacial Cu-O and Pt-O bonding with a very small contribution from the interface Ta atoms as can be seen from Fig. 10. In the central region, on the other hand, both O and Ta atoms have no contribution around Fermi level (see Fig. 10), indicating that the central region of the a-$Ta_2O_5$ is insulating. Thus, the Cu/a-$Ta_2O_5$/Pt interface forms a Schottky contact.

### 3.2.3 Schottky Barrier Height of Cu/a-$TaO_x$/Pt Structure

As we know, the Schottky barrier height (SBH) is the important quantity that should be considered in controlling the initial set bias voltage of Cu/a-$Ta_2O_5$/Pt atomic

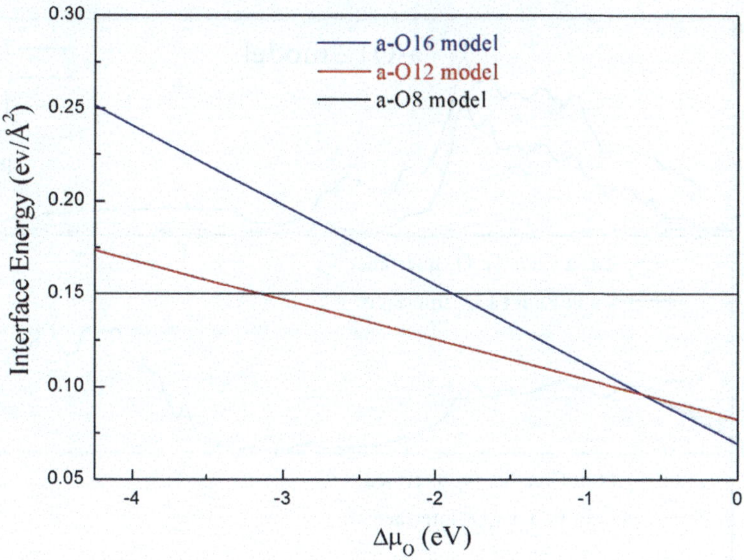


**Fig. 9** Interface energies of a-O8, a-O12 and a-O16 models. Adapted from Ref. [8] with permission from American Chemical Society

switches. In general, SBH could be estimated from the partial density of states (PDOS) as the difference between the Fermi level and valence band maximum (VBM) of the bulk like layer. However, it is difficult to estimate the SBH for our a-$Ta_2O_5$ heterostructures because of their relatively small thickness. In addition, the distribution of macroscopically averaged electrostatic potentials is fluctuant in these heterostructures due to the disordered arrangement of atoms.

Considering the above, we constructed the Cu/λ-$Ta_2O_5$/Pt model with well-defined interfaces. Here the λ-$Ta_2O_5$ is the most stable structure of crystal $Ta_2O_5$ at low temperature [26]. To construct the Cu/λ-$Ta_2O_5$/Pt heterostructure, Cu and Pt (111) surfaces were connected to the (001) surface of λ-$Ta_2O_5$ with O-termination. The lattice mismatch is less than 1%. The corresponding chemical composition of this model is $Cu_{69}/Ta_{36}O_{102}/Pt_{63}$, which is labeled as c-O12 since the λ-$Ta_2O_5$ slab cleaved contains 12 O atoms in the interface layer. Figure 11 shows the PDOS of the Cu/λ-$Ta_2O_5$/Pt models with various O concentrations at the interfaces. From this figure, the SBHs can be obtained as the difference between the Fermi level and VBM in the bulk-like layer. It should be noted that VBM can be more accurately computed than the conduction band minimum within DFT [27]. The estimated p-type SBH is 1.1 eV for the c-O12 case. The large values of p- and n-type SBHs reveal that when the voltage applied to the Cu or Pt electrode is not so large, no conduction channel for the electron flow is formed in this heterostructure.

Next, the SBH in the Cu/λ-$Ta_2O_5$/Pt was calculated as a function of the interface O concentration. In doing so, the corresponding O atoms (from 0 to 50%) were removed from both c-O12 interfaces. As seen in Fig. 11, the PDOS analysis reveals

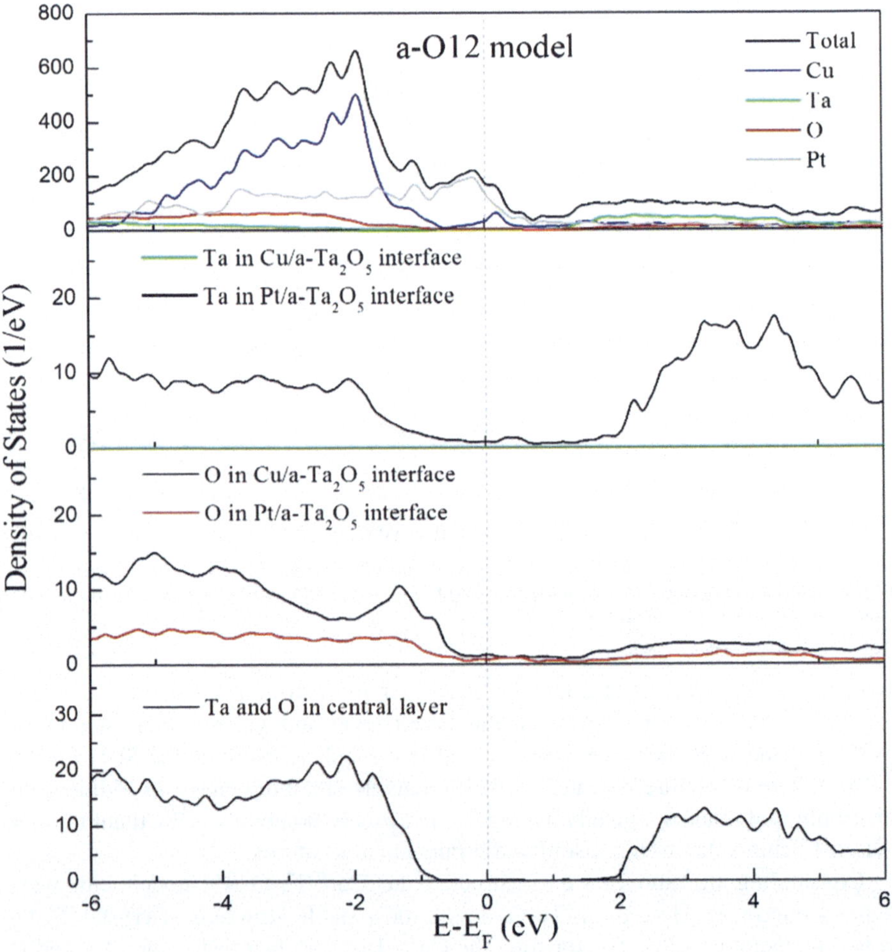


**Fig. 10** Density of states of a-O12 model. Adapted from Ref. [8] with permission from American Chemical Society

that O atoms mainly contribute to the VBM of λ-$Ta_2O_5$, and thus the amount of interface O atoms has a great impact on the VBM of metal oxides. As a consequence, the valance band offset (VBO) increases with the reduction of the amount of interface O atoms, and the VBO can increase up to 1.5 eV by gradually reducing the interface O atoms from 0 to 50%. This increase is attributed to the downward shift of the VBM of $Ta_2O_5$ due to the reduced electronegativity of the total interfacial oxygen as compared to that of the c-O12. Thus, the initial SBH of the Cu/$Ta_2O_5$/Pt atomic switch can be experimentally controlled by tuning the interface O concentration, a property directly related to the ambient oxygen pressure.

In summary of this subsection, we have discussed the interface structures of Cu/a-$Ta_2O_5$/Pt atomic switch. Our results reveal that the Cu atoms tend to be ionized at the

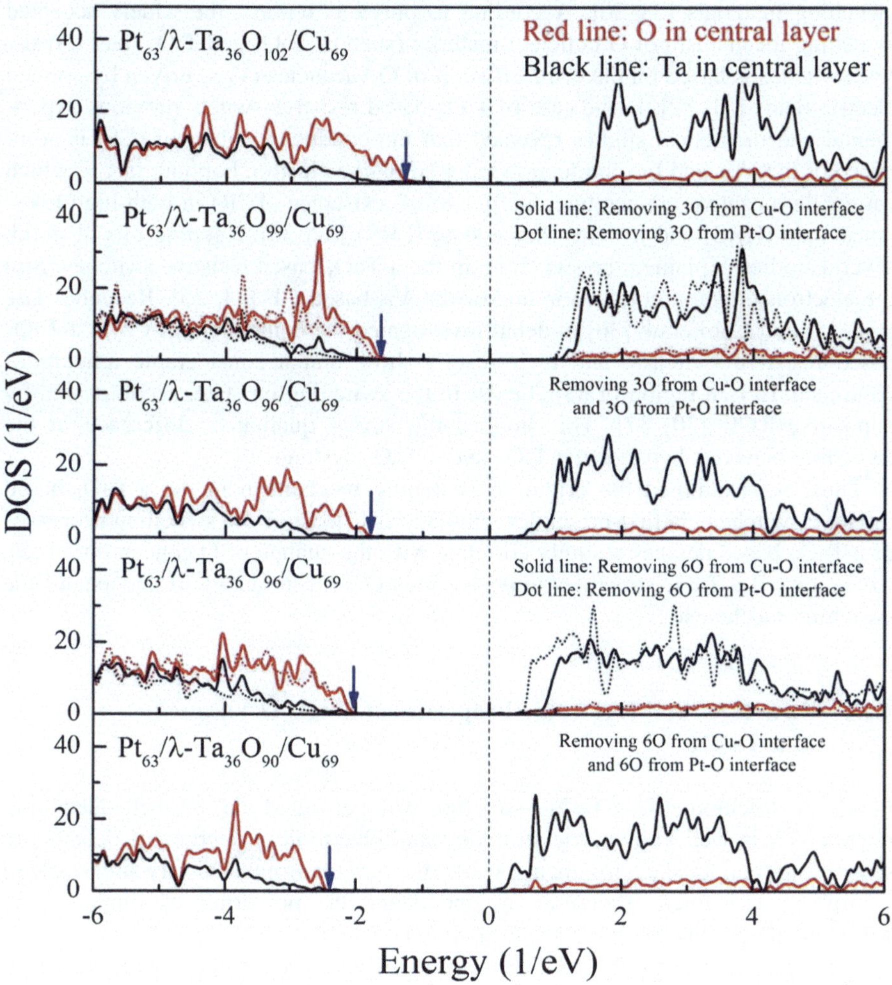


**Fig. 11** Density of States of Pt/c-$Ta_2O_5$/Pt with various interface O contents

interface of Cu/a-$Ta_2O_5$, and the ionization behavior becomes more obvious as the interface O content and/or temperature increases.

# 4 Switching Mechanism of Pt/a-$TaO_x$/Pt Resistive Switch

## *4.1 Conduction Path in Pt/a-$TaO_x$/Pt Resistive Switch*

So far, a-$TaO_x$ ($x < 2.5$) based resistive switches (i.e., Pt/$TaO_{2.5-x}$/$TaO_{2-x}$/Pt) have been extensively studied in the literature and shown superior performance to other switching materials [28–30]. According to previous reports, the widely accepted switching mechanism of O-deficient-material (such as $TaO_x$ and $TiO_x$) based resistive switches is based on the drift/diffusion of O vacancies ($V_O$s) driven by applied electric fields [31, 32]. In the case of $TiO_x$-based resistive switch, previous experimental and theoretical studies revealed that the conduction filament (CF) is composed of the $V_O$s and the conduction is based on the electron hopping in CF, which causes the negative temperature coefficient of resistance (TCR) in both high resistance state (HRS) and low resistance state (LRS) [32, 33]. Keeping this in mind, several studies explained the switching in the a-$TaO_x$ based resistive switches from the electron hopping conduction and/or the $V_O$-based CF [34, 35]. Recently, Lee et al. [5] and Choi et al. [36] in detail investigated the component of CF in a-$TaO_x$ based resistive switches, and their results show that a considerable amount of continued Ta-rich region ($TaO_{1-x}$) exist in the switching-on state, accompanied by a positive TCR [30, 37]. This implies the strong qualitative difference in the switching behavior between the $TiO_x$ and a-$TaO_x$ systems.

Thus, we examined the origin of switching mechanism in the a-$TaO_x$-based resistive switches via first-principles simulations [7]. Since the switching processes of a-$TaO_x$ based devices strongly correlate with the change of O concentration [5], we examined a-$TaO_x$ structures under various O concentrations to understand the switching mechanism.

### 4.1.1 Structures and Electronic Properties of Single O Vacancies in a-$Ta_2O_5$

Firstly, a stoichiometric a-$TaO_{2.5}$ structure was generated via the melt quenching method. As is well known, the structural and electronic properties of defects are sensitive to their local environments, while the local environments vary site by site in amorphous structures. Therefore, to understand the properties of single $V_O$ in a-$TaO_{2.5}$, all the possible vacancy sites were considered.

As can be seen from Fig. 12a, the stability of vacancy correlates with the Ta-Ta bond length, that is, a structure with a shorter bond length has a lower energy [7]. The length of the Ta-Ta dimer bond formed at the vacancy site is shorter than that in a-$TaO_{2.5}$ (3.20 Å). In particular, the bond length of Ta-Ta dimer (2.84 Å) in the most stable $V_O$ structure is comparable with the one (2.86 Å) in Ta metal. The short bond length suggests good stability of the Ta dimer at $V_O$. To further confirm this, MD simulations were performed at different temperature as shown in Fig. 12b, c. The results reveal that the Ta dimer is very stable at room temperature, while the position of such Ta dimer structure migrates into its adjacent sites at high temperature (i.e., 873 K). Thus, the motion of such a single Ta dimer structure is likely to occur under external electric fields in real devices.

To understand the electronic properties of the $V_O$ in a-$TaO_{2.5}$, we have calculated the DOS for the energetically most stable $V_O$ structure with HSE06 hybrid functional for the exchange-correlation term [38]. The results show that the defect state (mainly coming from $V_O$) appears near the Fermi level, and its location (0.8 eV and 2.5 eV

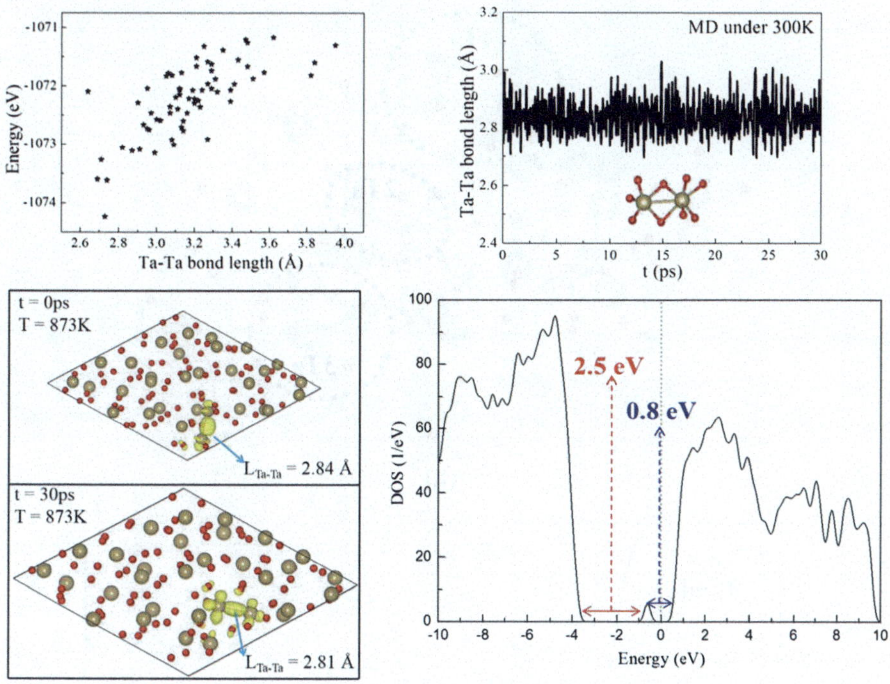


**Fig. 12** (**a**) Relationship between the stability of single O vacancy ($V_O$) and the Ta-Ta bond length at the vacancy site in amorphous $Ta_2O_5$. The change of (**b**) Ta-Ta bond lengths and (**c**) the position of Ta dimer during 30 ps molecular dynamics (MD) simulation. (**d**) DOS of the most stable single $V_O$ structure. Adapted from Ref. [7] with permission from The Royal Society of Chemistry

away from the conduction and valence bands, respectively) agrees with previous experiments [39, 40]. Note that several types of $V_O$ states were found among the 80 single $V_O$ structures in a-$TaO_{2.5}$, which may correspond to the existence of several kinds of current leakage sources in a-$TaO_{2.5}$ capacitor in experiment [41].

#### 4.1.2 Structures and Electronic Properties of a-$TaO_{2.5}$ with High $V_O$ Concentration

Next, we consider the a-$TaO_{2.5}$ with high $V_O$ concentration (such as a-$TaO_{2.25}$). The a-$TaO_{2.25}$ structure was generated via melt-quenching method after randomly removing 8 O atoms from the original a-$TaO_{2.5}$ (with 32 Ta and 80 O atoms). In the a-$TaO_{2.25}$, Ta dimer, trimer and tetramer structures coexist as seen in Fig. 13a [7]. The DOS analysis (Fig. 13b) shows the existence of defect states in the band gap region, which consist of mainly the components of these Ta clusters. The energy positions of the defect states become deeper with the increase of the size of Ta cluster, indicating the good stability of big Ta clusters. Subsequent MD simulation at

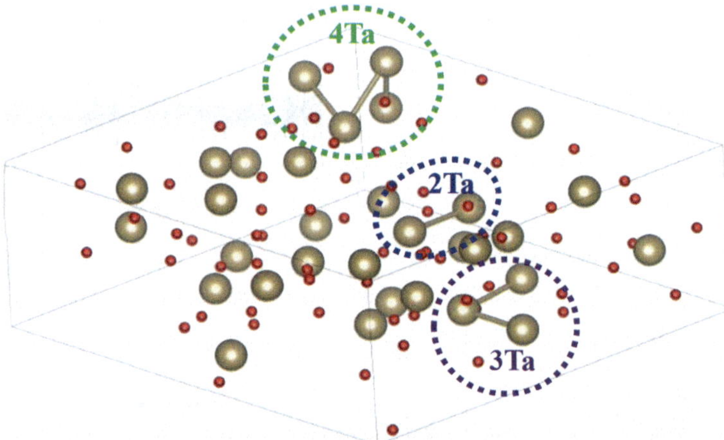


(a)

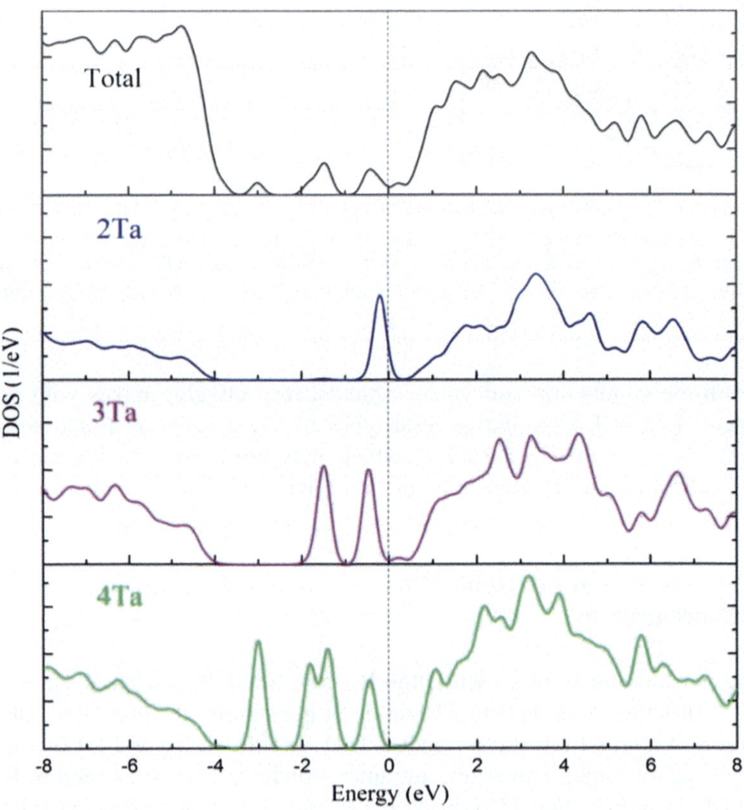


(b)

**Fig. 13** (**a**) Structure and (**b**) density of states of a-$TaO_{2.25}$

1073 K shows that the Ta dimer structures tend to merge into the Ta tetramer, which further confirms that the formation of Ta clusters is energetically preferable in the O-deficiency state of a-$Ta_2O_5$. This could be probably attributed to the simple equilibrium solid-phase diagram for the Ta-O system: there are only two stable phases below 1273 K: the single-metal-valence compound $Ta_2O_5$ and Ta metal [42].

Subsequently, in order to understand the structural evolution of Ta-O systems during operation, various a-$TaO_x$ ($x = 2.85, 2.67, 2.50, 2.25, 2, 1.5, 1$ and $0.75$) structures were examined [7]. The values of Ta-O/Ta-Ta bond lengths and Ta(O)/O (Ta) coordination numbers (CNs) as the function of O contents are shown in Fig. 14a,b, respectively. Note that Ta(O) CN denotes the number of O atoms surrounding a Ta atom. From the figure, we can see that: (1) with the increase of x values, the Ta-O bond length gradually decreases; (2) the Ta(O) CN values increase with the increase of x values until it keeps almost a constant value (corresponding to $TaO_6$) in the region of $x > 2.50$. All the above features agree well with experimental data [43], supporting the reliability of our simulation results. In addition, other features of the structural evolution can also be seen: (3) the O(Ta) CN increases with the increase of $V_O$ concentrations; (4) there is the close relationship between the O(Ta) CN and Ta-O bond length, and between the Ta(O) CN and Ta-Ta bond length; (5) considerable amount of Ta-Ta metallic bonds are formed in the case of O-deficient a-$TaO_x$, which could dramatically enhance the electronic conductivity of a-$TaO_x$ and corresponds to LRS of the a-$TaO_x$ based device [5, 28].

To understand the correlation between electronic properties of a-$TaO_x$ and O contents, the calculated DOS of all the considered a-$TaO_x$($x = 2.85, 2.67, 2.50, 2.25, 2, 1.5, 1$ and $0.75$) are shown in Fig. 15. This figure reveals that: (1) in the case of a-$TaO_x$ with $x > 2.50$, the defect states near the Fermi level mainly consist of the O-O bonds [43]; (2) a-$TaO_{2.5}$ is an insulator with a big band gap; (3) in the case of a-$TaO_x$ with $x < 2.50$, the Fermi level shift towards the conduction band due to the formation of Ta-Ta bonds, and the intensity of defect states around the Fermi level becomes stronger with the decrease of O content concentration.

As revealed in the previous experiment [28], the chemical compositions of Ta-O system for HRS and LRS in a-$TaO_x$ based device correspond to the values $x > 2$ and $x < 1$, respectively. Accordingly, we chose the a-$TaO_{2.25}$ and a-$TaO_{0.75}$ to represent of the HRS and LRS, respectively, and studied their structures and electronic properties in detail (Fig. 16). In the case of a-$TaO_{2.25}$, several Ta clusters can be found which are apart from each other, and the corresponding local density of states (LDOS) around the Fermi level (Fig. 16a) shows that the electrons are localized on these Ta clusters. Accordingly, no connected conduction path is formed. It is noted that the isolated Ta-rich regions were observed in the experiment in the HRS of a-$TaO_x$ based device [4]. On the other hand, in a-$TaO_{0.75}$ (Fig. 16b), a considerable amount of Ta-Ta bonds can be found which are connected with each other. The corresponding LDOS (Fig. 16b) reveals that electrons are delocalized on these Ta-Ta bonds, which results in the formation of conduction path. The results agree well with experiments showing the formation of a connected Ta-rich region in LRS of such device [4, 28]. In addition, the volume of a-$TaO_{2.25}$ structure is about twice of a-$TaO_{0.75}$, which agree with the experimental finding that the thickness of a-$TaO_x$

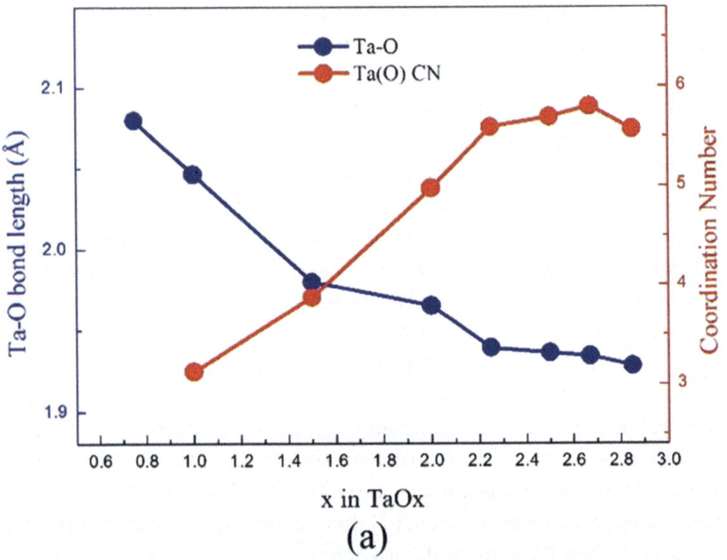


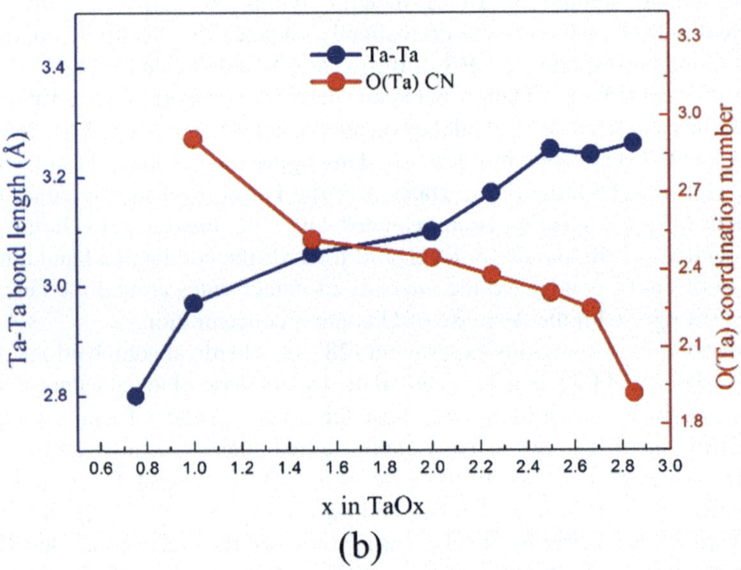


**Fig. 14** Changes of Ta-O/Ta-Ta bond lengths and Ta(O)/O(Ta) CNs dependent on x value in a-$TaO_x$. Adapted from Ref. [7] with permission from The Royal Society of Chemistry

was decreased by half after the electrical operations [37]. On the basis of these results, we conclude that the formation of Ta-Ta bonds, but not $V_O$s, is responsible for the LRS in a-$TaO_x$ based resistive switch.

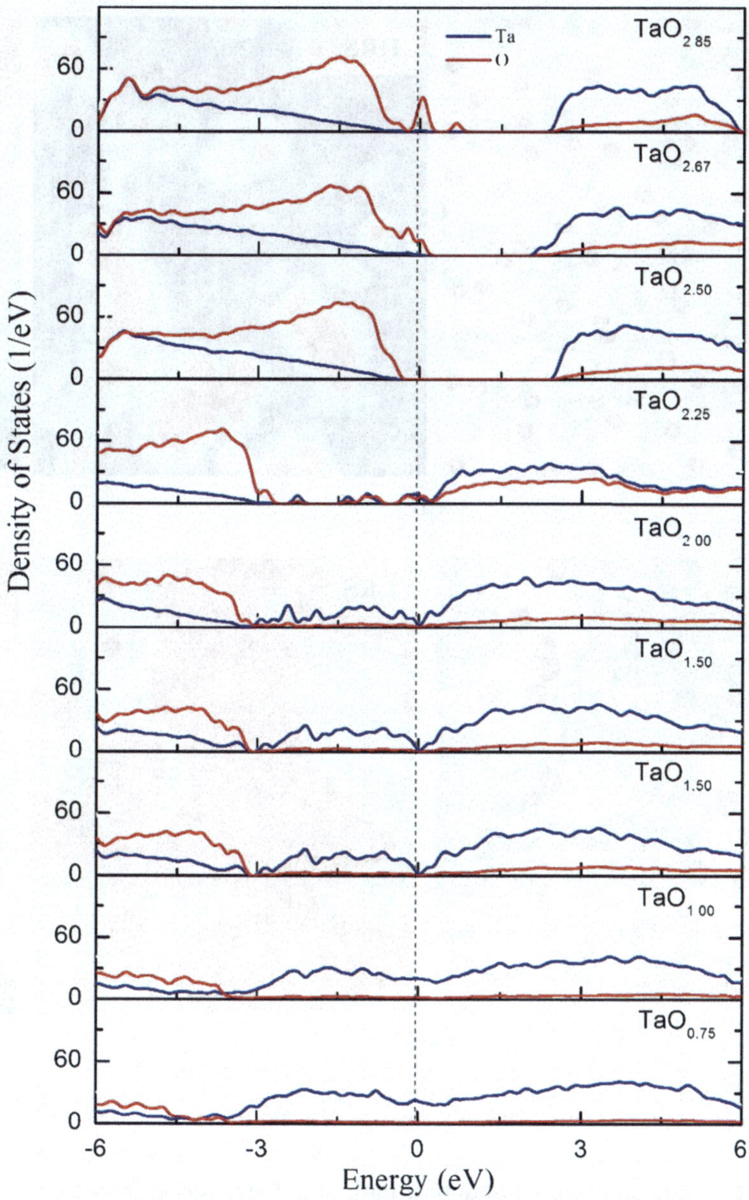


**Fig. 15** DOS of various a-$TaO_x$. Adapted from Ref. [7] with permission from The Royal Society of Chemistry

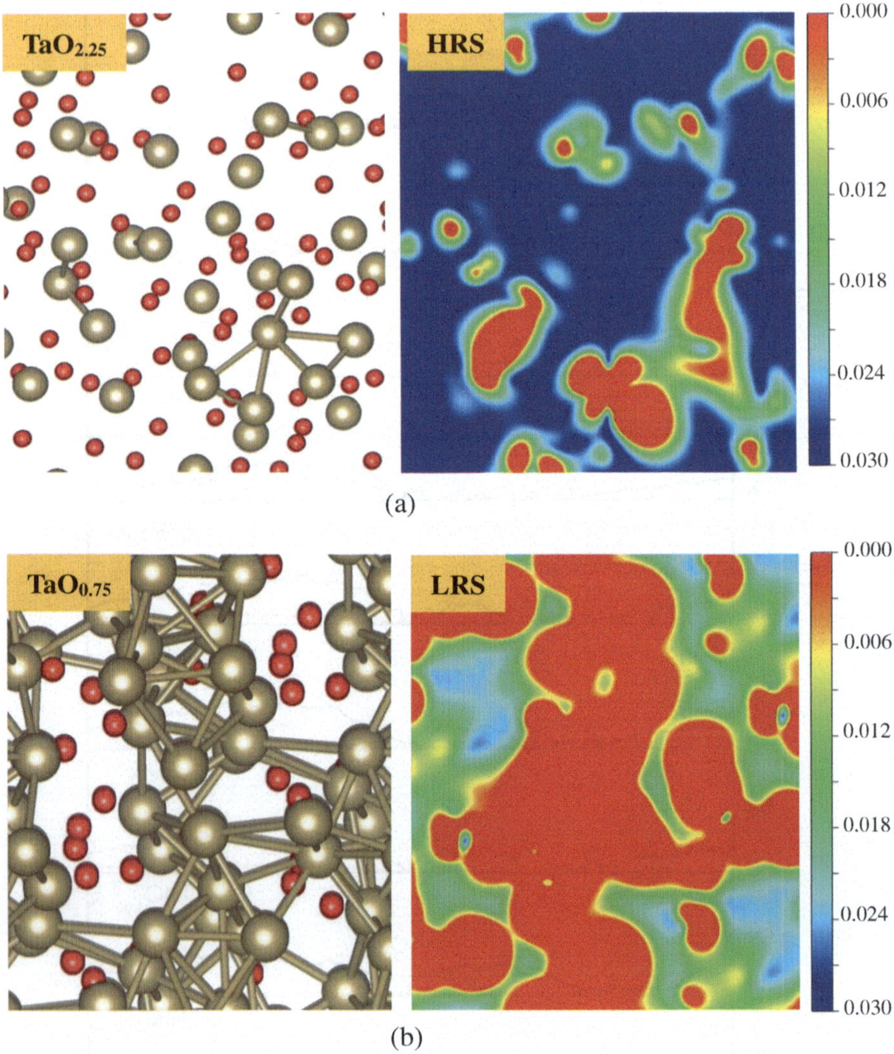


**Fig. 16** Structure and local density of states near the Fermi level of a-$TaO_{2.25}$ and a-$TaO_{0.75}$, respectively. Adapted from Ref. [7] with permission from The Royal Society of Chemistry

### 4.1.3 Crystallization of Conduction Path in a-$TaO_x$ Based Resistive Switch

Even though the Ta-rich region in CF of a-$TaO_x$-based resistive switch had been identified in experiment, its atomic detail was still unknown due to the experimental limitations. Therefore we examined the structure of CF (a-$TaO_{0.75}$) on atomic level from theoretical point of view [7]. As shown in Fig. 17, the atomic arrangement in

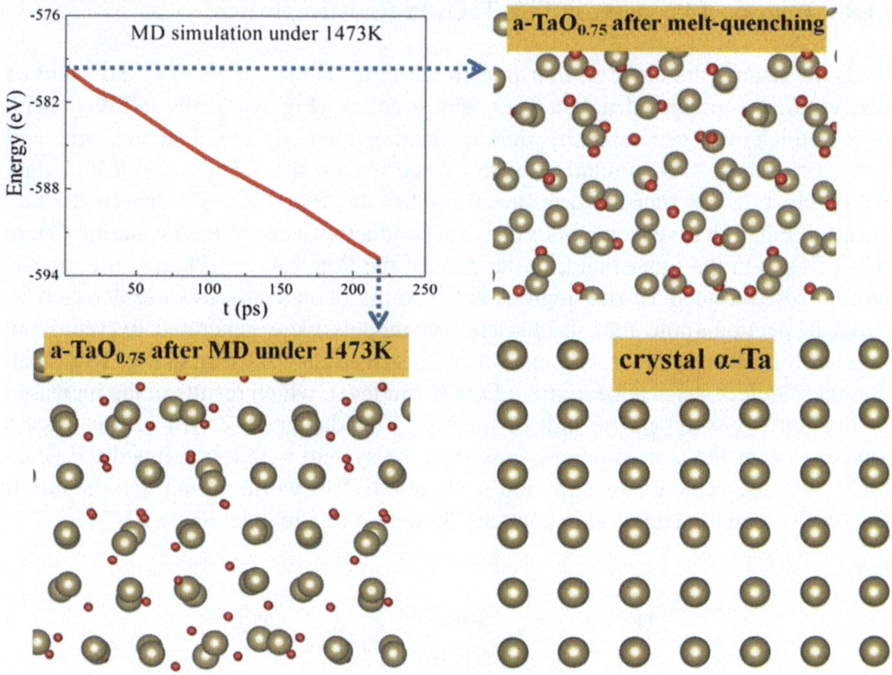


**Fig. 17** Change of a-$TaO_{0.75}$ structure during MD simulation. Structure of α-Ta is shown for comparison. Adapted from Ref. [7] with permission from The Royal Society of Chemistry

the a-$TaO_{0.75}$ structure generated by the melt-quenching method is disordered (Fig. 17). However, it becomes ordered after the MD simulation at 1473 K, accompanied by the decrease of the total energy. As can be seen in Fig. 17, the arrangement of Ta atoms in such ordered structure is similar to that of the crystalline α-Ta, which is the most stable phase of Ta [44]. In order to further explore the relationship between the a-$TaO_{0.75}$ and crystalline α-Ta, the structure relaxation was performed after removing all the O atoms in a-$TaO_{0.75}$. As expected, the generated structure is identical to the crystal α-Ta, and thus the structure of a-$TaO_{0.75}$ can be viewed as the α-Ta with interstitial O atoms. On the Basis of these results, we propose that the phase transition from a-$TaO_{0.75}$ to α-Ta with interstitial O atoms is likely to occur at high temperature.

In real devices, the CF region is considered to become very hot (from 873 K to 1633 K) during the electrical operation probably due to Joule heating. Furthermore, the crystallization of CF in a-$TaO_x$-based resistive switch has been observed during a long time electrical operation on LRS [28, 29], which supports our results.

### 4.1.4 Transport Property of Pt/a-$TaO_x$/Pt Resistive Switch

Next, we discuss the electron transport in the Pt/a-$TaO_x$/Pt device [7]. To examine this, we first constructed a Pt/a-$TaO_{2.5}$/Pt structure (Fig. 18a): the a-$TaO_{2.5}$ with 16.5 Å thick was generated by melt-quenching method, and then the structural optimization and MD simulation were carried out on the a-$TaO_{2.5}$ with including the Pt layers. The transmission spectrum for the Pt/a-$TaO_{2.5}$/Pt heterostructure shown in Fig. 18b reveals the poor electron conductivity due to the insulating nature of a-$Ta_2O_5$. On the other hand, in the case of the Pt/a-$TaO_{2.5-x}$/Pt structure, generation of discontinued Ta-rich regions between the two electrodes were detected by visual inspection (note that the O-deficient models were generated by removing oxygen atoms that were located near other oxygen atoms). As seen in Fig. 18b, small amount of defect states appear around the Fermi level, which results in the increased conduction (to ~0.03 in the unit of quantized conductance, $2e^2/h$) in this system compared with the stoichiometric case. For the system with a continued CF (Pt/a-$TaO_{2-x}$/Pt), the value of transmission is about 0.6. It is worth noting that the on/off ratio of 20 in our calculation is comparable to that in a real device (~10) [4].

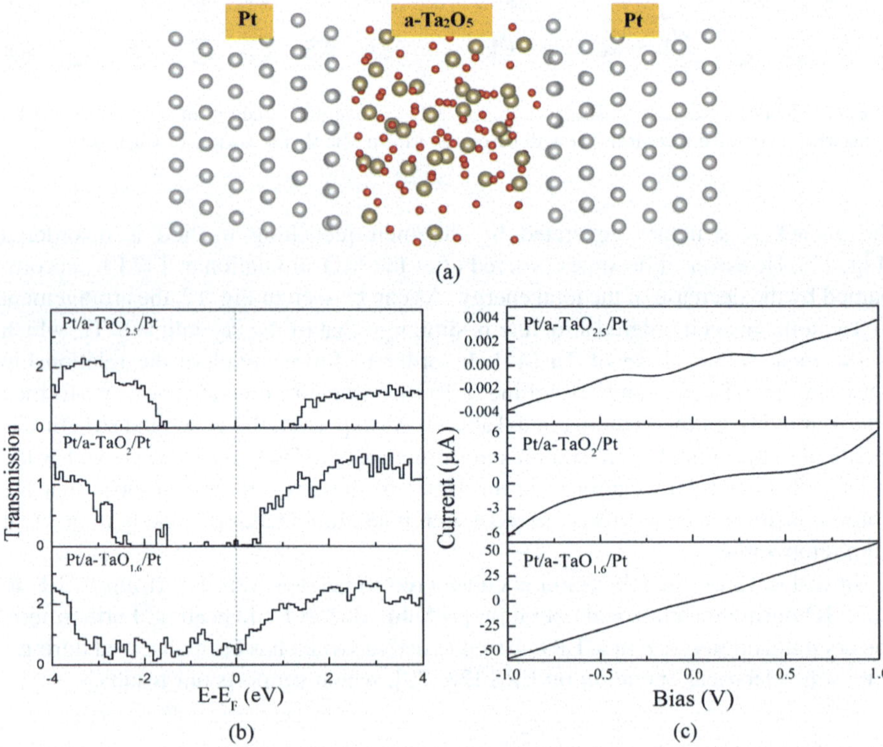


**Fig. 18** (**a**) Structure of Pt/a-$Ta_2O_5$/Pt, (**b**) transmission spectra and (**c**) current-voltage curves of the Pt/a-$TaO_x$/Pt. Adapted from Ref. [7] with permission from The Royal Society of Chemistry

Next, we calculated the current-voltage (I-V) curves of these three heterostructures, which are shown in Fig. 18c. The results further confirm the above features of conduction seen in the transmission spectra. For example, at the bias voltage of 1.0 V, a much larger current (~57 μA) flows in the heterostructure with a continued CF, as compared to the case with a discontinued CF (6 μA). The current value (~57 μA) is comparable to that in the real device [4], though this agreement may be accidental.

In summary of this subsection, the structures and electronic properties of a-$TaO_x$ ($x = 2.85, 2.67, 2.50, 2.25, 2, 1.5, 1$ and $0.75$) were examined via first-principles calculations. Our results reveal that there is a strong correlation between the Ta (O) coordination number and the O-Ta (Ta-Ta) bond length. More importantly, we suggested that the formation of Ta-Ta bonding structures, but not $V_O$, is mainly responsible for the LRS in the Pt/a-$TaO_x$/Pt resistive switches.

## 4.2 *Diffusion of Metal and Oxygen Ions in a-$TaO_x$ Based Resistive Switch*

Almost all the models proposed so far for the switching mechanism of a-$TaO_x$-based atomic switches are based on the diffusion of O ions or vacancies in a-$TaO_x$ [45]. Very recently, it was reported that the diffusion of Ta ions could also play an important role in the switching process of a-$TaO_x$-based resistive switches [46]. It should be noted that the diffusion of both Ta and O ions have already been observed during the growth of a-$TaO_x$ ($x < 2.5$) film [47, 48]. However, details of the behaviors of Ta and O ion diffusion in a-$TaO_x$ during the switching process are still unclear. Thus, to deepen the understanding on the switching mechanism of a-$TaO_x$-based resistive switches, we performed atomistic simulations to clarify the diffusion behaviors of both Ta and O ions during the switching processes [10].

### 4.2.1 Diffusion Coefficients and Barriers of Ta and O Ions in a-$TaO_x$

To examine the diffusion behaviors of Ta and O ions in a-$TaO_x$ ($x = 2$, 1.5 and 1), we calculated the time-average mean square displacement (MSD) at the temperatures of 873 K, 1073 K, 1273 K, 1473 K and 1673 K. From the time evolution of MSDs of Ta and O ions, we extracted the corresponding diffusion coefficients (D). Then from the temperature dependence of D shown in Fig. 19, we have evaluated the activation energies ($E_a$) according to Arrhenius equation. Our results revealed that D and $E_a$ of Ta and O ions have strong correlation with the O concentration in a-$TaO_x$: D ($E_a$) values of Ta and O ions increase (decrease) with the decrease of O content in a-$TaO_x$.

Our calculated values of $E_a$ for O ions in a-$TaO_x$ ($x = 2$, 1.5 and 1) are 0.59, 0.41 and 0.31 eV, respectively, which agrees with the experimental results (from 0.29 to

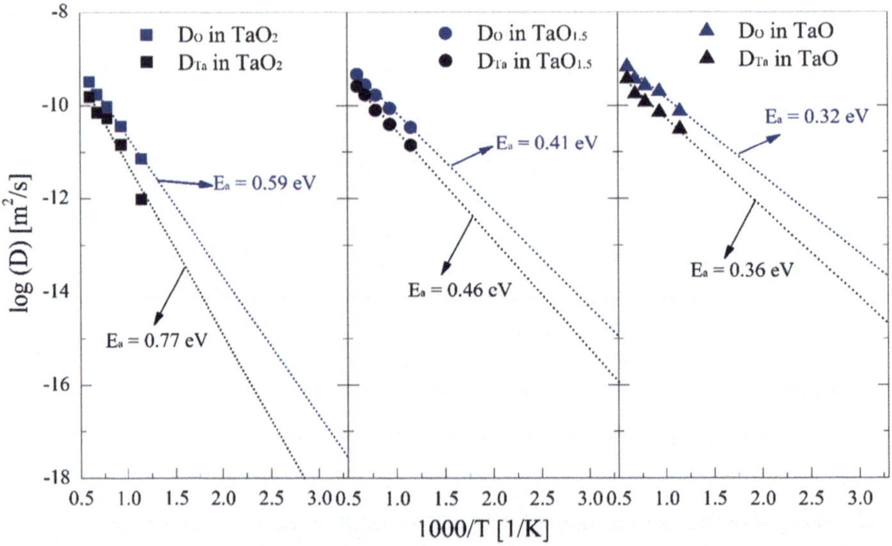


**Fig. 19** Diffusion coefficients and energy barriers of Ta and O ions in a-$TaO_x$ with various O concentrations. Adapted from Ref. [10] with permission from American Chemical Society

0.45 eV) [49]. As can be seen in Fig. 19, the values of D of Ta ions ($D_{Ta}$) are always smaller than those of O ions in a-$TaO_x$ irrespective of the O contents. Interestingly, as the O vacancy concentration increases, $D_{Ta}$ approaches to $D_O$, which means that the diffusion of Ta is more significant in the conductive filament (O-poor region) than in the O-rich region.

As mentioned in the previous experiment [50], in the LRS of the a-$TaO_x$-based atomic switches, the CF is composed of a-TaO. During the switching process, the local temperature of filament rises sharply due to the Joule heating, especially during the RESET process in the bipolar device (from 300 to 600 K) and unipolar device (from 300 to 1200 K) [51–53]. As seen in Fig. 19, $D_{Ta}$ in a-TaO is four times smaller than $D_O$ at 300 K, but the ratio is reduced to 2 at 1200 K. Our results agree with the experimental observation that $D_{Ta}$ is two to three times lower than $D_O$ during the growth of a-$TaO_x$ [47]. On the basis of these results, we can say that the diffusion of Ta ions is non-negligible in the CF, and thus responsible for the switching of a-$TaO_x$-based atomic switches, especially for unipolar switching. It is noted that the value of $D_{Ta}$ in a-$TaO_{1.5}$ is only three to ten times smaller than $D_O$ in the temperature range from 300 to 1200, which suggests that the diffusion of Ta ions may occur during the whole switching process of a-$TaO_x$-based devices.

On the basis of the above, we proposed a schematic model to explain the switching process of a-$TaO_x$ based devices [10], that is, the diffusion of O ions is predominant in the initial forming process of a-$TaO_x$; as the O concentration decreases, the Ta ion diffusion becomes relatively more significant, especially in the LRS region.

### 4.2.2 Diffusion Mechanism of Ta and O Ions in a-$TaO_x$

To deepen our understanding on the diffusion behaviors of Ta and O ions in a-$TaO_x$, here we discuss them in the light of the dependence of the structural parameters (such as CN and bond length) on x value of a-$TaO_x$ ($x = 2.85$, 2.67, 2.50, 2.25, 2, 1.5, 1 and 0.75) [10]. The x dependence of the Ta-O bond lengths and CNs, both Ta (O) and O(Ta), are plotted in Fig. 14. The results show that the Ta-O bond length increases with the decrease of x, which expands the free atomic volume around both metal (Ta) and O atoms, and consequently enhances their mobility. In particular, the CN of O(Ta) increases by ~1 as x decreases from 2.85 to 0.75 as seen in Fig. 14. This corresponds to the change of the component unit from $OTa_2$ to $OTa_3$, and as a result suppresses the mobility of O ions in some degree due to the increased steric hindrance around themselves. In contrast, the CN of Ta(O) shows obvious decrease (by ~3) with the decrease of x as seen in Fig. 14. This corresponds to the change of the component unit from octahedral $TaO_6$ to $TaO_3$, and will dramatically enhance the mobility of Ta ions due to the decreased steric hindrance around themselves. Thus, the enhancement of Ta ion mobility in a-$TaO_x$ is more obvious than that of O ions as the decrease of O concentration, and consequently the mobility of Ta and O ions is close to each other under low O concentration.

It is noted that our theoretical results agree with the previous experiments on the growth of $TaO_x$ ($x < 2.5$), which shows that the diffusion of Ta ions in a-$TaO_x$ layer is non-negligible [47]. Accordingly, from the diffusion behaviors of metal ions during the growth of metal oxide films, we may be able to obtain useful information to understand the switching mechanism of metal oxide based resistive switches. In experiments, the metal ion diffusion has already been observed in metal sub-oxide layer during oxide film growth for Ti, Al, Nb and W [47]. On the basis of these results, we can speculate that the metal ion diffusion in $TaO_x$, $TiO_x$, $AlO_x$, $NbO_x$ and $WO_x$ could occur during the switching of resistive switches based on these materials.

In summary of this subsection, we investigated the diffusion behavior of Ta and O ions in a-$TaO_x$ with various O concentrations. Our results reveal that the diffusion of Ta ions is enhanced with the decrease of O concentration, and approaches to that of O ions at the extremely low O concentration cases. Such phenomenon can be attributed to the combined effect of the changes in the Ta/O coordination numbers and the Ta-O/Ta-Ta bond strengths with the charge of O concentration in a-$TaO_x$.

# 5 Concluding Remarks

In this chapter, we have discussed several issues related to the switching mechanism of amorphous $TaO_x$ (a-$TaO_x$) based resistive switches on the basis of our simulations within the density functional theory. Our results reveal that (1) Cu nanowires with a diameter of three atoms or larger can work as conduction filaments (CFs) in the Cu/a-$Ta_2O_5$/Pt atomic switch; (2) Cu atoms tend to be ionized at the Cu/a-$Ta_2O_5$ interface, and the ionization behavior becomes obvious with the increase of interface O concentration and/or temperature; (3) the CF in the Pt/a-$TaO_x$/Pt resistive switch consists of Ta-Ta bonding structures, but not O vacancies; (4) the mobility of Ta ions in the CF region of Pt/a-$TaO_x$/Pt device is comparable to that of O ions, and thus could actively participate in the switching process of a-$TaO_x$ based devices.

For the atomistic understanding on the switching mechanism of the (a-$TaO_x$) based resistive switches, several issues still remain as future tasks. For example, the atomistic details on the effects of moisture on the Cu diffusion [5] have not been clarified yet. In fact, our preliminary simulation, where we compared the Cu diffusion behaviors between bare and water-adsorbed a-$Ta_2O_5$ surfaces, suggests that the Cu diffusion can be faster on the water-adsorbed surface than the bare one [11]. Theoretical studies based on a more realistic model, however, would be desirable.

For tackling with such issues, density functional calculations are still heavy even on the current supercomputers. So development of methods that have both reliability and computational efficiency is highly desired. For this purpose, the interatomic potentials constructed by combining the density functional calculations and machine-learning techniques have attracted much attention recently. We applied such approaches to Cu diffusion in the a-$Ta_2O_5$ [54] and Li diffusion in a-$Li_3PO_4$ [55], but extensive investigation using these approaches remains as a future task.

**Acknowledgments** We thank Prof. Tinkun Gu, Prof. Tomofumi Tada for collaboration in the early stage of the present work. SW also thanks Prof. Shu Yamaguchi, Prof. Tsuyoshi Hasegawa, Dr. Tohru Tsuruoka, Dr Toshi Sakamoto and Dr Naoki Banno for fruitful discussion. This work was partially supported by CREST-JST "Atom transistor", Low Power Electronics Association and Projects, the grant-in-aid for Innovation Area "Computics" by MEXT, Japan, and Global COE program "Global COE for Mechanical Systems Innovation" by MEXT, Japan.

## References

1. Wang, J.Y., Li, L.Z., Huyan, H., Pan, X.Q., Nonnenmann, S.: Highly uniform resistive switching in $HfO_2$ films embedded with ordered metal nanoisland arrays. Adv. Funct. Mater. **29**, 1808430 (2019)
2. Zhou, G.D., Duan, S.K., Li, P., Sun, B., Wu, B., Yao, Y.Q., Yang, X.D., Han, J.J., Wu, J.G., Wang, G., Liao, L.P., Lin, C.Y., Hu, W., Xu, C.Y., Liu, D.B., Chen, T., Chen, L.J., Zhou, A.K., Song, Q.L.: Coexistence of negative differential resistance and resistive switching memory at room temperature in $TiO_x$ modulated by moisture. Adv. Electron. Mater. **4**, 1700567 (2018)
3. Prakash, A., Jana, D., Maikap, S.: $TaO_x$-based resistive switching memories: Prospective and challenges. Nanoscale Res. Lett. **8**, 418 (2013)
4. Lee, M.J., Lee, C.B., Lee, D., Lee, S.R., Chang, M., Hur, J.H., Kim, Y.B., Kim, C.J., Seo, D.H., Seo, S., Chung, U., Kim, K.: A fast, high-endurance and scalable non-volatile memory device made from asymmetric $Ta_2O_{5-x}/TaO_{2-x}$ bilayer structures. Nat. Mater. **10**, 625–630 (2011)
5. Valov, L., Tsuruoka, T.: Effects of moisture and redox reactions in VCM and ECM resistive switching memories. J. Phys. D. Appl. Phys. **51**, 413001 (2018)
6. Xiao, B., Gu, T.K., Tada, T., Watanabe, S.: Conduction paths in Cu/amorphous- $Ta_2O_5$/Pt atomic switch: First-principles studies. J. Appl. Phys. **115**, 034503 (2014)

7. Xiao, B., Watanabe, S.: Oxygen vacancy effects on an amorphous-$TaO_x$-based resistance switch: A first principles study. Nanoscale. **6**, 10169–10178 (2014)
8. Xiao, B., Watanabe, S.: Interface structure in Cu/$Ta_2O_5$/Pt resistance switch: A first-principles study. ACS Appl. Mater. Interfaces. **7**, 519–525 (2015)
9. Xiao, B., Yu, X.F., Cheng, J.B.: Atomic insight into the origin of various operation voltages of cation-based resistance switches. ACS Appl. Mater. Interfaces. **8**, 31978–31985 (2016)
10. Xiao, B., Yu, X.F., Watanabe, S.: A comparative study on the diffusion behaviors of metal and oxygen ions in metal-oxide-based resistance switches via ab initio molecular dynamics simulations. ACS Appl. Electron. Mater. **1**, 585–549 (2019)
11. Xiao, B., Watanabe, S.: Moisture effect on the diffusion of Cu ions in Cu/$Ta_2O_5$/Pt and Cu/$SiO_2$/Pt resistances switches: A first-principle study. Sci. Technol. Adv. Mat. **20**, 580–588 (2019). https://doi.org/10.1080/14686996.2019.1616222
12. Kresse, G., Furthmüller, J.: Efficiency of ab-initio total energy calculations for metals and semiconductors using a plane-wave basis set. Comput. Mater. Sci. **6**, 15–50 (1996)
13. Kresse, G., Furthmüller, J.: Efficient iterative schemes for ab initio total-energy calculations using a plane-wave basis set. Phys. Rev. B. **54**, 11169 (1996)
14. Kresse, G., Joubert, J.: From ultrasoft pseudopotentials to the projector augmented-wave method. Phys. Rev. B. **59**, 1758 (1999)
15. Wang, Y., Perdew, J.P.: Correlation hole of the spin-polarized electron gas, with exact small-wave-vector and high-density scaling. Phys. Rev. B. **44**, 13298 (1991)
16. Brandbyge, M., Mozos, J., Ordejon, P., Taylor, J., Stokbro, K.: Density-functional method for nonequilibrium electron transport. Phys. Rev. B. **65**, 165401 (2002)
17. Terabe, K., Hasegawa, T., Nakayama, T., Aono, M.: Quantized conductance atomic switch. Nature. **433**, 47–50 (2005)
18. Waser, R., Aono, M.: Nanoionics-based resistive switching memories. Nat. Mater. **6**, 833–840 (2007)
19. Sawa, A.: Resistive switching in transition metal oxides. Mater. Today. **11**, 28–36 (2008)
20. Gu, T.K., Wang, Z.C., Tada, T., Watanabe, S.: First-principles simulations on bulk $Ta_2O_5$ and Cu/$Ta_2O_5$/Pt heterojunction: Electronic structures and transport properties. J. Appl. Phys. **106**, 103713 (2009)
21. Gu, T.K., Tada, T., Watanabe, S.: Conductive path formation in the Ta2O5 atomic switch: First-principles analyses. ACS Nano. **4**, 6477–6482 (2010)
22. Kim, M.J., Seo, Y., Cruz, M.A., Wiley, B.J.: Metal nanowire felt as a flow-through electrode for high-productivity electrochemistry. ACS Nano. **13**, 6998–7009 (2019). https://doi.org/10.1021/acsnano.9b02058
23. Gonzalez, J.C., Rodrigues, V., Bettini, J., Rego, L.G.C., Rocha, A.R., Coura, P.Z., Dantas, S. O., Sato, F., Galvao, D.S., Ugarte, D.: Indication of unusual pentagonal structures in atomic-size Cu nanowires. Phys. Rev. Lett. **93**, 126103 (2004)
24. Zhou, Z., Zhao, J.J., Chen, Z.F., Gao, X.P., Lu, J.P., Schleyer, P.V.R., Yang, C.K.: True nanocable assemblies with insulating BN nanotube sheaths and conducting Cu nanowire cores. J. Phys. Chem. B. **110**, 2529–2532 (2006)
25. Banno, N., Sakamoto, T., Iguchi, N., Matsumoto, M., Imai, H., Ichihashi, T., Fujieda, S., Tanaka, K., Watanabe, S., Yamaguchi, S., Hasegawa, T., Aono, M.: Structural characterization of amorphous $Ta_2O_5$ and $SO_2$-$Ta_2O_5$used as solid electrolyte for nonvolatile switches. Appl. Phys. Lett. **97**, 113507 (2010)
26. Lee, S., Kim, J., Kim, S., Kim, S., Park, G.: Hidden structural order in orthorhombic. Phys. Rev. Lett. **110**, 235502 (2013)
27. KC, S., Dong, H., Longo, R.C., Wang, W.C., Xiong, K., Wallace, R.M., Cho, K.: Electronic properties of InP (001)/$HfO_2$ (001) interface: Band offsets and oxygen dependence. J. Appl. Phys. **115**, 023703 (2014)
28. Park, G.S., Kim, Y.B., Park, S.Y., Li, X.S., Heo, S., Lee, M.J., Chang, M., Kwon, J.H., Kim, M., Chung, U.I., Dittmann, R., Waser, R., Kim, K.: In situ observation of filamentary

conducting channels in an asymmetric $Ta_2O_{5-x}/TaO_{2-x}$ bilayer structure. Nat. Commun. **4**, 2382 (2013)
29. Miao, F., Strachan, J.P., Yang, J.J., Zhang, M.X., Goldfarb, I., Torrezan, A.C., Eschbach, P., Kelley, R.D., Medeiros-Ribeiro, G., Williams, R.S.: Anatomy of a nanoscale conduction channel reveals the mechanism of a high-performance memristor. Adv. Mater. **23**, 5633–5640 (2011)
30. Jagath, A.L., Kumar, T.N., Almurib, H.A. Modeling of current conduction during RESET phase of $Pt/Ta_2O_5/TaO_x/Pt$ bipolar resistive RAM devices. In: IEEE 7$^{th}$ Non-Volatile Memory Systems and Applications Symposium (NVMSA). (2018). https://doi.org/10.1109/NVMSA.2018.00014
31. Yang, J.J., Kobayashi, N.P., Strachan, J.P., Zhang, M.X., Ohlberg, D.A.A., Pickett, M.D., Li, Z., Medeiros-Ribeiro, G., Williams, R.S.: Dopant control by atomic layer deposition in oxide films for memristive switches. Chem. Mater. **23**, 123–125 (2011)
32. Yang, J.J., Strachan, J.P., Miao, F., Zhang, M.X., Pickett, M., Yi, W., Ohlberg, D., Medeiros-Ribeiro, G., Williams, R.: Metal/$TiO_2$ interfaces for memristive switches. Appl. Phys. A Mater. Sci. Process. **102**, 785–789 (2011)
33. Xu, N., Gao, B., Liu, L.F., Sun, B., Liu, X.Y., Han, R.Q., Kang, J.F., Yu, B.: A unified physical model of switching behavior in oxide-based RRAM. Symposium on VLSI Technology Digest of technical papers, pp. 100–101 (2008)
34. Hur, J.H., Lee, M.J., Lee, C.B., Kim, Y.B., Kim, C.J.: Modeling for bipolar resistive memory switching in transition-metal oxides. Phys. Rev. B. **82**, 155321 (2010)
35. Wei, Z., Takagi, T., Kanzawa, Y., Katoh, Y., Ninomiya, T., Kawai, K., Muraoka, S., Mitani, S., Katayama, K., Fujii, S., Miyanaga, R., Kawashima, Y., Mikawa, T., Shimakawa, K., Aono, K.: Demonstration of high-DENSITY ReRAM ensuring 10-year retention at 85°C. Based on a newly developed reliability model. IEDM11–721, 31.4.1–31.1.4 (2011)
36. Choi, S., Yang, Y.C., Lu, W.: Random telegraph noise and resistance switching analysis of oxide based resistive memory. Nanoscale. **7**, 400–404 (2014)
37. Yang, J.J., Zhang, M.X., Strachan, J.P., Miao, F., Pickett, M.D., Kelley, R.D., Medeiros-Ribeiro, G., Williams, R.S.: High switching endurance in $TaO_x$ memristive devices. Appl. Phys. Lett. **97**, 232102 (2010)
38. Heyd, J., Scuseria, G.E., Ernzerhof, M.: Hybrid functionals based on a screened coulomb potential. J. Chem. Phys. **118**, 8207 (2003)
39. Lau, W.S., Leong, L.L., Han, T., Sandler, N.P.: Detection of oxygen vacancy defect states in capacitors with ultrathin $Ta_2O_5$ films by zero-bias thermally stimulated current spectroscopy. Appl. Phys. Lett. **83**, 2835 (2003)
40. Ivanov, M.V., Perevalov, T.V., Aliev, V.S., Gritsenko, V.A., Kaichev, V.V.: Effect of dielectric stoichiometry and interface chemical state on band alignment between tantalum oxide and platinum. J. Appl. Phys. **110**, 024115 (2011)
41. Lau, W.S.: Engineering implication of the correlation between the leakage current in high-k dielectric materials and the electronic defect states detected by zero-bias thermally simulated current spectroscopy. China Semiconductor Technology International Conference (CSTIC). (2018). https://doi.org/10.1109/CSTIC.2018.8369248
42. Garg, S.P., Krishnamurthy, N., Awasthi, A., Venkatraman, M.: The O-Ta (oxygen-tantalum) system. J. Phase Equilib. **17**, 63–77 (1996)
43. Alderman, O.L.G., Benmore, C.J., Neuefeind, J., Coillet, E., Mermet, A., Martinez, V., Tamalonis, A., Weber, R.: Amorphous tantala and its relationship with the molten state. Phys. Rev. Mater. **2**, 043602 (2018)
44. Jiang, A.Q., Tyson, T.A., Axe, L.: The structure of small Ta clusters. J. Phys. Condens. Matter. **17**, 1841 (2005)
45. Park, T.H., Kwon, Y.J., Kim, H.J., Woo, H.C., Kim, G.S., An, C.H., Kim, C.H., Kwon, D.E., Hwang, C.S.: Balancing the source and sink of oxygen vacancies for the resistive switching memory. ACS Appl. Mater. Interfaces. **10**, 21445–21450 (2018)

46. Wedig, A., Luebben, M., Cho, D.Y., Moors, M., Skaja, K., Rana, V., Hasegawa, T., Adepalli, K.K., Yildiz, B., Waser, R., Valov, L.: Nanoscale cation motion in $TaO_x$, $HfO_x$ and $TiO_x$ memristive systems. Nat. Nanotechnol. **11**, 67–74 (2016)
47. Davies, J.A., Domeij, B., Pringle, J.P.S., Brown, F.: The migration of metal and oxygen during anodic film formation. J. Electrochem. Soc. **112**, 675–680 (1965)
48. Khalil, N., Leach, J.S.L.: The anodic oxidation of valve metals—I. determination of ionic transport numbers by a-spectrometry. Electrochim. Acta. **31**, 1279–1285 (1986)
49. Chandrasekharan, R., Park, I., Masel, R.I., Shannon, M.A.: Thermal oxidation of tantalum films at various oxidation states from 300 to 700°C. J. Appl. Phys. **98**, 114908 (2005)
50. Oshima, M., Toyoda, S., Horiba, K., Yasuhara, R., Kumigashira, H.: Synchrotron radiation nano-spectroscopy of dielectrics for LSI and ReRAM. ECS Trans. **41**, 453 (2011)
51. Yang, J.J., Strukov, D.B., Stewart, D.R.: Memristive devices for computing. Nat. Nanotechnol. **8**, 13–24 (2013)
52. Yalon, E., Gavrilov, A., Cohen, S., Ritter, D.: Validation and extension of local temperature evaluation of conductive filaments in RRAM devices. IEEE Trans. Electron Devices. **62**, 3671–3677 (2015)
53. Sharma, A.A., Noman, M., Abdelmoula, M., Skowronski, M., Bain, J.A.: Electronic instabilities leading to electroformation of binary metal oxide-based resistive switches. Adv. Funct. Mater. **24**, 5522–5529 (2014)
54. Li, W., Ando, Y., Watanabe, S.: Cu diffusion in amorphous $Ta_2O_5$ studied with a simplified neural network potential. J. Phys. Soc. Jpn. **86**, 104004 (2017)
55. Li, W., Ando, Y., Minamitani, E., Watanabe, S.: Study of Li atom diffusion in amorphous $Li_3PO_4$ with neural network potential. J. Chem. Phys. **147**, 214106 (2017)

# Development of Three-Terminal Atomic Switches and Related Topics

**T. Hasegawa, T. Tsuruoka, Y. Itoh, T. Sakamoto, and M. Aono**

**Abstract** Three-terminal atomic switches control formation and annihilation of a conductive path between a source and a drain by the electric field applied through a gate electrode. Advantage of the three-terminal atomic switches to the two-terminal atomic switches is that a signal line of the three-terminal atomic switches is separated from a signal line, which increases their controllability especially when used as logic devices. Three-terminal atomic switches are classified into two types by a species of controlled ions that form a conductive path. One type controls metal cations, so that a conductive path is a metal filament. The other type controls oxygen ions, so that a conductive path is an oxygen deficient conductive region. Both types show nonvolatile switching with a high on/off ratio over five orders of magnitude.

## 1 Introduction

Nobody doubts that a representative of three-terminal devices is a semiconductor transistor, that has enabled present-day highly sophisticated information society. Metal-oxide-semiconductor field-effect transistor (MOSFET) and its complementary system (CMOS) have drastically reduced a device size and power consumption of electronic devices. The miniaturization should be the major cause of the successful development of high performance computing systems. If there were disadvantage in MOSFETs, it must be its volatile operation requiring continuous bias application to keep its on-state. Actually, the volatility has become a big issue in terms of energy consumption in the last decade, inducing developments of nonvolatile devices such as storage class memories for replacing certain parts of electronic devices in CMOS

T. Hasegawa (✉)
Faculty of Science and Engineering, Waseda University, Tokyo, Japan
e-mail: thasega@waseda.jp

T. Tsuruoka · Y. Itoh · M. Aono
International Center for Materials Nanoarchitectonics (MANA), National Institute for Materials Science (NIMS), Tsukuba, Ibaraki, Japan

T. Sakamoto
System Platform Research Laboratories, NEC, Ibaraki, Japan



M. Aono (ed.), *Atomic Switch*, Advances in Atom and Single Molecule Machines,
https://doi.org/10.1007/978-3-030-34875-5_7

circuits to reduce the growing power consumption. Although there are various nonvolatile devices such as used in memories, most of them are two-terminal devices that cannot replace CMOS.

Nonvolatile three-terminal devices that controls ionic movement in solid electrolytes were developed and commercialized in 1970s [1]. Since the size of the devices was in the order of 10 mm, the early-developed nonvolatile three-terminal devices did not have chance to be used in information processing systems. In 1990s, 'memistor', the three-terminal device aiming to be used in neural computing systems, was developed [2]. The memistor moves $H^+$ ions to a channel region of $WO_3$, which becomes conductive by being doped with $H^+$ ions. Namely, a source and a drain are electrically connected. After stopping bias application, $H^+$ ions gradually leave from the channel region typically in several minutes. The gradual decrease in conductivity was expected to emulate a neural function. Therefore, the memistor is a volatile device although it has a certain decaying time.

The nonvolatile three-terminal operation truly in a nanoscale was demonstrated in 2004 by electrochemical deposition of metal atoms on two Au electrodes dipped in an electrolyte [3]. The electrochemical deposition of metal (Ag) atoms onto the two Au electrodes thickened the two electrodes, making a gap between the two electrodes smaller and finally bridging them. Bias application in an opposite polarity dissolved the deposited metal atoms into an electrolyte, resulting in a disconnection of the two electrodes. Precise control of a gate bias achieved switching between quantized conductance states.

## 2 Metal Cation-Controlled Three-Terminal Atomic Switches

Metal cation-controlled type is developed based on the operating mechanism of the two-terminal atomic switches. It is classified into two types by a rate limiting process in the metal filament formation in their switching on process. In the first type, drift of metal cations is a rate limiting process, where a conductive filament gradually grows by incorporating metal atoms drifting from a gate electrode. In the second type, nucleation of metal atoms is a rate limiting process, where nucleus is formed when a concentration of metal cations at a channel region reaches a certain value. Depending on applications, one can choose either of the two types.

### *2.1 Filament Growth Controlled Type*

As mentioned in the former section, deposition and dissolution of metal atoms between two electrodes in an electrolyte (liquid), such as $AgNO_3 + HNO_3$ [3] and $CuSO_4$ [4], can be used for nonvolatile three-terminal operations, as shown in

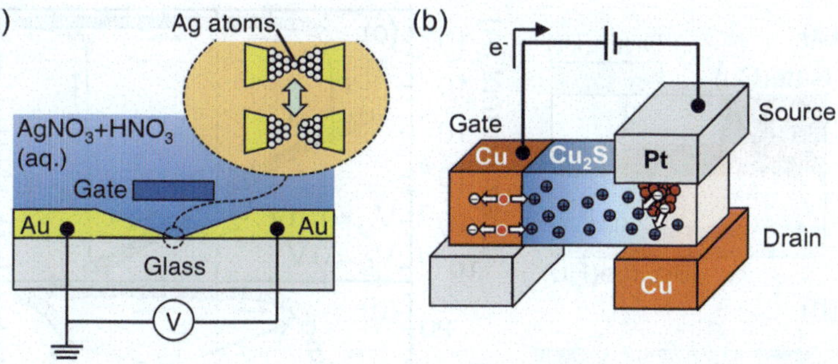


**Fig. 1** Schematics of three-terminal switches where formation/annihilation of a metal filament is achieved by a gate bias controlled electrochemical deposition/dissolution in (**a**) an electrolyte [3] and (**b**) a solid electrolyte [4]

Fig. 1a. The similar operation is available in a solid electrolyte. Such an operation was first demonstrated using $Cu_2S$ as a solid electrolyte, in which migration of $Cu^{2+}$ cations is controlled by a gate bias [4]. Namely, application of a positive gate bias moves $Cu^{2+}$ cations towards a channel region, where $Cu^{2+}$ cations are reduced and deposited on a source and a drain, as shown in Fig. 1b. Thickening of the source and the drain results in their electrical connection by a Cu filament bridge between them, similar to the phenomena controlled in electrolytes (liquid). Application of a negative gate bias re-oxidizes Cu atoms deposited on a source and a drain. Dissolving of re-oxidized Cu atoms, i.e., $Cu^{2+}$, into a solid electrolyte ($Cu_2S$) annihilates a Cu filament, resulting in disconnection of the source and the drain. In this type of the three-terminal atomic switch, initialization that forms a Cu filament by supplying $Cu^{2+}$ cations from a drain electrode made of Cu, using the operating mechanism of a gapless-type atomic switch, is very effective for reducing the filament size (length) in the switching controlled by the gate electrode. The initialization ensures the electrical isolation of a gate electrode from a source and a drain, resulting in the high reliability as well as the short switching time in the operation.

After the first development, the device structure that does not require the initialization was also developed by using Cu for the gate electrode and Pt for both the drain and source electrodes [5]. In the three-terminal atomic switches, a gate electrode should be far enough in order to avoid a Cu filament formation between a gate electrode and source/drain electrodes, i.e., short-circuit. At the same time, a gate electrode has a function of applying an electric field in a layer of a solid electrolyte to move $Cu^{2+}$ cations. Because of this reason, $Cu_2S$ that has both electronic conductivity and ionic conductivity was employed in the development. An on/off ratio of higher than $10^4$ and a cyclic endurance of $10^2$ times have been achieved using the new structure. The device structure and the operating results are shown in Fig. 2.

Although the use of $Cu_2S$ resulted in the gate current in the order of μA, it is acceptable when used as reconfigurable switches in programmable devices because

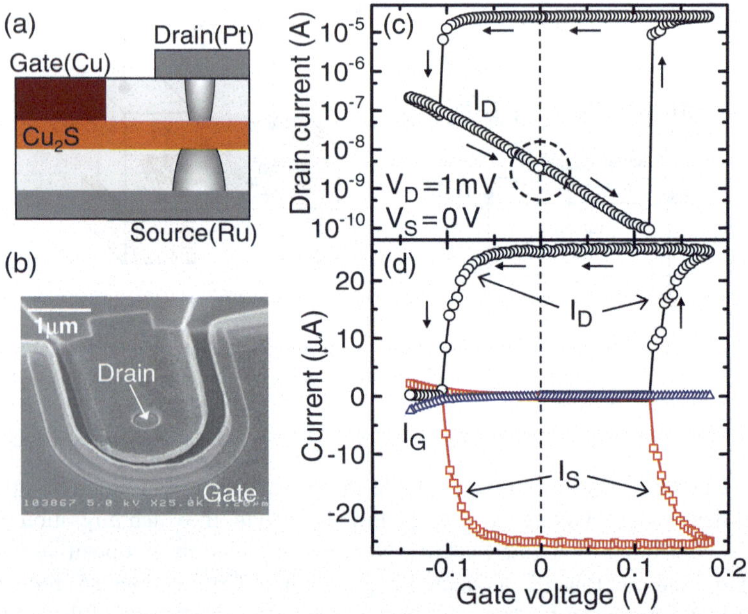


**Fig. 2** Initialization free three-terminal atomic switch formed in a metal wiring layer. (**a**) Schematic cross section of the switch, (**b**) A scanning electron microscopic image of the switch, (**c**) Change in drain current ($I_D$) on a log scale and (**d**) change in drain ($I_D$), source ($I_S$) and gate ($I_G$) currents while sweeping the gate bias. (Reproduced with permission from Ref. [5])

they are not turned on/off so often. That the new structure can be easily formed in metal wiring layers is also advantageous for the application.

Although the deposition and dissolution of Cu atoms occur with the smaller gate bias such as 0.1 V, this does not cause any degradation in the nonvolatility and reliability of the device. This is because a Cu filament is thick enough to be stable and a bias application is needed to dissolve a filament.

## 2.2 *Nucleation Controlled Type*

Another type of three-terminal atomic switch has been developed based on the nucleation control of metal clusters. In the gapless-type atomic switches, there are four elemental processes in their SET process [6]. (1) Oxidation of metal atoms of an active electrode, (2) Drift of metal cations introduced from an active electrode, (3) Nucleation of a metal cluster at a counter inert electrode, and (4) filament growth of the nucleus. When we use sulfide, e.g., $Cu_2S$, as an ionic transferring material, oxidation and reduction of metal atom easily occur with a smaller bias, as shown in the former sections and chapters, resulting in that the rate limiting process is the drift of metal cations traveling in an ionic transfer layer. Since metal cation's drift is an activation process, traveling time of metal cations can be expressed using Arrhenius equation as follows

$$t \approx \sum_j A_j \cdot exp\left(\frac{E(x_j) - e\Delta V(x_j)}{kT}\right),$$

where $E(x_j)$ and $\Delta V(x_j)$ are a diffusion barrier at a position of $x_j$ without bias application and a potential drop at $x_j$ due to a bias application. From the equation, it is understood that metal cation's drift occurs even with an application of smaller bias although the traveling time exponentially becomes longer. As a result, there is no threshold bias for the switching although switching time changes over several orders of magnitude depending on an applied bias. Therefore, SET bias depends on the speed of bias sweeping. This isn't a matter when atomic switches are used such as a memory device and a programmable switch. Only when used as logic devices, it becomes a matter that is the reason why we developed another type of the three-terminal atomic switches.

In order to develop a three-terminal atomic switch that has a clear threshold bias for SET and RESET, we employed $Ta_2O_5$, which is an insulator in terms of electrical conductivity, as the ionic transferring material. This is because the rate limiting process of $Ta_2O_5$-based gapless atomic switches is nucleation of metal atoms in a $Ta_2O_5$ layer at the counter electrode side. Since the nucleation occurs when a concentration of metal cations reaches a certain value, i.e., super-saturation value, nucleation controlled three-terminal atomic switch should have a clear threshold bias.

Figure 3 shows the operating concept of the three-terminal atomic switch. In the SET process, metal cations supplied from the gate electrode migrate towards the

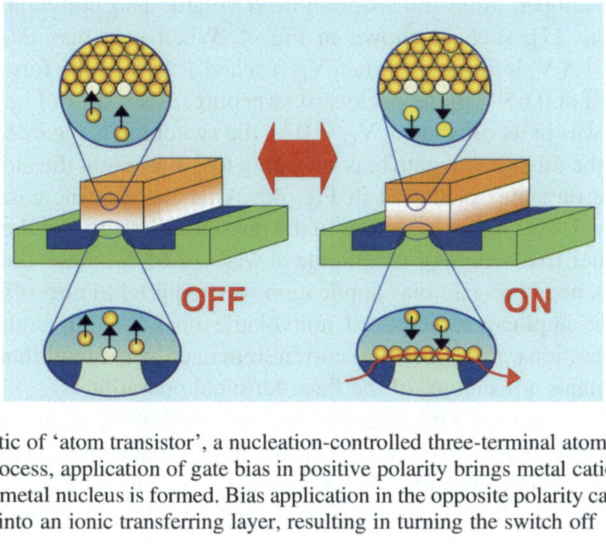


**Fig. 3** Schematic of 'atom transistor', a nucleation-controlled three-terminal atomic switch. In the switching-on process, application of gate bias in positive polarity brings metal cations to a channel region, where a metal nucleus is formed. Bias application in the opposite polarity causes dissolution of metal atoms into an ionic transferring layer, resulting in turning the switch off

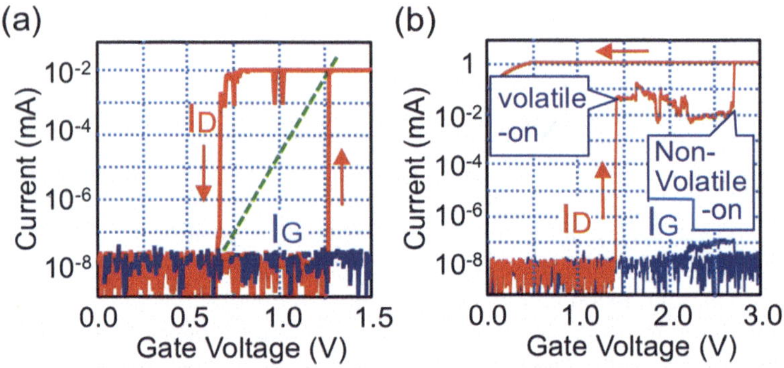


**Fig. 4** Operating results of atom transistor consisting of a Cu(gate)/$Ta_2O_5$/Pt(source), Pt(drain) structure. (**a**) Change in drain (red) and gate (blue) currents while gate bias sweeping from 0 to 1.5 V, and vice versa. (**b**) Change in drain (red) and gate (blue) currents while gate bias sweeping from 0 to 3 V, and vice versa. Both in (**a**) and (**b**), 5 mV was applied between a drain and a source to measure the drain current. (Reproduced with permission from Ref. [7])

source and drain electrodes. When the concentration of metal cations at the channel region reaches a value of super-saturation, metal nucleus is formed that connects the source and the drain. Since the concentration is a function of the gate electric field, there should be a clear threshold bias in the SET process. The bias application in the opposite polarity oxidizes metal atoms of a nucleus, resulting in dissolution of a nucleus. Namely, the device turns off. As such, on-state is made by a metal nucleus that is highly conductive, while off-state is made by $Ta_2O_5$ that is highly insulator. Since the nucleation controlled three-terminal atomic switch works as a circuit element in a similar way to a silicon transistor, i.e., the state variable is gate voltage ($V_G$), it is referred to as the 'Atom Transistor' [7].

The first developed atom transistor showed volatile and nonvolatile dual functional operations [7], such as shown in Fig. 4. When gate bias ($V_G$) was swept between 0 and 1.5 V, it turned on when $V_G$ reached 1.25 V in the forward sweeping and it turned off at 0.65 V in the backward sweeping, as shown in Fig. 4a. Since the atomic switch was in its off state at $V_G = 0$ V, the switching is regarded as a volatile operation. On the other hand, gate bias sweeping to 3 V brought the atomic switch to the nonvolatile on state, as shown in Fig. 4b. After the first increase in the drain current at $V_G = 1.4$ V, which is similar to the one shown in Fig. 4a, the drain current increases another two orders of magnitude at $V_G = 2.65$ V. Since the on-state was kept at $V_G = 0$, negative gate bias application was required to turn off. Namely, the larger gate bias application achieved nonvolatile operation. In both volatile and nonvolatile operations, gate (leakage) current remained very small that we expected as one of the major advantages of the three-terminal operation.

The dual functionality, i.e., volatility and non-volatility, should be caused by stability of a metal nucleus in $Ta_2O_5$. Critical size of a metal nucleus to be stabilized is a function of a concentration of metal cations in the around [8]. In the case of a volatile operation, the lower gate bias application could form a metal nucleus that

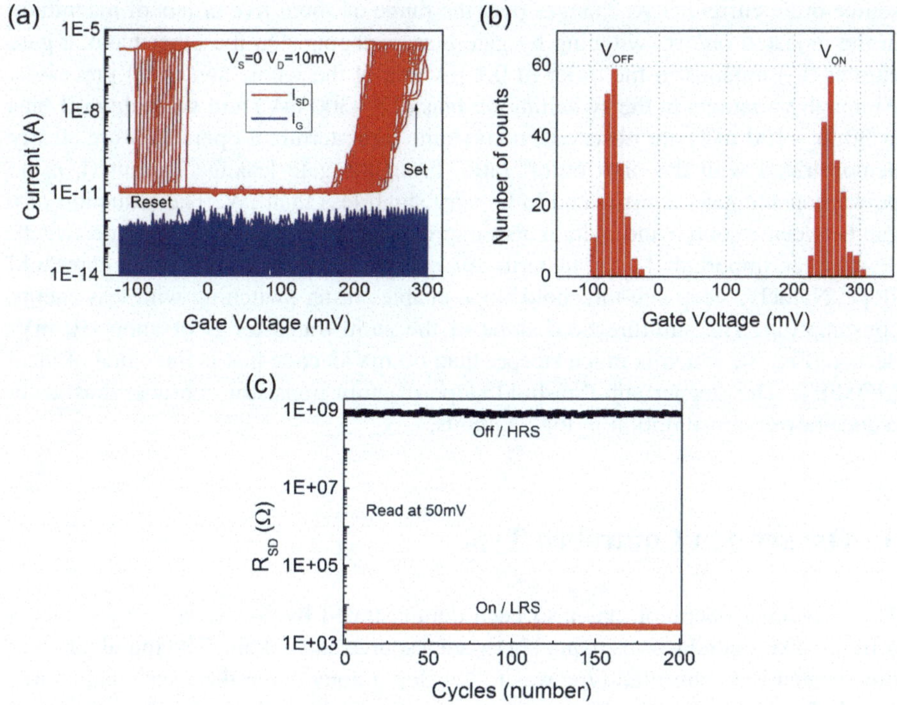


**Fig. 5** Operating results of a Ag(gate)/$Ta_2O_5$/Pt(source), Pt(drain) atom transistor, in which a 3.5 nm-thick $SiO_2$ layer separated a source and a drain. (**a**) Change in a current flowing between a source and a drain ($I_{SD}$) and a gate current ($I_G$) in a 200 times sequential switching. 10 mV was applied between a source and a drain in the operation for measuring $I_{SD}$. (**b**) Distribution of $V_{ON}$ and $V_{OFF}$ in the 200 times sequential switching. (**c**) Distribution of $R_{ON}$ and $R_{OFF}$ for those read at 50 mV. (Reproduced with permission from Ref. [9])

can exist while gate bias application. However, the decrease in a concentration of metal cations in the around due to the backward sweeping of the gate bias dissolved a nucleus. Namely, the size of a nucleus was not big enough. On the other hand, the larger bias application made a size of a nucleus large enough, resulting in the nonvolatile operation.

The gate bias used in the first demonstration of atom transistor was larger than that of the conventional semiconductor transistors. Further reduction in the gate bias was a big issue of the atom transistor to be overcome in order to compete with semiconductor transistors. Our strategy was to reduce a gap size between a source and a drain that should enable switching with the formation and annihilation of a smaller metal nucleus, which is easily formed and dissolved with a smaller gate bias. The gap size in the first demonstration was 10 nm.

Figure 5 shows the operating result of Ag(gate)/$Ta_2O_5$/Pt(source), Pt(drain) atom transistor, in which a 3.5 nm-thick $SiO_2$ layer separates the source and the drain [9]. As we expected, switch was achieved with a smaller gate bias. In the operations, source-drain current ($I_{SD}$) changes over the range of about five orders of magnitude in the repeated on/off switching by gate bias sweeping. On the other hand, a gate current ($I_G$) remains in the order of 0.1 pA during the whole switching processes. Although variations in the switching-on bias (220–300 mV) and switching off bias (−30 to −100 mV) are observed, nonvolatile three-terminal operations are clearly demonstrated with the high on/off ratio. From the gate leakage current ($I_G$), the power consumption is estimated to be very small less than fW. The reduction of a gap between a source and a drain also improved repeatability of the atomic switch.

Another important factor in terms of energy consumption is a sub-threshold slope. Namely, steep sub-threshold slope enables faster switching with less energy consumption. The sub-threshold slope of the atom transistor is of about 10 mV/decade (Fig. 5), which is much steeper than 60 mV/decade that is the value of ideal MOSFET. The steeper sub-threshold slope of atom transistor can also drastically reduce power consumption in logic circuits.

## 3 Oxygen Ion Controlled Type

Three-terminal operation has also been demonstrated by controlling oxygen ion's drift to make a conductive channel between a source and a drain. The initial target of this oxygen ion controlled type was to develop a nonvolatile three-terminal atomic switch that turns on with a negative gate bias application. Since the atom transistor mentioned in the former section turns on with a positive gate bias application, the development of the three-terminal atomic switch that turns on with a negative gate bias would have enabled a nonvolatile complementally operating system that can fully replace CMOS systems.

For that purpose, we decided to use the phenomena that some metal oxides become conductive with increase in the concentration of oxygen ions from the stoichiometric condition. Since a negative gate bias application brings oxygen ions towards a channel region, the phenomena should turn on an oxygen ion controlled three-terminal switch.

In order to confirm the mechanism, we fabricated a sidewall gate type device, as shown in Fig. 6. We employed TaOx as a channel material instead $Ta_2O_5$ in order to easily move oxygen ions. Another difference from the atom transistor is that a gate electrode consists of Pt so that ions that can move in a metal oxide layer is only oxygen ions. The operating result is shown in Fig. 7. As can be seen in Fig. 7a, resistance between a source and a drain decreased to the order of 10 kΩ from the order of 10 GΩ, while resistance between a gate and source/drain was kept at around the order of GΩ, suggesting a three-terminal switching due to the increase in oxygen ions occurred. Figure 7b shows I/V characteristics between a source and a drain measured in the off-state and the on-state. The I/V characteristics in the on state shows nonlinearity, suggesting that a conductive path is semiconductive. As such we succeeded in turning on by increasing oxygen ions at a channel region. However, the repeatability was not good because the increasing oxygen ions more than the

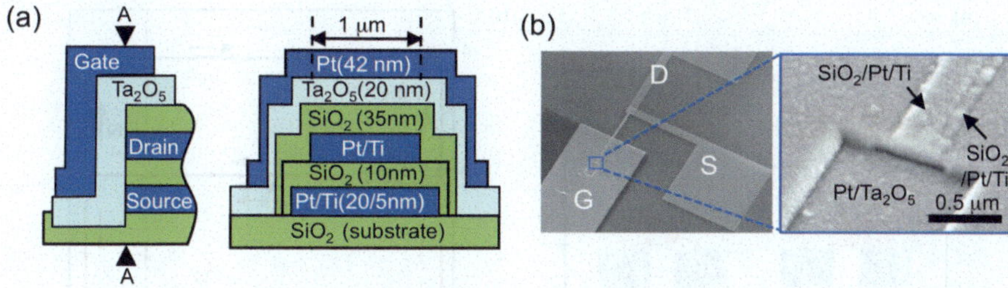


**Fig. 6** Oxygen ion controlled three-terminal switch. (**a**) Schematic of the fabricated device. (**b**) A scanning electron microscopic image of the device. (Reproduced with permission from Ref. [10])

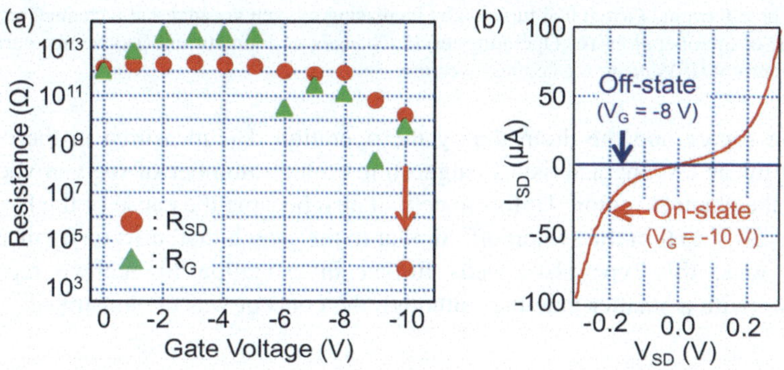


**Fig. 7** Result of the oxygen controlled operation. (**a**) Change in resistance between a source and a drain ($R_{SD}$) and that between a gate and source/drain ($R_G$) as a function of a gate bias. (**b**) $I_{SD}$ vs. $V_D$ measured at $V_G = -8$ V and $-10$ V. (Reproduced with permission from Ref. [10])

stoichiometric value often gives a severe damage in a TaOx layer. Namely, a hard breakdown occurs. Exploring of metal oxide that is not damaged by the increase in oxygen ions is needed.

Three-terminal operation is also achievable by decreasing oxygen ion concentration at a channel region [11], which is the method to make a conductive channel used in the conventional two-terminal metal oxide resistive switches [12]. Its operating concept is shown in Fig. 8a, where an oxygen vacancy $V_O^{++}$ are drawn instead of an oxygen ion. The application of a positive gate bias brings oxygen vacancies towards a source and a drain. When a gap between a source and a drain is sufficiently narrow, oxygen vacancies also migrate to the channel region, making the channel conductive. Application of a negative gate bias brings oxygen vacancies back from the channel region, returning the channel insulative. Figure 8b shows an example of operating results of the oxygen ion-controlled three-terminal switch. Gate bias sweeping in the positive polarity steeply increased source current ($I_S$) and drain current ($I_D$) at 8 V. The large currents of a source ($I_S$) and a drain ($I_D$) were kept until a negative gate bias of $-38$ V was applied. The gate current ($I_G$) remained at a value much smaller than $I_S$ and $I_D$, indicating the gate electrode was electrically isolated

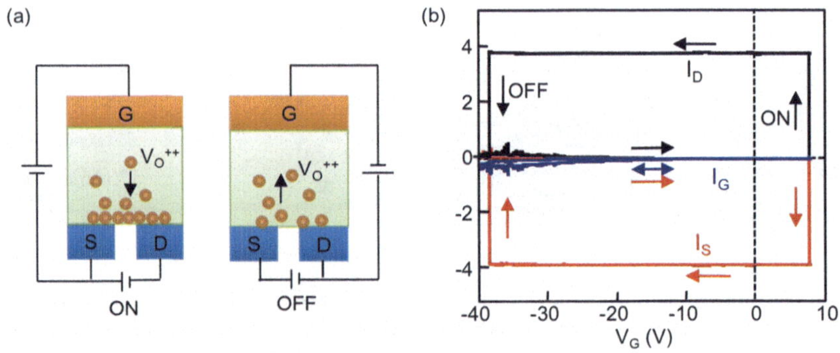


**Fig. 8** Three-terminal switch that turns on by increasing oxygen vacancies at a channel region. (**a**) Schematic of the operation. (**b**) Operating result. Changes in drain current ($I_D$), source current ($I_S$) and gate current ($I_G$) during a gate bias sweeping

from the source and the drain during the operation. $I_{SD}$ measured in the on-state showed linear I/V characteristics, suggesting enough number of oxygen vacancies were at the channel region. Unfortunately, it also became the cause of the large gate bias of −38 V to turn the switch off. Similar to the switch that turns on by increasing oxygen ions, this type also needs further investigation to achieve repeatable switching with a smaller gate bias although the concept was demonstrated.

## 4 Summary

A three-terminal operation has an advantage to a two-terminal operation in terms of controllability and energy efficiency. This is because the control line separated from a signal line can drastically reduce switching current, i.e., energy consumption. Both cation-controlled type and anion-controlled type have been developed. Cation controlled type operates by forming a metal filament/nucleus that connects a source and a drain. It shows high enough on/off ratio up to the six orders of magnitude and repeated operations. Oxygen controlled type can be turned on both by increasing oxygen ions and by decreasing oxygen ions at a channel. However, repeatability should be improved by optimizing materials and structures.

## References

1. Ikeda, H., Tada, K., Narukawa, S., Ooe, Y.: Denki Kagaku. **44**, 535 (1976)
2. Thakoor, S., Moopenn, A., Daud, T., Thakoor, A.P.: J. Appl. Phys. **67**(6), 3132 (1990)
3. Xie, F.-Q., Nittler, L., Obermair, C., Schimmel, T.: Phys. Rev. Lett. **93**, 128303 (2004)
4. Banno, N., Sakamoto, T., Iguchi, N., Kawaura, H., Kaeriyama, S., Mizuno, M., Terabe, K., Hasegawa, T., Aono, M.: IEICE Trans. Electron. **E89-C**, 1492 (2006)

5. Sakamoto, T., Iguchi, N., Aono, M.: Appl. Phys. Lett. **96**, 252104 (2010)
6. Tsuruoka, T., Terabe, K., Hasegawa, T., Aono, M.: Nanotechnology. **21**, 425205 (2010)
7. Hasegawa, T., Itoh, Y., Tanaka, H., Hino, T., Tsuruoka, T., Terabe, K., Miyazaki, H., Tsukagoshi, K., Ogawa, T., Yamaguchi, S., Aono, M.: Appl. Phys. Express. **4**, 15204 (2011)
8. Tsuruoka, T., Terabe, K., Hasegawa, T., Aono, M.: Nanotechnology. **22**, 2534013 (2011)
9. Wang, Q., Itoh, Y., Tsuruoka, T., Aono, M., Hasegawa, T.: Adv. Mater. **27**, 6029 (2015)
10. Wang, Q., Itoh, Y., Tsuruoka, T., Hasegawa, T., Watanabe, S., Yamaguchi, S., Hiramoto, T., Aono, M.: Key Eng. Mater. **596**, 111 (2014)
11. Wang, Q., Itoh, Y., Hasegawa, T., Tsuruoka, T., Yamaguchi, S., Watanabe, S., Hiramoto, T., Aono, M.: Appl. Phys. Lett. **102**, 233508 (2013)
12. Akinaga, H., Shima, H.: Proc. IEEE. **98**, 2237 (2010)

# Solid-Polymer-Electrolyte-Based Atomic Switches

**Tohru Tsuruoka, Karthik Krishnan, Saumya R. Mohapatra, Shouming Wu, and Masakazu Aono**

**Abstract** The atomic switch operation is demonstrated using a solid polymer electrolyte (SPE) as a matrix material. Similar to inorganic electrolyte-based atomic switches, SPE-based atomic switches exhibit not only bi-stable resistive switching but also quantized conductance. The high ionic conductivity of SPE enables us to directly observe filament growth behaviors even in micrometer-scaled devices, and to reveal the kinetic factors determining the filament growth processes. We also succeed in fabricating devices on a plastic substrate using an ink-jet technique, and to demonstrate stable resistive switching under substrate bending. All the results indicate that the SPE-based atomic switch has great potential for the development flexible switch/memory devices as well as new types of atomic-scale devices with high-speed operation and ultra-low power consumption.

## 1 Introduction

Over the last decades, organic and polymer switching memories have become attractive as an emerging research topic in electronics, because they become alternatives to, or supplement, current memory technologies based on inorganic semiconductor materials [1]. In general, organic materials exhibit significant advantages over inorganic materials in that their dimensions are easily scalable, they are readily

T. Tsuruoka (✉) · M. Aono
International Center for Materials Nanoarchitectonics (MANA), National Institute for Materials Science (NIMS), Tsukuba, Ibaraki, Japan
e-mail: TSURUOKA.Tohru@nims.go.jp

K. Krishnan
Council of Scientific and Industrial Research, Central Electrochemical Research Institute, Karaikudi, India

S. R. Mohapatra
Department of Physics, National Institute of Technology, Cachar Silchar, India

S. Wu
Zhejiang Fluoride and Silicon Research Institute, Quzhou National Hi-Tech Industrial Development Zone, Quzhou, Zhejiang, China

 
M. Aono (ed.), *Atomic Switch*, Advances in Atom and Single Molecule Machines,
https://doi.org/10.1007/978-3-030-34875-5_8

processed into a variety of shapes and geometries, and the use of molecular design is possible through chemical synthesis [2]. Polymer materials possess additional favorable properties, such as good mechanical stability, flexibility, and processability onto various substrates [3]. Therefore, they should be a promising candidate for use in future organic device applications. Many research groups have reported resistive switching in polymer-based devices, and several charge-trapping mechanisms were proposed to explain the observed switching behaviors [1, 3–5]. Despite that resistive switching memories based on electrochemical reactions in polymer materials are very attractive, a limited number of studies reported this type of polymer memory in the second half of the 2000's [6, 7]. Even in the research of atomic switches, only inorganic solid electrolytes were used during that time period.

In 2008, we considered the possibility of realizing atomic switches using solid polymer electrolytes (SPEs), which have used as the base materials for rechargeable batteries, electrochromic windows, and light-emitting devices [8]. SPEs exhibited high mechanical flexibility and stabilities up to their melting temperatures. Earlier studies on SPEs were mostly related to alkali-metal (such as $Li^+$ and $Na^+$) salt complexes, which are developed for battery applications. However, alkali-metal salt complexes are not suitable for electronic devices, because they easily react with water under ambient conditions. In contrast, $Ag^+$-salt complexes are expected to be more stable under the same conditions. Thus, in 2009, we started development of atomic switches using a $Ag^+$-conductive SPE, and have attempted to demonstrate resistive switching and the unique functions of these switches. In this chapter, we describe the basic characteristics, the fundamental mechanism, and the novel functions of SPE-based atomic switches.

## 2 Invention of SPE-Based Atomic Switch

We selected polyethylene oxide (PEO) as the first polymer material. PEO was discovered in 1973 by Wright and co-workers who studied the ionic conductivity of alkali-metal salt complexes using PEO [9]. After the discovery, PEO came to be recognized as the base material of rechargeable batteries, and the properties of PEO and its derivatives have since been extensively investigated. In this section, we describe the switching characteristics, operating mechanism, and memory properties of an atomic switch based on a $Ag^+$-conductive PEO.

### 2.1 *Typical Switching Characteristics*

The initial devices were fabricated with SPE films prepared by drop-casting from PEO ($M_W = 2 \times 10^6$) and silver perchlorate ($AgClO_4$), which were dissolved in water [10, 11]. The device has a cross-point structure with top Ag and bottom Pt electrodes, as illustrated schematically in Fig. 1a. The Ag salt concentrations of the

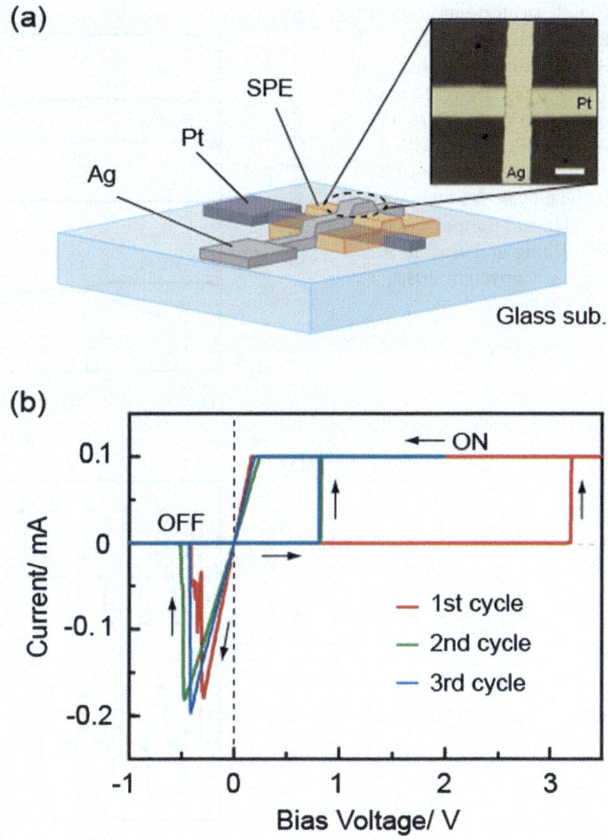


**Fig. 1** (**a**) Schematic and optical microscope image (top view) of the fabricated device. The scale bar is 50 μm. (**b**) Typical *I–V* curves of a Ag/Ag-PEO/Pt device with a Ag salt concentration of 3 wt%, measured for three consecutive sweep cycles. Copyright 2011, John Wiley & Sons

Ag-included PEO (AG-PEO) film ranged from 1 to 8 wt%. Thicknesses of SPE films were on the order of 1 μm. Figure 1b shows typical *I–V* curves measured for a Ag/Ag-PEO/Pt device with a Ag salt concentration of 3 wt%. A current compliance $I_{CC}$ was set only for positive bias to regulate a low resistance (on) state current. The devices initially exhibited a high resistance (off) state with greater than 10 GΩ. When the bias voltage was applied to the Ag electrode, the current suddenly jumped up to the compliance level at a relatively higher bias voltage, which turns on the device with an on state of 3 kΩ. Subsequent negative bias sweep sharply switched the device back to the off state of ~1 GΩ, while the device kept a high resistance to 0 V, indicating bipolar nonvolatile switching behavior. A higher positive voltage is always required to turn on the device on the first sweep cycle, corresponding to a forming process. The device showed high on/off resistance ratio of ~$10^5$, and the on resistance was found to decrease linearly with a decrease of the ambient temperature, suggesting metallic transport in the on state.

The turn-on and turn-off voltages significantly changed with an increase of the Ag salt concentration. The turn-on voltage gradually increases up to 4 wt% and then distributes over a wide bias voltage range at 5 wt%. The turn-off voltages decrease in

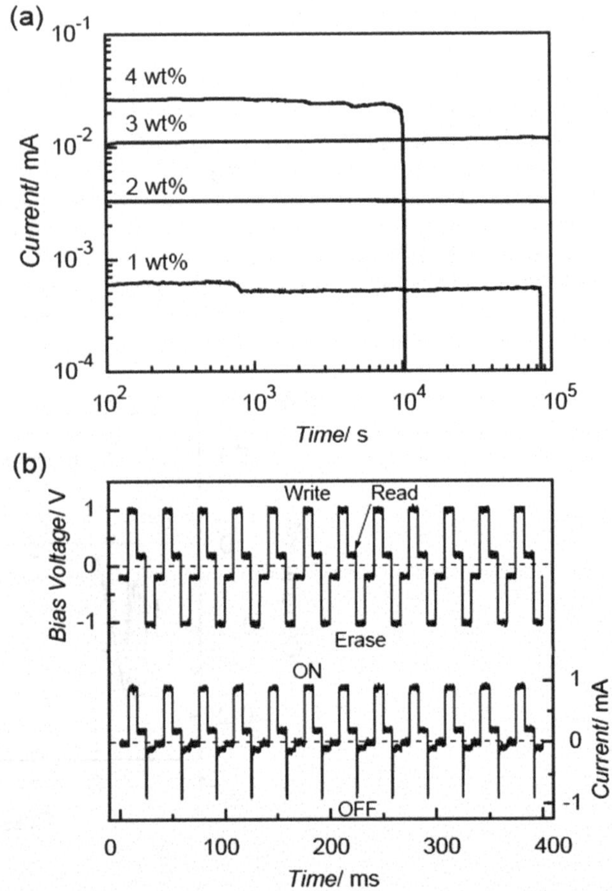


**Fig. 2** (**a**) Retention test of the on state for devices with Ag salt concentrations from 1 to 4 wt%. (**b**) Write-read-erase-read cycle test of a Ag/Ag-PEO/Pt device with a Ag salt concentration of 2 wt %. The cycle frequency is 30 kHz and the pulse width for writing and erasing is 8.3 ms. Copyright 2011, John Wiley & Sons

magnitude in the negative bias range with increasing salt concentration. However, most of them appear in the positive bias range at 5 wt%, suggesting a transition to volatile switching behavior. For higher than 6 wt%, the device retains the pristine off state and no longer shows any resistive switching behavior.

Typical retention characteristics for the on state are shown in Fig. 2a for the devices with salt concentrations of from 1 to 4 wt%. All the devices exhibit long retention times of over $10^4$ s, without any degradation. In particular, the devices with 2 and 3 wt% can retain the on state for more than 1 week. The off state can be maintained for over 1 week for all the devices. These results indicate excellent retention properties of the SPE-based atomic switch.

Figure 2b shows a write-read-erase-read cycle test conducted on the device with a Ag salt concentration of 2 wt%. The writing and erasing pulses were $\pm 1$ V respectively, and the device resistance was probed with small voltages of $\pm 0.2$ V, as shown in the upper part of Fig. 2b. The lower part of Fig. 2b represents the current responses, showing that the device is turned on and off by the consecutive writing and erasing pulses. Switching speed changes from several hundred nanoseconds to a few hundred microseconds, depending on the voltage pulse condition, the thickness of the Ag-PEO film, and the fabrication conditions of the devices. Even at this early stage, the SPE-based atomic switch exhibits high programing speeds ($<1\ \mu s$) and long retention times ($>1$ week), demonstrating as appropriate capabilities for non-volatile memories.

## 2.2 Switching Mechanism

To understand the observed switching behavior, the thermal and transport properties of the Ag-PEO film were investigated as a function of the Ag salt concentration [11]. The pure PEO film shows a low conductivity of ~$10^{-9}$ S cm$^{-1}$ and a glass transition temperature $T_g = -55$ °C. With increasing Ag salt concentration, conductivity rises by two orders of magnitude up to 3 wt% and then decreases slightly for higher concentrations. $T_g$ increases monotonically and reaches −25 °C at 6 wt%. The electronic transport number is estimated to be less than $3 \times 10^{-3}$ and is almost independent of the salt concentration. In contrast, the cation transport number increases up to $7 \times 10^{-2}$ at 2 wt%, and then decreases for higher Ag salt concentrations. This behavior can be explained by initial increase of carrier ion numbers with introducing Ag salts and subsequent decreased number of cations due to formation of ionic aggregates such as ion pairs and triple ions [12].

The standard redox potentials for Ag in aqueous solutions are given as $Ag^{2+}/Ag^{+}$ ($E^{\phi,1} = 1.98$ V) and $Ag^{+}/Ag$ ($E^{\phi,2} = 0.8$ V) [13]. Since water is absorbed from the atmosphere into the PEO matrix [14], the residual water can involve redox reactions at the interface and the general trend of redox potentials holds in our devices. According to the standard redox potentials, oxidation from Ag to $Ag^{+}$ is thermodynamically the most favorable. Cyclic voltammetry measurements were performed for Ag/Ag-PEO/Pt devices with different salt concentrations to investigate redox reactions at the Ag/PEO interface. At 3 wt%, current peaks were observed at around $\pm 1$ V, which can be attributed to oxidation from Ag to $Ag^{+}$ and reduction from $Ag^{+}$ to Ag respectively. The redox current was reduced for higher salt concentrations, although the oxidation and reduction peaks were still observed. This current reduction is considered to come from a decrease in both the mobility and number of free carriers, due to the formation of ionic aggregates.

Based on the results obtained, we inferred the switching mechanism that is illustrated schematically in Fig. 3. When a positive bias voltage is applied to the Ag electrode at the first sweep cycle (Fig. 3a), $Ag^{+}$ ions oxidized at the Ag/PEO interface and initially dissolved in the PEO matrix transport to the Pt electrode. At the same time, $ClO_4^{-}$ ions move toward the Ag electrode. Then, a metal filament is formed between the Ag and Pt electrodes by heterogeneous nucleation of Ag on Pt followed by the Ag nucleus growth, which turns on the device (Fig. 3b). When a negative bias voltage is subsequently applied, the oxidation reaction takes place at surfaces of the metal filament, and the thinnest part of the metal filament is dissolved,

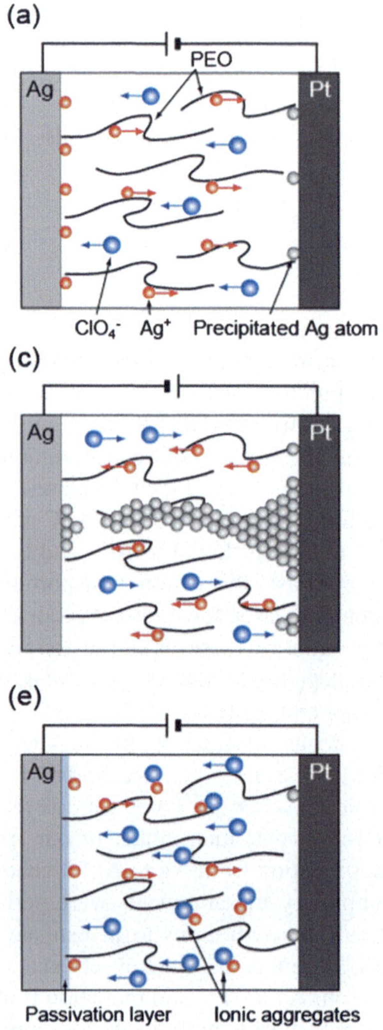


**Fig. 3** Switching mechanism of the Ag/Ag-PEO/Pt atomic switch. (**a**) Transport of $Ag^+$ and $ClO_4^-$ ions under positive bias. (**b**) Formation of a metal filament by inhomogeneous nucleation and subsequent growth of Ag on Pt. (**c**) Local dissolution of the metal filament at the thinnest part under negative bias. (**d**) Reformation of the metal filament under positive bias. (**e**) Formation of ionic aggregates and a passivation layer and (**f**) incomplete formation of the metal filament at higher salt concentrations. Copyright 2011, John Wiley & Sons

disconnecting the metal filament locally (Fig. 3c). The device is switched back to the off state. After that, resistive switching occurs by alternative positive and negative biasing, based on the reformation and dissolution at the thinnest part of the metal filament (Fig. 3d and c). Because reformation and dissolution of the filament occur in a small gap, the turn-on voltage is lowered after the second sweep cycle.

When the Ag salt concentration increases, the number of free $Ag^+$ ion decreases by the formation of ionic aggregates (Fig. 3e). In addition, $ClO_4^-$ anions can form radicals by decomposition, giving rise to the degradation of polymer molecules to produce a passivation layer at the electrode interface. This may form higher-ordered ionic associations. These species are immobile in the polymer matrix and contribute to the further build-up of the passivation layer, which prevents the supply of $Ag^+$ ions into the polymer matrix. As a result, metal filament formation becomes increasingly difficult and higher bias voltages are needed to turn on the device. For Ag salt concentrations higher than 5 wt%, metal filaments are no longer formed (Fig. 3f) and resistive switching never take place. This model explains the experimental observations well and indicates the importance of the salt concentration so as to obtain the desired characteristics.

# 3 Kinetic Factors Determining Filament Formation

To realize stable resistive switching, an understanding of the filament growth kinetics under voltage biasing is very important. The structural characteristics of SPE films are considered to play a major role in both ion transport and subsequent filament growth behavior [15, 16]. One can expect that the morphology of thin SPE films has a major impact on ion transport as well as on the formation of conducting filaments in the atomic switch. We have systematically studied the filament growth behavior of SPE-based atomic switches by means of in situ and ex situ microscopy [17–19]. In this section, we discuss how switching behavior is controlled by kinetic factors, which dominates the filament formation kinetics in SPE-based atomic switches.

## *3.1 Direct Observation of Filament Growth Processes*

Planar devices were fabricated on a $SiO_2$-covered Si substrate, in which opposing Ag (or Pt) and Pt (or Ag) electrodes were formed with different gaps of 0.5–8 μm [17, 18]. An SPE film with a Ag salt concentration of 3 wt% was formed on the electrodes by a spin coating method. Thickness of the coated SPE films was ~200 nm. Under voltage biasing, direct observation and video capture of the device were performed using an optical microscope equipped with a CMOS camera.

Figure 4a shows a typical current-time (*I–t*) plot obtained for a Ag/Ag-PEO/Ag planar device with a gap of ~8 μm under constant voltage application of +1 V. A low $I_{CC}$ of 11 nA was preset to regulate the initial stage of filament growth. The corresponding optical microscope images, taken at the selected times, are presented in Fig. 4b1–b8. After biasing, the current gradually decreased with time. This initial current originates from the oxidation of Ag at the biased electrode interface. Small voids, appeared in the biased electrode, evidence the dissolution of Ag atoms. These

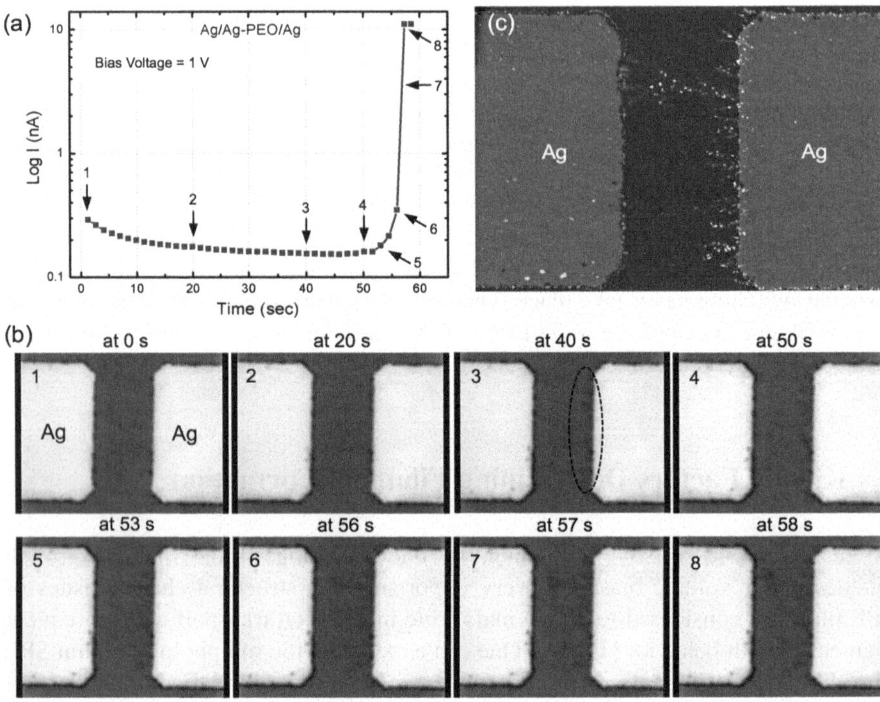


**Fig. 4** (**a**) Typical *I–t* curve measured under of a bias voltage of 1 V for a symmetric Ag/Ag-PEO/Ag planar device with a gap of 8 μm. (**b**) In-situ optical microscope snapshots measured at the selected times indicated in (**a**). (**c**) SEM image taken after the bias voltage application. Copyright 2016, IOP Publishing

voids became larger with longer biasing times. The current continued to decrease up to a certain time (~50 s) and then jumped to the compliance level, indicating the formation of conducting filaments between the electrodes. However, it can be observed that Ag precipitations started to appear from the grounded electrode at an earlier time (~40 s), as indicated by the dotted circle in Fig. 4b3. These precipitations are attributed to the reduction of $Ag^+$ ions on the grounded electrode. With increasing biasing time, filaments grew toward the biased electrode from many precipitation sites, and finally one of the filaments made a connection between the electrodes.

The morphology of the formed filaments was examined by SEM. Figure 4c shows an SEM image taken after biasing. Many voids are created in the biased (left) electrode and a number of filaments protrude from the grounded (right) electrode. The image also shows that the filaments contain a lot of Ag clusters as well as very thin filament structures, and that the filaments grow in a unidirectional manner near the grounded electrode. On the other hand, away from the grounded electrode and beyond a certain critical length, the growth morphology changes to a random and dendritic growth behavior.

The same measurements were performed for Ag/Ag-PEO/Pt, Ag/PEO/Pt, and Pt/Ag-PEO/Pt planar devices with gaps of 6–9 μm. Depending on the electrode material and salt inclusion in the PEO matrix, different filament growth behaviors as well as different current responses were observed. Ag included PEO-based devices exhibited similar current behavior as the Ag/Ag-PEO/Ag device (Fig. 4a), whereas PEO-based devices showed that the current never reach the compliance level due to the large gap. The symmetric electrode configuration (Ag/Ag) induced gradual and random growth of conducting filaments, but the asymmetric electrode configuration (Ag/Pt) promoted unidirectional filament growth.

## *3.2 Impacts of Device Configuration and Experimental Parameters*

The observation of a planar device enabled us to investigate how redox reactions and subsequent filament formations take place under certain bias conditions. However, for gaps of μm scale, the devices exhibit no switching behavior even after the forming process is clearly observed. This indicates that planar devices with μm gaps are difficult to correlate the filament growth processes to the actual switching behavior. Hence, we used EB lithography processes to fabricate planar devices. Owing to the decreased distances between the electrodes, stable switching behavior could be observed, and the subsequent SEM observations allowed us to investigate the filament growth processes affected by the device and experimental parameters such as the gap between electrodes, salt inclusion, and $I_{CC}$ [19]. Most of this subsection is reproduced with permission from Ref. 19. Copyright 2016 the Royal Society of Chemistry.

Figure 5a–c show SEM images of Ag/Ag-PEO/Pt devices with gaps of 2, 1, and 0.5 μm, taken after the forming process was done under positive bias sweeping with a constant sweep rate and a certain $I_{CC}$. All the SEM images show dendritic filament morphologies, in which one of the filaments connects between the electrodes. For the smallest gap (0.5 μm), narrower filaments grow randomly from the Pt electrode. As the gap increases, the filaments grow further with changing to dendritic morphologies. This indicates that the dendritic growth can be enhanced in devices with larger gaps, although the filament growth is driven by the same dynamics. The forming voltage is reduced for decreased gaps, suggesting that the number of $Ag^+$ ions required to form a filament is reduced for smaller gaps. This is evidenced by the decreased void area in the Ag electrode for reduced gaps.

As shown in Fig. 5d–f, the impact of the gap between electrodes on the filament growth processes is more pronounced for Ag/PEO/Pt devices. No completed filament formation occurred for the largest gap (2 μm), even after biasing up to ~9 V, although some precipitations appeared on the Pt electrode (Fig. 6d). This means that the device cannot be turned on for such a gap range under the bias condition used. However, when the gap was reduced, very narrow filaments, consisting of small Ag

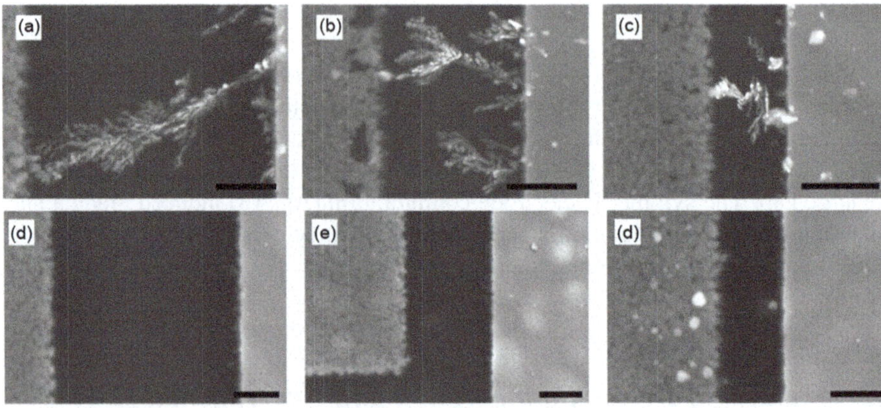


**Fig. 5** SEM images of the filament growth morphology between Ag (left side) and Pt (right side) electrodes, after the forming process, taken for Ag/Ag-PEO/Pt and Ag/PEO/Pt devices with gaps of 2 (**a**, **d**), 1 (**b**, **e**), and 0.5 μm (**c**, **f**). The scale bar is 500 nm. Copyright 2016, the Royal Society of Chemistry

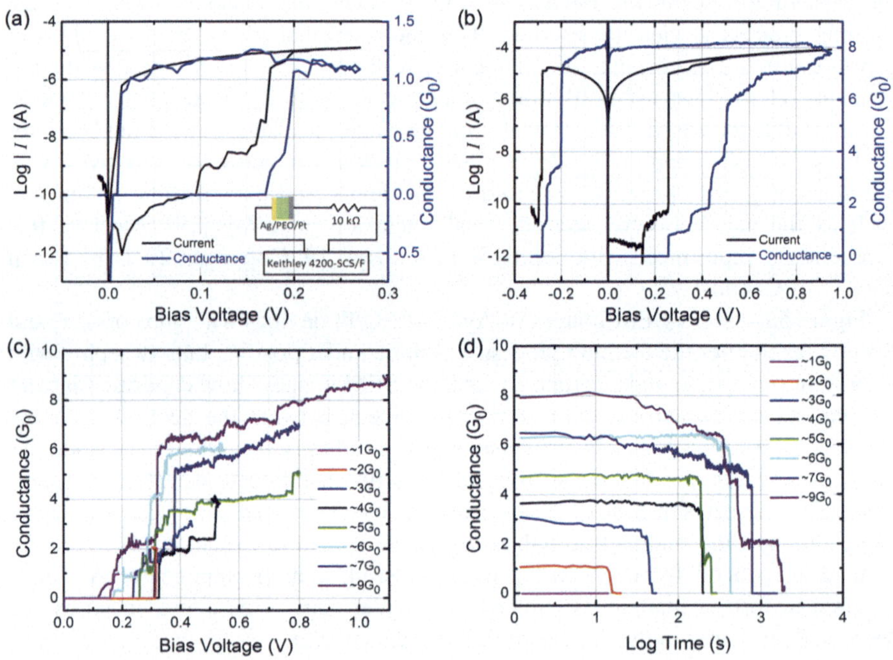


**Fig. 6** *I*–*V* and the corresponding conductance curves measured with different $V_S$, resulting in maximum quantized steps of ~1$G_0$ (**a**) and ~8$G_0$ (**b**) in a Ag/PEO/Pt device. To control the conductance level, a current-limiting resistor of 10 kΩ was inserted in series with the device, as illustrated in the inset of (**a**). (**c**) Stimulated multilevel conductance quantization with various $V_S$. (**d**) Spontaneous conductance decay behavior measured as a function of time after realizing each conductance state shown in (**c**). Copyright 2017, John Wiley & Sons

clusters, were formed between the electrodes (Fig. 6e and f) after forming process. These results show that the gap between electrodes plays a crucial role in the filament formation in the PEO-based device.

The effects of the salt inclusion in the polymer matrix are found in the comparison of the observations between Ag/Ag-PEO/Pt and Ag/PEO/Pt devices. Since pure PEO is a highly crystalline polymer [20], redox reactions at the Ag/PEO interface and $Ag^+$ ion mobility are significantly hindered in the Ag/PEO/Pt device. This results in almost no filament formation and only precipitations on the Pt electrode for a gap of 2 μm (Fig. 6d). With decreasing gap, precipitated Ag atoms tend to grow along sites where $Ag^+$ ions transport, inducing unidirectional filament growth, as seen in Fig. 5e and f. The observed filaments consist of small Ag clusters rather than being connected continuously. This situation changes drastically in the Ag/Ag-PEO/Pt device. The salt inclusion of 3 wt% reduces the crystalline degree of the PEO matrix, giving rise to an increased ionic conductivity of more than two orders of magnitude [11]. The oxidation rate at the Ag/Ag-PEO interfaces is also enhanced due to an increased amorphous phase. Therefore, both the $Ag^+$ ions dissolved from the Ag electrode and the pre-existing $Ag^+$ ions can contribute to ion transport. Once precipitations take place on the Pt electrode, the Ag nucleus grow by incorporating $Ag^+$ ions from the surrounding area. As a result, the precipitated atoms form dendritic morphologies with tree-like branching structures, as seen in Fig. 5a–c.

If the $I_{CC}$ is increased, the filament morphologies are retained in both the Ag/Ag-PEO/Pt and Ag/PEO/Pt devices, but the density and size of the filament structures are increased. The $I_{CC}$ level also determines whether the switching behavior is volatile or nonvolatile. The results obtained indicate that the filament formation and switching behavior depend strongly on several kinetic factors, such as the redox reaction at the electrode/polymer interfaces, ion mobility and electric field strength in the polymer matrix, and reduction sites for precipitations. The different filament formations, between unidirectional and dendritic growth, can be controlled by tuning specific parameters, which can improve the stability and performance of SPE-based atomic switches.

## 4 Highly Reproducible Conductance Quantization

Besides having bi-stable switching ability, one of the characteristic features of atomic switches is conductance quantization. This quantized conductance was first realized in a $Ag_2S$-based atomic switch [21, 22], and was subsequently demonstrated in a $Cu_2S$-based atomic switch [23] and a $Ta_2O_5$-based atomic switch [24]. We found that it is also possible to observe highly reproducible conductance quantization in Ag/PEO/Pt devices. In this section, we first present experimental results on the quantized conductance of Ag/PEO/Pt devices, and then discuss, using first-principles density functional theory (DFT) simulations, how the structural variation of atomic point contacts influences the stability of the quantized conductance. Most of this section is reproduced with permissions from Ref. 25. Copyright 2017 John Wiley & Sons, and from Ref. 26. Copyright 2017 IOP Publishing.

## 4.1 I–V *Characteristics in the Atomic Contact Regime*

*I–V* measurements were performed for a Ag/PEO/Pt device with a 40 nm-thick PEO film, at room temperature in air [25, 26]. Quantized conductance can be observed by carefully tuning the stop voltage ($V_S$) and voltage sweep rate in positive bias. Figure 6a and b depict the typical *I–V* and corresponding conductance curves at different $V_S$ with a voltage sweep rate of 3.5 mV $s^{-1}$. For these measurements, a 10 kΩ resistor was inserted in series with the device to regulate the on state current, as illustrated in the inset of Fig. 7a. With $V_S = 0.27$ V, the device conductance jumped from zero to $1G_0$ (Fig. 6a). This conductance jump to $1G_0$ represents the formation of a single atom point contact in the narrowest region of the metal filament formed. As the bias voltage was subsequently swept back, the conductance state was initially retained at the ~$1G_0$ level, but dropped to zero before the bias voltage reached 0 V, indicating volatile switching behavior. On the other hand, with $V_S = 0.95$ V, the device conductance increased in a stepwise fashion and finally reached $8G_0$ (Fig. 6b). The conductance state changed not only to integer multiples but also to half-integer multiples. In addition, the conductance state held its level even if the bias voltage was swept back to 0 V and dropped to zero in negative bias, indicating nonvolatile behavior.

The retention characteristics of different conductance values were also investigated. Figure 6c plots *I–V* curves measured with a constant sweep rate (3.5 mV $s^{-1}$) and various $V_S$ (0.18–1.1 V) to realize quantized conductance states with a specific value. Then, the spontaneous conductance decay (retention) was measured as a function of time for each quantized conductance state (Fig. 6d). Conductance states of $\leq 1G_0$ dropped to zero before starting the retention measurement and thus did not show any retention characteristics. In contrast, conductance states of $\geq 2G_0$ exhibited different retention time depending on the conductance value. The retention time became longer and increased exponentially for higher conductance values.

The conductance state distribution was evaluated from the *I–V* data with different $V_S$. Figure 7 depicts conductance-state histograms by counting the number of states for different $V_S$. The Ag/PEO/Pt device exhibited large peaks at integer multiples of $G_0$, clearly demonstrating quantized conductance behavior. The peaks of the state distributions shift to the higher conductance side with increasing $V_S$. In addition, distinct peaks appear at half-integer multiples of $G_0$, and fractional conductance variations are also observed between peaks.

## 4.2 *Transport Simulations for Atomic Point Contacts*

Detailed atomistic simulations were performed to investigate the correlation between the observed quantized conductance and the atomic point contact structure [25, 26]. The quantized conductance originates from the narrowest region of a metal filament between electrodes, which consists of the number of atoms

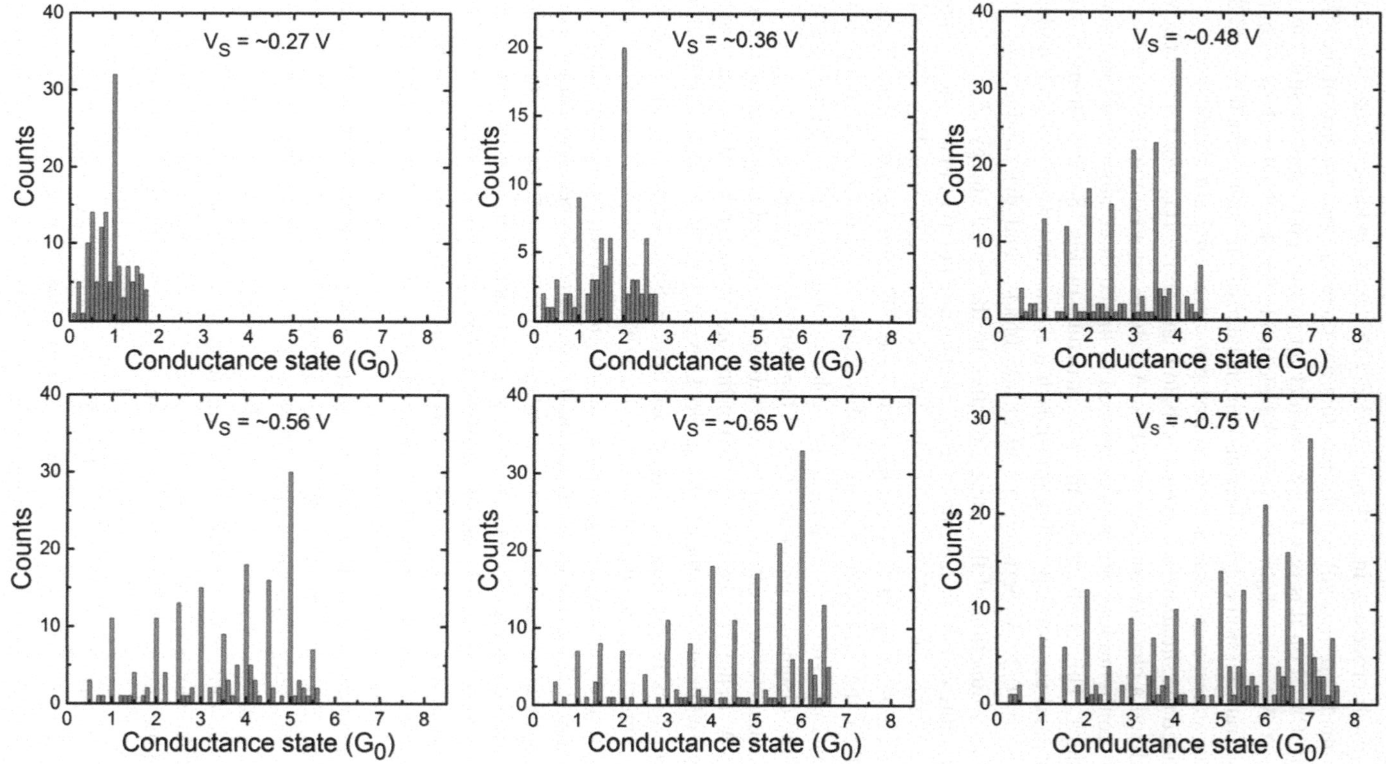


**Fig. 7** Conductance state histograms in $G_0$ evaluated from conductance curves with different $V_S$ for a Ag/PEO/Pt atomic switch. Copyright 2017, John Wiley & Sons

$n$ (where, $n = 1, 2, 3, \ldots$) participating in an atomic point contact. Observations of filament growth processes in a planar Ag/PEO/Pt device evidenced that the metal filament is made up of small Ag clusters that are unevenly precipitated in the PEO matrix [20]. Based on this result, we consider an atomic point contact structure as consisting of Ag atom chains between Ag electrode blocks. Under these circumstances, the transmission eigenvalues and eigenstates of the entire structure, and the conductance states, were evaluated from first-principles DFT simulations.

We divided the situation into two parts; below and above $1G_0$. For below $1G_0$, we first considered a channel consisting of two Ag atoms in a symmetric three-to-one atom chain configuration, which is connected to linked three-atom chain structures on both sides. This is shown as the top structure in Fig. 8a. The transmission eigenvalues and the corresponding transmission eigenstates were analyzed under a bias voltage of 0.25 V between the electrode blocks, and the conductance was calculated as a function of the energy relative to the Fermi energy ($E_F$) at 0 eV. The top structure of Fig. 8a has one transmission channel of $1G_0$, in the energy range of between −0.5 and 0.5 V, as shown by the black curve in Fig. 8b, and a transmission eigenvalue at $E_F$ is estimated to be 0.999. This suggests that electrons incoming from the left electrode are negligibly scattered and reach the right electrode without any loss. The channel length was calculated to be 7.58 Å, which length is defined as the distance between the tips of the two electrode blocks.

The conductance was calculated by extending the channel length up to 2.5 Å in steps of 0.5 Å. In the calculations, the two in-line atoms were relaxed to minimize the total energy, while atoms in the electrode blocks were fixed in position. The resulting relaxed structures are presented in Fig. 8a for three extended distances of 0.5, 1, and 2 Å. The channel lengths are estimated to be 8.08, 8.58, and 9.08 Å, respectively. With extending the channel length, the center region breaks easily and a small gap appears, even at the extended distance of 0.5 Å. This gap becomes larger as the channel length is increased, and the corresponding conductance is substantially reduced, as shown in Fig. 8b. The result shows that such a single-atom point contact is unstable and breaks easily under small or no bias conditions. Thus, the volatile switching behavior that occurs below $1G_0$ originates from the change of the gap distance at the atomic point contact, where electron tunneling through Ag clusters dominates.

For conductance states above $1G_0$, we next considered the addition of Ag atoms into the channel with the same electrode configuration. DFT calculations were performed by adding different numbers of Ag atoms ($n = 2$–$5$) to the atom chains, as shown by the dotted circles in Fig. 9a. The channel length was fixed at 7.58 Å in all the calculations. Figure 9a represents the final (relaxed) structures for the given number of Ag atoms in the channel. The corresponding conductance is plotted as a function of the energy in Fig. 9b. The three-atom chain exhibits conductance slightly higher than $1G_0$ at $E_F$. This suggests that a three-atom chain forms almost a single-atom point contact after structural relaxation. In contrast, the four- and five-atom chains exhibit conductance values slightly lower than $2G_0$ at $E_F$, indicating the formation of a two-atom point contact. Atomic contacts that have more than three atoms are deformed and twisted, and lose translational invariance along the channel.

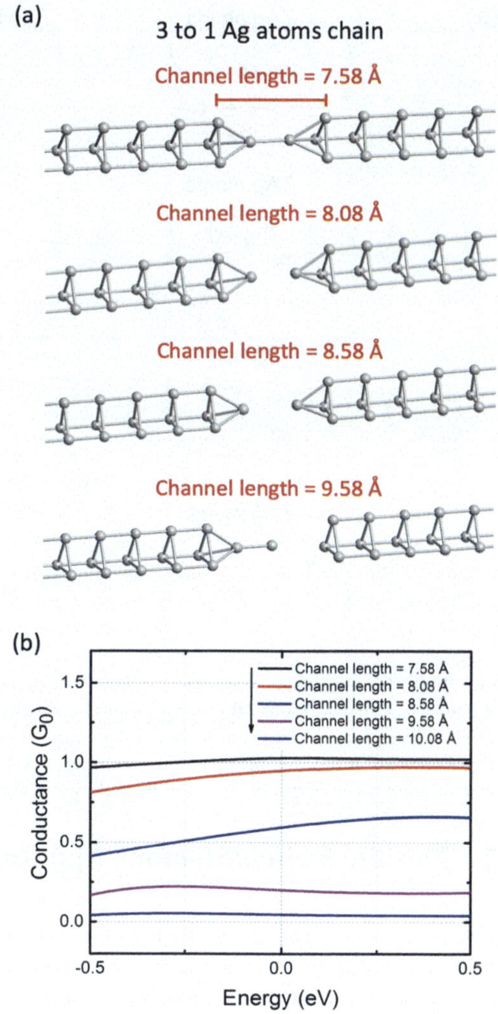


**Fig. 8** (**a**) Atomic configuration of an in-line two Ag atom chain connected to linked three Ag atom chain structures with different channel lengths. (**b**) Energy-dependent conductance plot for different channel lengths. Copyright 2017, IOP Publishing

Therefore, the electrons in the channels suffer from electron scattering, resulting in reduction of the transmission rates (conductance).

Transmission channel analysis also indicates that three conducting channels at $E_F$ (indicated by the arrow in Fig. 9b) can be realized for the three-atom chain configuration, as shown in Fig. 9c. The first two transmission channels have transmission eigenvalues higher than 0.1 and exhibit wave functions that reach the right electrode block, whereas the remaining channel has a very low transmission eigenvalue and its wave function does not reach the right electrode. Therefore, in this configuration, two partial transmission channels contribute to the single-atom point contact. A similar situation is confirmed for the four- and five-atom chain configurations. These results indicate how the atomic configuration is important in determining conductance states of atomic switches.

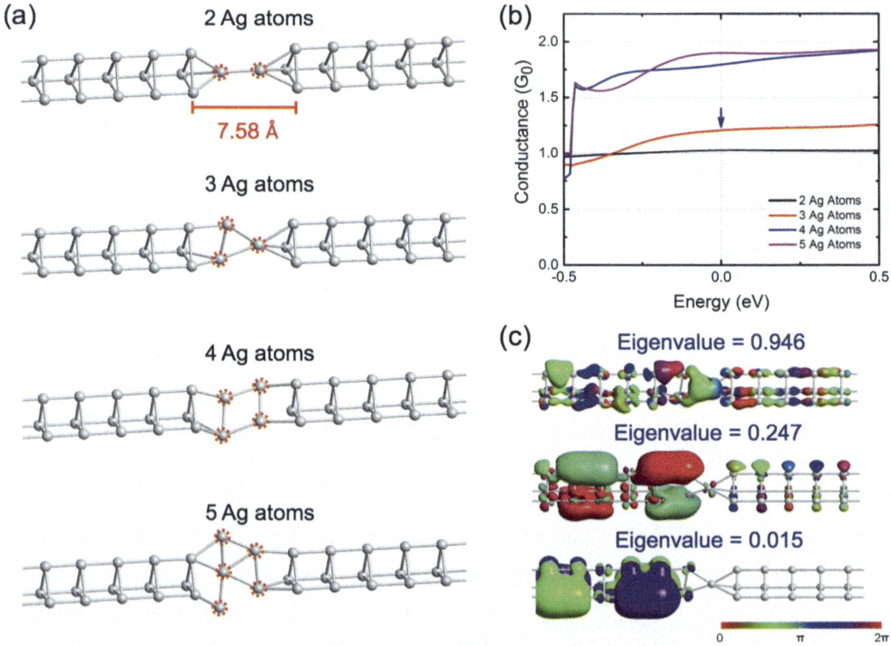


**Fig. 9** (**a**) Geometrically optimized (relaxed) structures of different numbers (2–5) of Ag atom chains connected to linked three Ag atom chain structures on both sides. (**b**) Energy-dependent conductance plot for various numbers of Ag atoms in the channel as shown in (**a**). (**c**) Transmission eigenstates of three Ag atom chains forming a single-atom point contact, calculated for the transmission eigenvalues at $E_F$, as indicated by the arrow in (**b**). Copyright 2017, IOP Publishing

## 5 Flexible Switch/Memory Applications

The process of fabricating device components with solution drop-casting or spin-coating techniques are considered to be more cost effective and easier than expensive lithographic techniques. Specifically, for polymer materials, solution-based deposition techniques are widely accepted. However, these solution casting methods sometimes lack the ability to control the thickness of SPE layers and pin-pointing of the deposition area. To overcome these problems, we tried a drop-on-demand ink-jet printing (IJP) method for deposition of SPE solutions. The advantage of IJP is that the required amount of the target solution is deposited onto specified positions only when needed; thus, it is more economical than other deposition methods [27]. In this section, we present the fabrication and characterization of atomic switches using an IJP SPE film. Most of this section is reproduced with permissions from Ref. 28. Copyright 2012 AIP Publishing, and from Ref. 29. Copyright 2012 Cambridge University Press.

A Ag/Ag-PEO/Pt device was fabricated on a polyethylene naphthalate (PEN) substrate with a thickness of 0.2 mm, on which 5-nm-thick Ti and 30-nm thick Pt were formed as the adhesion layer and bottom electrode [28]. The step-by-step

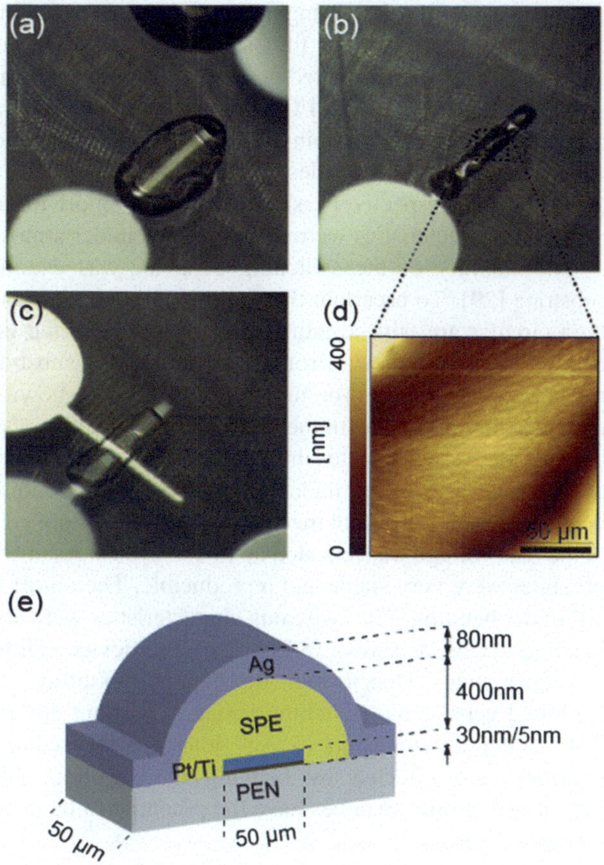


**Fig. 10** Optical images of a Ag-PEO film deposited on a Pt electrode immediately after IJP (**a**) and after drying for 24 h (**b**). (**c**) The device after deposition of a Ag electrode. (**d**) An AFM image of an inkjet-printed film on the Pt electrode. (**e**) Schematic illustration of the cross-section of the fabricated Ag/Ag-PEO/Pt device. Copyright 2012, AIP Publishing

process for the device fabrication by IJP of the Ag-PEO solution is depicted in Fig. 10. The SPE film was deposited on the Pt electrode by dropping an acetonitrile solution, mixed with PEO ($M_W = 20{,}000$) and $AgClO_4$ (3 wt%), from an ink-jet head with an 80 μm nozzle. The deposition of SPE solution was quite tricky, requiring lots of optimization of various parameters such as the PEO molecular weight, solution concentration, nozzle size, pulse width and height of the voltages applied to the ink-jet head. The deposited solution initially spread out around the Pt electrode to form a coffee-ring pattern, as shown in Fig. 10a, which pattern is often observed in IJP under certain conditions [30]. Then, during evaporation of the solvent, the solution gradually coagulated on the Pt electrode, because of higher surface energy of Pt (~1000 dyn/cm) than PEN (~44 dyn/cm). This high contrast of the surface energy guided the deposition behavior of the solution. As a result, the solution covered most of the Pt electrode (50 μm width) after drying, as shown in Fig. 10b. AFM measurements revealed that the deposited Ag-PEO film is cylindrical, with a maximum thickness of ~400 nm and a base length of ~150 μm. The AFM image (Fig. 10d) also shows a complex lamellar morphology in the Ag-PEO film, suggesting that the ink-jet printed film has a highly crystalline nature. Finally, 80-nm-thick Ag was deposited on the Ag-PEO film as the top electrode, as shown in Fig. 10c. The cross-section of the fabricated device is illustrated schematically in Fig. 10e.

The fabricated Ag/Ag-PEO/Pt device exhibited bipolar resistive switching characteristics under bias sweeping. Resistive switching could be repeated for more than 100 consecutive sweep cycles. The on and off resistances were of the order of 1 kΩ and 100 MΩ, respectively, showing a high on/off resistance ratio of $\sim 10^5$. Good retention characteristics were also observed under atmospheric conditions.

The stability of the switching behavior was examined by bending the PEN substrate [29]. To ascertain the bending amount, the bending radius was estimated by a circular arc with a radius *r*, as illustrated in the inset of Fig. 11a. A smaller bending radius implies a stronger bending of the substrate. Variations in the on/off resistances and the turn-on/turn-off voltages are shown in Fig. 11a and b, respectively. In comparison with the flat substrate, there is no significant change in on and on and off resistance states under substrate bending. The retention properties of the device were also tested under substrate bending conditions. By keeping a certain bending radius, a constant mechanical stress was applied to the device and sweeping of the bias voltage was repeated up to 100 cycles. As observed in Fig. 11c, the on and off states were very stable and reproducible. The on/off resistance ratio remained at $10^5$ under bending. The switching characteristics were also examined with respect to bending cycles, as shown in Fig. 11d. The device exhibited very stable on and off resistance states. Due to high mechanical flexibility of the PEO film, the device exhibited very stable switching behavior without any deterioration. This confirms that neither the transport of $Ag^+$ ions nor the redox reactions at the electrode interfaces are affected by bending the substrate. These results indicate that SPE-based atomic switches have the potential to be developed into flexible switch/memory devices.

## 6 Summary

In this chapter, we described the fabrication, characterization, and various functions of SPE-based atomic switches. Similar to atomic switches based on inorganic solid electrolytes, SPE-based atomic switches exhibit not only bi-stable resistive switching but also quantized conductance. This is achieved by optimizing the device structure, the fabrication method, and the measurement conditions. It is significant that the high ionic conductivity of SPE enables us to directly observe filament growth behaviors even in micrometer-scaled devices. Thus, we could investigate how the filament growth processes are determined by kinetic factors such as the redox reaction rates at interfaces, electric field strength, and ion mobility and reduction sites in the polymer matrix. By combining with first-principles DFT simulations, it was revealed that the observed quantized conductance originates from the existence of a tunneling gap and the atomic rearrangement at an atomic point contact. Furthermore, we succeeded in fabricating devices on a plastic

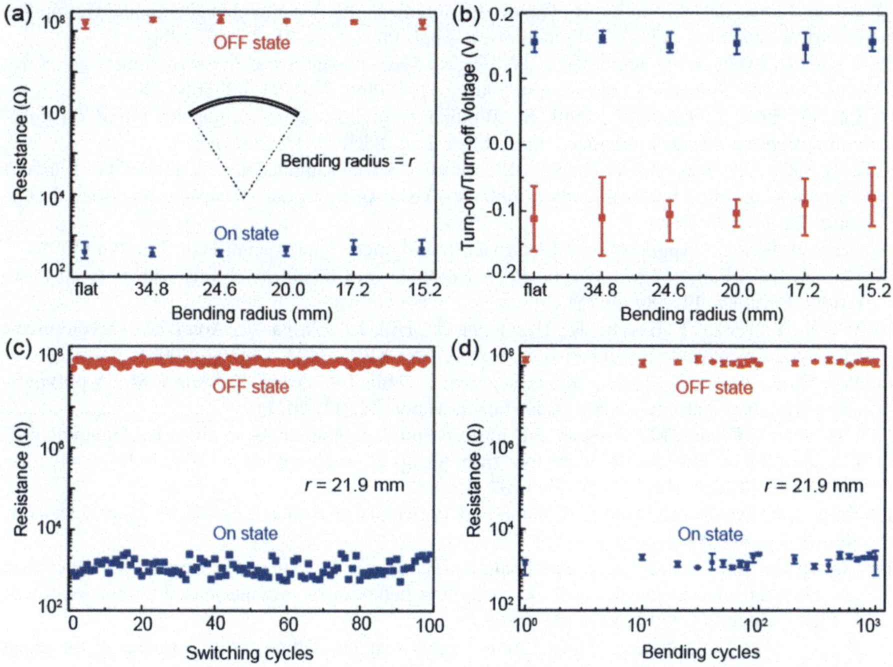


**Fig. 11** Switching properties of an IJP-based Ag/Ag-PEO/Pt device under substrate bending. (**a**) Variation in the on and off resistances and (**b**) the turn-on and turn-off voltages. (**c**) and (**d**) Show the retention properties with respect to sweeping and bending cycles. Copyright 2012, Cambridge University Press

substrate using a SPE solution drop-on-demand IJP technique, and to demonstrate stable resistive switching behavior under substrate bending. The SPE-based atomic switch shows promise for the development not only of flexible switch/memory devices but also of new types of atomic-scale devices with high-speed operation and ultra-low power consumption.

**Acknowledgments** We would like to thank to our collaborators, T. Hasegawa, K. Terabe, J. P. Hill, K. Ariga, M. Muruganathan, and H. Mizuta.

## References

1. Scott, J.C., Bozano, L.D.: Nonvolatile memory elements based on organic materials. Adv. Mater. **19**, 1452 (2007)
2. Altebaeumer, T., Gotsmann, B., Pozidis, H., Knoll, A., Duerig, U.: Nanoscale shape-memory function in highly cross-liked polymers. Nano Lett. **8**, 4398 (2008)
3. Ling, Q.D., Liaw, D.J., Zhu, C., Chen, D.S., Kang, E., Neoh, K.: Polymer electronic memories: materials, devices and mechanisms. Prog. Polym. Sci. **33**, 917 (2008)

4. Billen, J., Steudel, S., Müller, R., Genoe, J., Heremans, P.: A comprehensive model for bipolar electrical switching of CuTCNQ memories. Appl. Phys. Lett. **91**, 263507 (2007)
5. Kever, T., Böttiger, U., Schindler, C., Waser, R.: Mechanism for resistive switching in an oxide-based electrochemical metallization memory. Appl. Phys. Lett. **91**, 083506 (2007)
6. Lai, Q., Shu, Z., Chen, Y., Patil, S., Wudi, F.: Analog memory capacitor based on field-configurable ion-doped polymers. Appl. Phys. Lett. **88**, 133515 (2006)
7. Ssenyange, S., Yan, H., MaCreery, R.L.: Redox-driven conductance switching via filament formation and dissolution in carbon/molecule/$TiO_2$/Ag molecular electronic junctions. Langmuir. **22**, 10689 (2006)
8. Scrosati, B. (ed.): Application of Electroactive Polymers. Chapman & Hall, London (1993)
9. Fenton, D.E., Parker, J.M., Wright, P.V.: Complexes of alkali metal ions with poly(ethylene oxide). Polymer. **14**, 589 (1973)
10. Wu, S., Tsuruoka, T., Terabe, K., Hasegawa, T., Hill, J.P., Ariga, K., Aono, M.: Development of polymer electrolyte based resistive switch. Proc. SPIE. **7493**, 749364 (2009)
11. Wu, S., Tsuruoka, T., Terabe, K., Hasegawa, T., Hill, J.P., Ariga, K., Aono, M.: A polymer-electrolyte-based atomic switch. Adv. Funct. Mater. **21**, 93 (2011)
12. Lascaud, S., Perrier, M., Armand, M., Prud'homme, J., Kapfer, B., Vallée, A., Gauthier, M.: Evidence for ion pairs and/or triple ions from transport measurements in mixed-alkali polyether electrolytes. Electrochim. Acta. **43**, 1407 (1998)
13. Bard, A.J., Parson, R., Jordan, J.: Standard Potentials in Aqueous Solution. Marcel Dekker, New York (1985)
14. Mohapatra, S.R., Tsuruoka, T., Krishnan, K., Hasegawa, T., Aono, M.: Effects of temperature and ambient pressure on the resistive switching behavior of polymer-based atomic switches. J. Mater. Chem. C. **3**, 5715 (2015)
15. Bruce, P.G.: Ion-polyether coordination complexes: crystalline ionic conductors for clean energy storage. Dalton Trans. **11**, 1365 (2006)
16. Christie, A.M., Lilley, S.J., Staunton, E., Andreev, Y.G., Bruce, P.G.: Increasing the conductivity of crystalline polymer electrolytes. Nature. **433**, 50 (2005)
17. Krishnan, K., Tsuruoka, T., Mannequin, C., Aono, M.: Mechanism for conducting filament growth in self-assembled polymer thin films for redox-based atomic switches. Adv. Mater. **28**, 640 (2016)
18. Krishnan, K., Tsuruoka, T., Aono, M.: Direct observation of anodic dissolution and filament growth behavior in polyethylene-oxide-based atomic switch structures. Jpn. J. Appl. Phys. **55**, 06GK02 (2016)
19. Krishnan, K., Aono, M., Tsuruoka, T.: Kinetic factors determining conducting filament formation in solid polymer electrolyte based planar devices. Nanoscale. **8**, 13976 (2016)
20. Ratner, M.A., Shriver, D.F.: Ion transport in solvent-free polymers. Chem. Rev. **88**, 109 (1988)
21. Terabe, K., Hasegawa, T., Nakayama, T., Aono, M.: Quantized conductance atomic switch. Nature. **433**, 47 (2005)
22. Ohno, T., Hasegawa, T., Tsuruoka, T., Terabe, K., Jimzewski, J.K., Aono, M.: Short-term plasticity and long-term potentiation mimicked in single inorganic synapses. Nat. Mater. **10**, 591 (2011)
23. Nayak, A., Ohno, T., Tsuruoka, T., Terabe, K., Hasegawa, T., Aono, M.: Controlling the synaptic plasticity of a $Cu_2S$ gap-type atomic switch. Adv. Funct. Mater. **22**, 3606 (2012)
24. Tsuruoka, T., Hasegawa, T., Terabe, K., Aono, M.: Conductance quantization and synaptic behavior in a $Ta_2O_5$-based atomic switch. Nanotechnology. **23**, 435705 (2012)
25. Krishnan, K., Muruganathan, M., Tsuruoka, T., Mizuta, H., Aono, M.: Highly reproducible and regulated conductance quantization in a polymer-based atomic switch. Adv. Funct. Mater. **27**, 1605104 (2017)
26. Krishnan, K., Muruganathan, M., Tsuruoka, T., Mizuta, H., Aono, M.: Quantized conductance operation near a single-atom point contact in a polymer-based atomic switch. Jpn. J. Appl. Phys. **56**, 06GF02 (2017)

27. de Gans, B.J., Duineveld, P.C., Schubert, U.S.: Inkjet printing of polymers: state of the art and future developments. Adv. Mater. **16**, 203 (2004)
28. Mohapatra, S.R., Tsuruoka, T., Hasegawa, T., Terabe, K., Aono, M.: Flexible resistive switching memory using inkjet printing of a solid polymer electrolyte. AIP Adv. **2**, 022144 (2012)
29. Mohapatra, S.R., Tsuruoka, T., Hasegawa, T., Terabe, K., Aono, M.: Flexible polymer atomic switch using ink-jet printing technique. Mater. Res. Soc. Symp. Proc. **1430**, 69–74 (2012)
30. Shin, K.Y., Lee, S.H., Oh, J.H.: Solvent and substrate effects on inkjet-printed dots and lines of silver nanoparticle colloids. J. Micromech. Microeng. **21**, 045012 (2011)

# Nanoionic Devices for Physical Property Tuning and Enhancement

**Takashi Tsuchiya, Kazuya Terabe, and Masakazu Aono**

**Abstract** Nanoionic devices for physical property tuning and enhancement have been developed to generate novel functions overcoming limitations of conventional materials synthesis and semiconductor technology. Local ionic transport near the solid/solid interface enabled in-situ tuning and enhancement of various physical properties. Two electronic carrier doping methods can be used to achieve extremely high-density electronic carriers: one is electrochemical carrier doping using a redox reaction; the other is electrostatic carrier doping using an electric double layer (EDL). Optical bandgap and photoluminescence are tuned for various applications including smart windows and biosensors. Magnetization and magnetoresistance are tuned for low-power-consumption magnetic storage devices. Superconducting transition temperature is enhanced for exploring high temperature superconductivity. Nanoionic devices for physical property tuning and enhancement are promising derivative of atomic switch technology.

## 1 Introduction

Nanoionic devices are functional devices achieved by ionic transport and the resultant electrochemical processes in nanoscale including inside of solid, surface of solid, or solid/solid interface [1–24]. The concept of nanoionic devices originated from an invention of quantized conductance atomic switch by Terabe et al. [1, 2]. Early history of nanoionic devices research, which was mainly dedicated to development of atomic switch, is described elsewhere. For the atomic switch, electrochemical processes are used to achieve generation and annihilation of metallic nanofilament, resulting in the nonvolatile and quantized conductance switching. However, electrochemical processes can differently function in the vicinity of solid electrolyte/electronic materials, whose physical properties are strongly

T. Tsuchiya (✉) · K. Terabe · M. Aono
International Center for Materials Nanoarchitectonics (MANA), National Institute for Materials Science (NIMS), Tsukuba, Ibaraki, Japan
e-mail: TSUCHIYA.Takashi@nims.go.jp



M. Aono (ed.), *Atomic Switch*, Advances in Atom and Single Molecule Machines,
https://doi.org/10.1007/978-3-030-34875-5_9

dependent on the electronic carrier density. In this contribution, we focus on nanoionic devices enabling physical property tuning and enhancement which are promising derivative of the atomic switch technology.

The electronic structure near Fermi level is the origin of many physical properties of materials (e.g., electronic conductivity, superconductivity, thermal conductivity, magnetic susceptibility, color, optical reflectance) [25]. Control of the electronic structure, or more specifically electronic carrier density, has long been the most important issue in materials science because so many physical properties can be widely tuned by varying the electronic carrier density [26].

Two carrier doping methods have mainly been used for controlling the electronic carrier density in materials. One is chemical doping accompanied by generation of defects in materials (e.g., hetero-atoms, vacancies, and interstitials). Electronic (hole) carriers are added through substitution of the original atoms of a material by hetero-atoms and compensation of excess charge induced by the defects. Not only the hetero-atom doping but also nonstoichiometry, caused by atomic vacancies or interstitials, induces excess charge and the resultant electronic carrier addition. The chemical doping enables the control of electronic carrier density over a very wide range as long as the materials can accept such defects, and this method has been used by researchers to explore various physical properties.

The other method is electrostatic carrier doping using a dielectric that works as a capacitor. In this method, since electronic carriers are electrostatically added without generating defects, the electronic structure of the material is less affected than in the case of chemical doping: electrostatic carrier doping can be understood as a Fermi level shift in a rigid band model rather than that with defect level generation. This method has been exploited in field-effect-transistors (FETs), wherein the electronic carrier density is tunable in situ by gate voltage modulation, although in this case, the controllability of carrier density, e.g., $10^{13}$ $cm^{-2}$, is not as good as that by chemical doping because of the small capacitance of dielectric materials (e.g., 0.7 $\mu F/cm^2$ for $SiO_2$) [26–30].

While the ability to control electronic carrier density had produced advances in the study of material properties and in semiconductor technology, two goals have remained unrealized: (1) a high electronic carrier density cannot be tuned in situ, and (2) a high electronic carrier density cannot be achieved by doping without generating defects.

If a high electronic carrier density could be tuned in situ, a wider range of physical properties could be exploited for the development of novel functional devices beyond the reach of conventional semiconductor technology. Moreover, if high densities could be reached by doping without generating defects, it would lead to remarkable advances in exploiting physical properties that are sensitive to disturbance in the electronic structure, such as superconductivity. Therefore, the realization of these two goals would have a huge impact on both fundamental research and practical applications by overcoming limitations in conventional materials synthesis and semiconductor technology. Is this possible?

Nanoionic devices are promising candidates to overcome these limitations. In the devices, ionic transport is used to achieve (1) electrochemical carrier doping using a

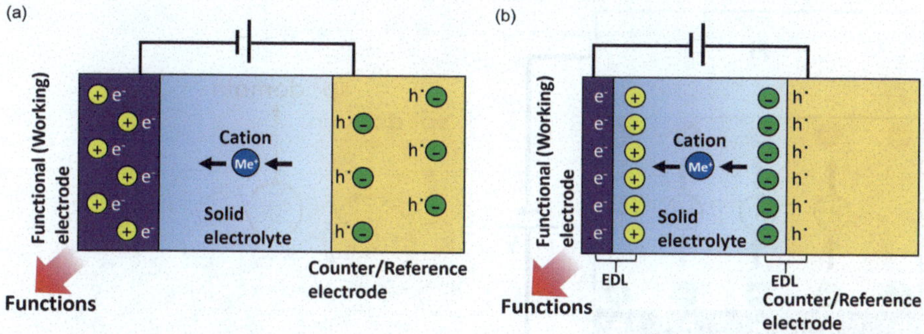


**Fig. 1** Illustrations of electrochemical carrier doping using redox reaction (**a**), and electrostatic carrier doping using EDL (**b**)

reduction and oxidation (redox) reaction or (2) electrostatic carrier doping using electric double layer (EDL), both of which can tune extremely high-density electronic carrier in electronic materials. Figure 1 shows illustrations of the two nanoionic devices. Since the definition for (1) is applicable also to conventional coulomb titration type electrochemical cells in which nonstoichiometry of electron-ion mixed conducting electrode is controlled by DC voltage application, apparently characteristic of nanoionic devices is not evident with respect to bulk-based conventional electrochemical devices. However, it can be clear given an operation at low temperature such as room temperature.

Regardless of the device size, effect of redox reaction should reach every part of functional working electrode within reasonable time range. This is not straightforward, however, for bulk-based conventional electrochemical devices because, at room temperature, ionic conductivity of solid materials is usually very low in spite of very long ionic transport distance. This makes materials selection for functional working electrode limited to a few electron-ion mixed conductors with exceptionally high ionic conductivity (e.g., halide, sulfide, tungstate) [31, 32]. Therefore, due to such a limitation, it is very difficult to develop bulk-based functional devices with various interesting physical properties. On the other hand, for nanoionic devices, usage of thin films or nanogap can significantly reduce ionic transport distance needed to tune carrier density of the functional working electrode entirely. This makes wide range of functional materials with low ionic conductivity possible choices for components of nanoionic devices.

Here, we review recent progress in nanoionic devices enabling physical property and tuning which are applied to optical bandgap, mobility, photoluminescence, magnetization, magnetoresistance, and superconductivity. With a help of nanoarchitectonics. The unique and excellent property of the devices indicates expansion of the paradigm that the atomic switch marked; nanoelectronics achieved by ions.

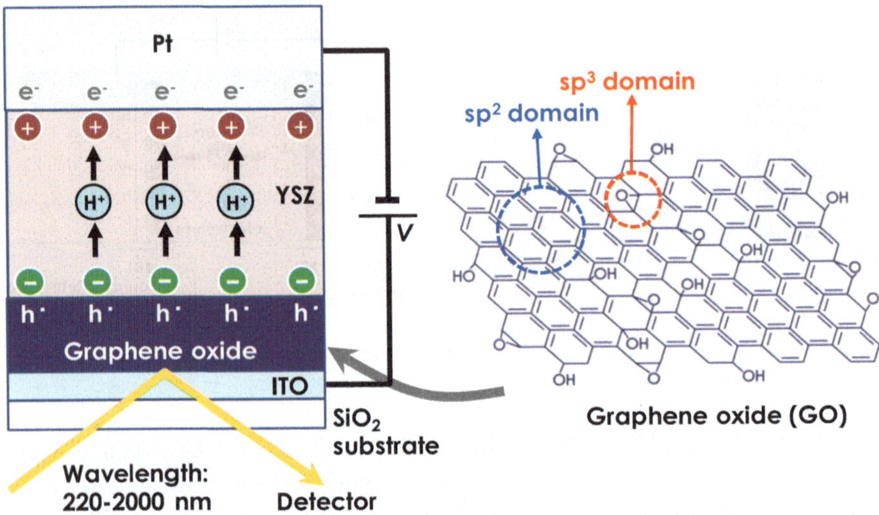


**Fig. 2** Schematic illustration of GO-based redox device with YSZ proton conductor. Brown and green circles represent positive and negative charges accumulated at interfaces owing to proton migration, respectively. $H^+$ represents positively charged protons [13]. Reprinted with permission from Ref. 13. Copyright (2014) John Willey and Sons

## 2 Bandgap Tuning of Graphene Oxide Achieved by Redox Reaction

Redox reaction is very useful to tune oxidation state of nonstoichiometric compounds. In particular, materials with both wide nonstoichiometry range and unique physical property related to oxidation state are well fit to redox reaction-based nanoionics devices. In such a view point, graphene oxide (GO), a derivative of graphene, is an idealistic material.

GO has been attracting much attention as a promising functional material. For instance, variable bandgap, wide energy range of photoluminescence, room-temperature ferromagnetism and other unique electronic properties are of great interest [33–37]. The unique functions of GO are caused by the electronic disorder due to $sp^3$ hybridization carbon in $sp^2$ conjugated networks. Because such $sp^3$ hybridization carbons are usually bonded with oxygen atoms or hydroxide ions, the $sp^2/sp^3$ fraction and the carbon/oxygen ratio (C/O) are in positive relationship, resulting in modulation of the band gap and other interesting properties [33–37]. We fabricated nanoionics devices to tune the $sp^2/sp^3$ fraction in situ by using redox reaction for development of multifunctional devices with merits of transparent, ultrathin, flexible, and low-cost [13–15].

Figure 2 illustrates an all-solid-state redox device composed of multilayer GO and an yttria-stabilized zirconia (YSZ) proton conducting thin film [13]. Proton is transported through the YSZ thin film under a DC voltage applied condition. The

**Fig. 3** DC bias dependence of optical band gap calculated from UV-Vis-NIR spectra of the cell shown in Fig. 2 [13]. Reprinted with permission from Ref. 13. Copyright (2014) John Willey and Sons

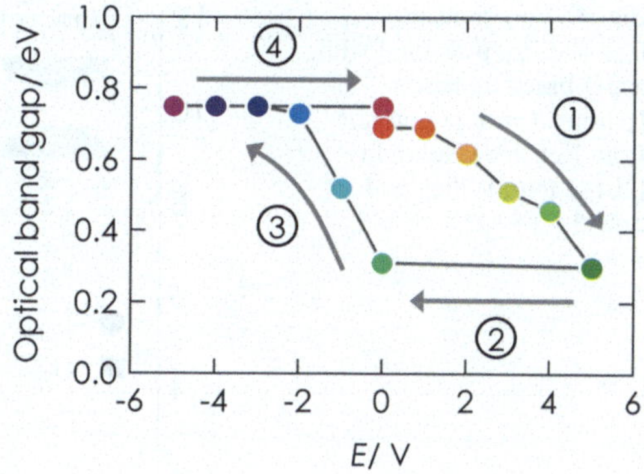


proton transport is accompanied by the redox reaction of GO in the vicinity of GO/YSZ interface.

$$C_xO_yH_z + 2H^+ + 2e' \leftrightarrow C_xO_{y-1}H_z + H_2O, \quad (1)$$

where $C_xO_yH_z$, $C_xO_{y-1}H_z$, and $H_2O$ are GO, reduced GO, and water molecules, respectively. The water molecules are known to adsorb on the surface of YSZ nano-grains, in which sufficient proton conduction is enabled even at room temperature [38, 39]. Figure 3 shows the variation in the band gap energy with various DC voltage application. It was obtained from UV-Vis-NIR reflectance measurements. As applied voltage was increased to positive value, the band gap decreased from $V = 2$ V and reached 0.30 eV at 5 V. On the other hand, the band gap showed an immediate increase when $-1$ V was applied. It was recovered at –3 V. This reversible bandgap tuning is due to the redox reaction of the GO on the basis of Eq. (1). This corresponds to the $sp^2/sp^3$ fraction tuning of in GO.

The present technique achieved the control of the ON/OFF ratio ($i_{ON}/i_{OFF}$) and field effect mobility ($\mu_{FE}$) in a GO-based transistor. Figure 4 shows the variations in $i_{ON}/i_{OFF}$ and $\mu_{FE}$ as functions of $i_{OFF}$. The $i_{ON}/i_{OFF}$ behavior is consistent with the bandgap variation shown in Fig. 8. The behavior of $\mu_{FE}$ is also consistent with the variable range hopping (VRH) mechanism for electronic transport in GO [35, 40].

Photoluminescence (PL) has been intensively studied among the unique property of GO due to its applications to nano-medicine. PL in GO are strongly affected by existence of various pathogens and neurotransmitters. GO-based biosensors use this effect to detect biomolecules [41–43]. The $sp^2/sp^3$ ratio tuning in GO is important because it directly affects the charge transfer between GO and biomolecule in the detection process because it changes the relative potential energies in GO and biomolecule. Figure 5 shows the normalized PL spectra tuned by various DC voltage application in an all-solid-state PL source using a mesoporous $SiO_2$ thin film as a proton conductor [15]. The mesoporous $SiO_2$ thin film has extremely low background PL, which is crucially important for the device. In the oxidation and reduction process, the peak wavelength was tuned from 712 nm (red) to 393 nm [near-ultraviolet (UV)]. The wavelength range covered that of chemically tuned GO,

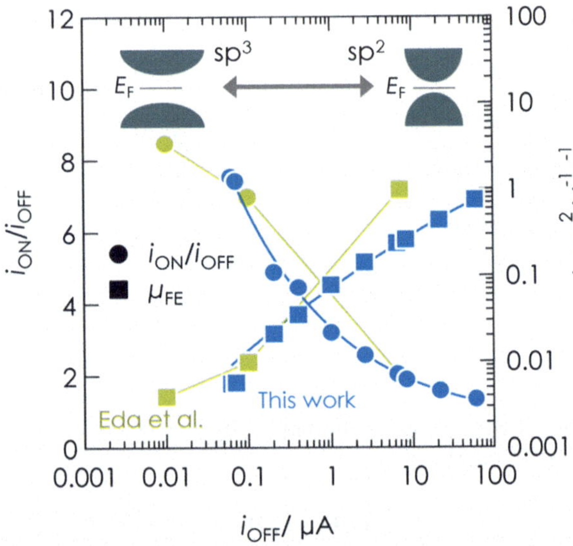


**Fig. 4** Variations in $i_{ON}/i_{OFF}$ and $\mu_{FE}$ measured with a GO-based transistor. Reprinted with permission from Ref. 13. Copyright (2014) John Willey and Sons [13]

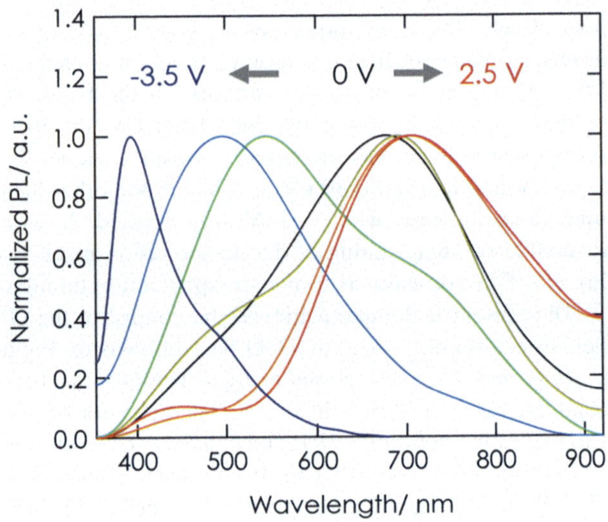


**Fig. 5** DC bias dependence of normalized PL spectra tuned by application of various DC bias voltages [15]. Reprinted with permission from Ref. 15. Copyright (2015) American Chemical Society

except at extremely long wavelengths (>712 nm). An alteration of the relative amounts of functional groups may be the operation principle here, rather than the $sp^2/sp^3$ variation [44].

An end member of GO is graphene and its bulk form is graphite. In contrast to that graphite is a typical inert material in electrochemistry, a composition from GO to graphene with a thickness of few nm is an active material for the redox reaction in the nanoionic devices shown above. This difference originates from the low out-of-plane proton conductivity in graphite and clearly illustrates an interesting characteristic of nanoionic devices: reducing the size to the nanoscale order leads to novel functions. In this sense, GO-based devices are typical nanoionic devices.

## 3 Magnetization and Magnetoresistance Tuning Achieved by Redox Reaction

Recently, magnetic storage devices [e.g., hard disks, magneto-optical disks, and magnetic random access memory (MRAM)] have been attracting much attention due to its high density. In particular, MRAM are of great importance to overcome limitation in conventional information and communication technology. One drawback for practical application of conventional MRAM is a relatively high energy consumption needed to switch ON and OFF, corresponding to tuning magnetization of ferromagnetic layer in the MRAM device. To the redox reaction occurring in nanoscale has been applied to in situ tuning of magnetic properties including magnetization and magnetoresistance in the quest to develop new forms of high-density magnetic storage.

Figure 6 shows an illustration of an all-solid-state redox device composed of magnetite ($Fe_3O_4$), which is a half-metallic ferromagnetic oxide at room temperature, and $Li^+$ conducting $Li_4SiO_4$ (LSO) and $Li^+$/electron mixed conducting $LiCoO_2$ (LCO) [18]. Tuning of magnetization and magnetoresistance can be achieved by extremely high carrier density doping of $Fe_3O_4$ by the electrochemical insertion of $Li^+$ [i.e., reaction (2) below]. This cannot be enabled with conventional electrostatic carrier doping using dielectric thin films due to its relatively poor electronic carrier controllability.

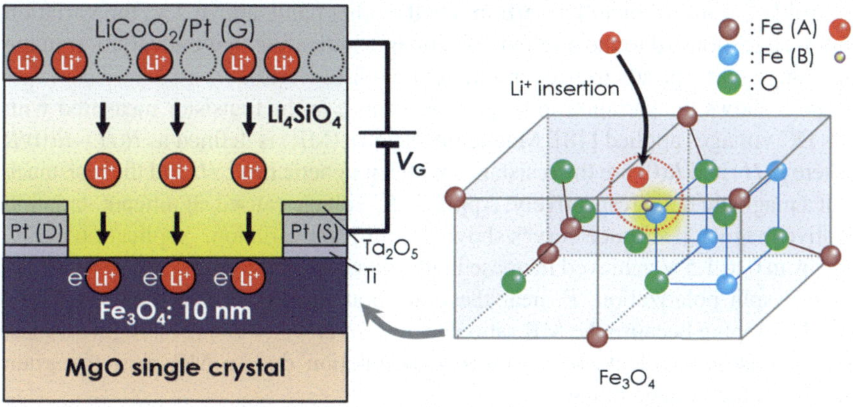


**Fig. 6** Schematic illustration of all-solid-state redox transistor with $Fe_3O_4$ and LSO lithium ion conductor. $Li^+$ represents positively charged lithium ions. Dotted circles in $LiCoO_2$ represent $Li^+$ vacancies [18]. Reprinted with permission from Ref. 18. Copyright (2016) American Chemical Society

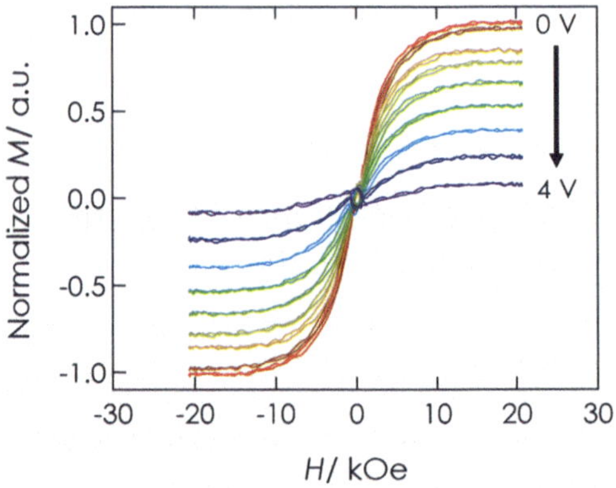


**Fig. 7** *M–H* loops of device measured at various DC voltages [18]. Reprinted with permission from Ref. 18. Copyright (2016) American Chemical Society

$$Fe_3O_4 + xLi^+ + xe^- \rightarrow Li_xFe_3O_4. \tag{2}$$

Figure 7 shows normalized magnetization-magnetic field (*M–H*) loops measured with various DC voltages applied [18]. Ferromagnetic hysteresis loop, which is typical behavior for $Fe_3O_4$ at room temperature, was observed under all voltage conditions. Saturation magnetization above 10 kOe gradually decreased as $Li^+$ was inserted in $Fe_3O_4$ by increasing positive voltage applied to the LCO. It was reported that electrostatic carrier doping in $Fe_3O_4$ using a solid dielectric enabled a similar decrease in saturation magnetization. However, the extent of the decrease in saturation magnetization was very small because electronic carrier density doped by a solid dielectric was low [45]. The decrease in saturation magnetization for our device is consistent with previous reports concerning $Fe_3O_4$ powder lithiated using liquid reagents [46–50]. The variation in saturation magnetization from 0 to 2 V was repeatable. This indicated that the magnetization tuning from 0 to 2 V was caused by reversible $Li^+$ insertion and desertion. On the other hand, above 2 V, the variation was not well responded to the applied DC voltage, indicating that the variation in the voltage range corresponds to irreversible reduction of $Fe_3O_4$.

Figure 8 shows the variation in magnetoresistance in the transistor measured with various DC voltages applied [18]. Magnetoresistance (MR) is defined as $R(H)–R(0)/R(0)$, where $R(H)$ and $R(0)$ are the resistance with a magnetic field, $H$, and the resistance without a magnetic field, respectively. Applied DC voltage caused significant variation in negative magnetoresistance curves showed significant variation. Application of DC voltage from 0 to 1.5 V achieved increase in negative MR value from 2.0 to 3.0%. This means that spin polarization, $P$, near the grain boundaries of $Fe_3O_4$ was enhanced through $Li^+$ doping because the MR ratio depends on $2P^2/(1 + P^2)$ [51]. Figure 8 inset shows an illustration of electronic current modulation due to MR near the grain boundaries in the voltage range.

For application to high-density magnetic storage devices, the *M* and MR tuning by using the present approach is quite useful. Furthermore, development of the technique toward magnetic vector manipulation should be promising for spintronics applications in which parallel and anti-parallel magnetization function to switch,

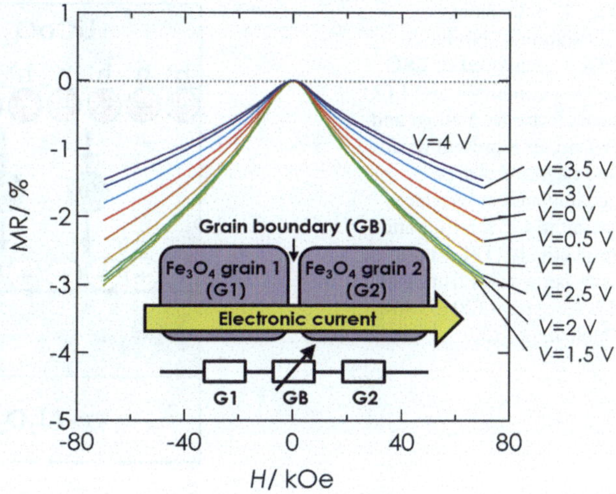


**Fig. 8** DC voltage dependence of MR in $Fe_3O_4$ thin film in device as a function of magnetic field [18]. Reprinted with permission from Ref. 18. Copyright (2016) American Chemical Society

because power consumption for ion insertion/desertion is very small. Nanoionics-based manipulation of magnetic property is an exciting subject for exploring novel and useful functions of the solid/solid interface.

## 4 Modulation of Superconducting Critical Temperature by All-Solid-State EDLT

Room-temperature superconductivity has long been a dream as it can solve serious energy and environmental problems. While synthesis of novel bulk materials and chemical doping have been performed to search for room-temperature superconductor, none of them were successful so far. One promising means is electrostatic carrier doping to potential superconductor, which cannot be superconductor due to lack of sufficient electronic carrier density, because of the freedom from structural disorder inherent in chemical doping.

Drozdov et al. reported that $H_2S$ solid showed superconducting transition at 200 K [52]. It strongly suggested that the possibility of high temperature superconductivity through electrostatic carrier doping in high Debye temperature solid, which is indicated to be potential high-temperature superconductor on the basis of Matthias' law for the BCS type superconductor [53],

$$T_c = 1.14\frac{\hbar\omega_D}{k_B}\exp\left(-\frac{1}{D(0)V}\right), \tag{3}$$

where $D(0)$ and $V$ are the density of states (DOS) near the Fermi level and the electron-phonon interaction, respectively [53]. Electrostatic carrier doping using

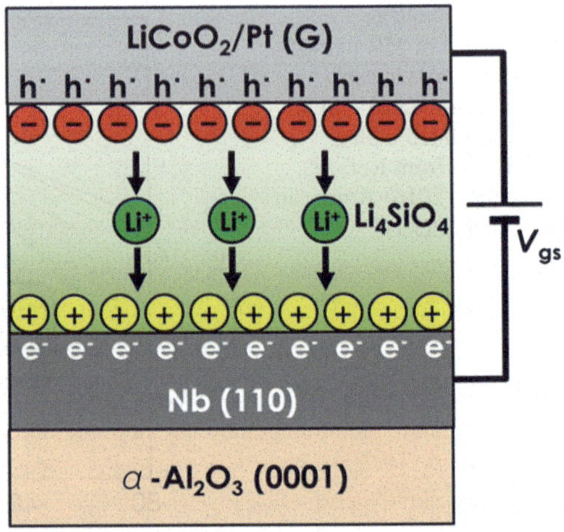


**Fig. 9** Schematic illustration of Nb-based EDLT composed of LSO $Li^+$ ion conductor and LCO gate electrode. Yellow and red circles represent positive and negative charges accumulated at interfaces owing to $Li^+$ ion migration, respectively [17]. Reprinted with permission from Ref. 17. Copyright (2015) AIP Publishing LLC

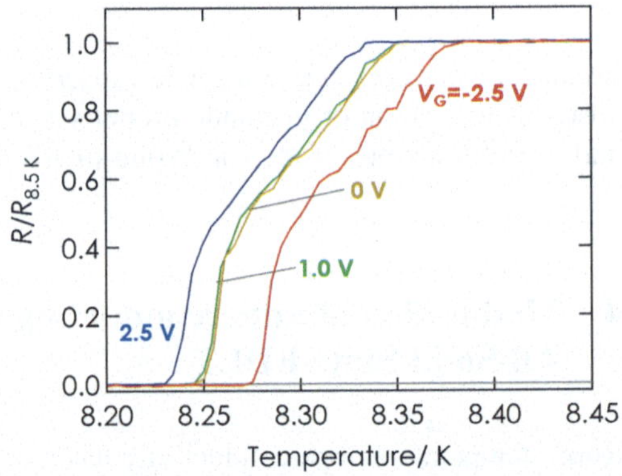


**Fig. 10** Temperature dependence of Nb channel resistance in Nb-based EDLT measured while applying various gate voltages near $T_c$ [17]. $T_c$ is defined as the onset of the transition from normal to superconductivity. Reprinted with permission from Ref. 17. Copyright (2015) AIP Publishing LLC

liquid type EDLTs has been applied to explore high temperature superconductivity, and it was accompanied by some interesting results. On the other hand, solid electrolyte type EDLTs, termed all-solid-state EDLT, as has not been used to search for superconductivity so far. Therefore, we studied all-solid-state EDLTs with superconductors to investigate the possibility of superconducting $T_c$ modulation using EDL carrier doping near superconductor/solid electrolyte interfaces [17].

Figure 9 shows an all-solid-state EDLT for modulation of superconducting critical temperature. It was composed of Nb thin film, which is well known as a typical metallic superconductor, and $Li_4SiO_4$ (LSO) $Li^+$ ion conducting amorphous oxide electrolyte, and a $LiCoO_2$ (LCO) mixed $Li^+$-hole conducting gate electrode [17]. The LCO gate electrode can supply a plenty of $Li^+$ ions to the Nb thin film/LSO electrolyte interface, in order to modulate charge density of an EDL at the interface. Figure 10 shows the temperature dependence of the Nb channel resistance, which was measured while applying various gate voltages (2.5 to −2.5 V). The Nb thin film showed superconductivity transition under all conditions. A monotonic variation in

$T_c$ was observed as the applied gate voltage was modulated. $T_c$ was enhanced from 8.33 to 8.39 K when the gate voltage was reduced from 2.5 to −2.5 V. These changes are almost three orders of magnitude larger than the values obtained using solid dielectric [54], indicating that the EDLT can modulate a much higher carrier density. In particular, modulation of $T_c$ in Nb film using an EDLT composed of an ionic liquid was recently reported [55]. The reported direction of the $T_c$ variation with respect to the gate voltage agrees with the result, but with a slightly smaller difference in $T_c$.

Superconducting $T_c$ of the BCS type superconductors is dependent on carrier density on the basis of Matthias' law shown in Eq. (3). Given that DOS curve and Fermi level of Nb film in the EDLT are similar to that without electrolyte, a positive gate voltage application and the resultant EDL charging causes decrease in $D(0)$, leading to decrease in $T_c$, and vice versa. The $T_c$ tuning function shown in Fig. 10 can be explained on the basis of the EDL mechanism.

Since chemical doping has been a sole mean to dope high density electronic carrier in superconductors and potential superconductors, development of high $T_c$ superconductor has been strictly limited by thermodynamic solubility of dopants so far. The electrostatic carrier doping using EDL in the vicinity of solid/solid interface should be applicable to potential superconductors which cannot sufficiently doped through chemical routes although redox reaction of mobile ions including decomposition of electrolyte (redox window) limits the extent of the carrier doping. The exploring room-temperature superconductivity in the vicinity of solid/solid electrolyte interface is underway.

## 5 Conclusions

Recent progress in nanoionic devices for various physical property tuning and enhancement has been reviewed. Nanoionic devices enabled electrochemical carrier doping using redox reaction and electrostatic carrier doping using EDL which are promising approaches to overcome the limitations in conventional material synthesis and semiconductor technology.

Nanoionic devices are novel approaches to enable various functions although they are on the basis of conventional electrochemical processes. Note that the approach applied here is not necessarily limited to the reduction of ionic transport distance due to usage of thin films. In the atomic switch as the first nanoionic device, quantized and high-speed conductance switching was achieved by generation and annihilation of silver nanofilament at a tunnel gap between a probe tip of scanning probe microscope (STM) and a counter platinum electrode. It was not electron-ion mixed conductor but the gap that was miniaturized to achieve the function. This indicates possible extension of the approach toward various directions.

A distinct advantage of nanoionic devices over electronic devices is their high density of electronic carriers achieved by use of EDLs and redox reactions. On the other hand, regarding conventional material synthesis, distinct advantages are in situ controllability of electronic carrier density and the freedom from structural disorders. In addition, nanoionic devices enable a high level of integration and a low power consumption owing to their simple structure and low-voltage operation. These advantages can be exploited for deriving various useful functions. Furthermore, the high compatibility of nanoionic devices with peripheral devices because of their all-solid-state structure would make them truly practical.

Compared with electronic devices, nanoionic devices have had a quite short history since the advent of the first atomic switch [1, 2], although conventional solid state ionic devices based on bulk have a comparably much longer history [31, 32]. As nanoionic devices are still at the dawn of their development, there is ample room to improve their characteristics (e.g., response time, retention time, and repeatability). Above all, the greatest challenge is high density carrier doping using all-solid-state EDLT for high temperature superconductivity. Here, one needs better understanding of the local ion behavior based on not only conventional electrochemistry but also operando approaches using transmission electron microscope (TEM) or X-ray spectroscopy in order to realize the full potential of the solid/solid interface [56, 57].

As clearly seen in the discovery of the atomic switch, which was born at the cross point of solid state ionics and scanning tunnel microscopy, which was a state-of-art approach in surface science at the time, the next horizon of nanoionic devices should lie also in unconventional meetings with other emerging fields. An atomic-switch-based decision maker (ASDM) is a good example of such a meeting with nanoionic devices and physics-based analog computing that may lead to a sophisticated means of decision making [58].

The solid/solid interface is one of the few great frontiers remaining in modern materials science and solid state physics, which have been intensively studied already. Physical property tuning and enhancement achieved by controlling ion transport in nanoscale is an exciting field to explore an unprecedented range of physical properties (e.g., room-temperature superconductivity) and functions (e.g., high-performance universal memories and artificial brains).

**Acknowledgments** The authors thank Dr. Tohru Tsuruoka, Dr. Satoshi Moriyama and Dr. Minoru Osada of the International Center for Materials Nanoarchitectonics, National Institute for Materials Science, and Dr. Tohru Higuchi of Tokyo University of Science for their assistance with PL, superconductivity, and magnetic property measurements.

## References

1. Terabe, K., Hasegawa, T., Nakayama, T., Aono, M.: RIKEN Rev. **37**, 7 (2001)
2. Terabe, K., Hasegawa, T., Nakayama, T., Aono, M.: Nature. **433**, 47 (2005)
3. Terabe, K., Hasegawa, T., Liang, C., Aono, M.: Sci. Technol. Adv. Mater. **8**, 536 (2007)
4. Sakamoto, T., Lister, K., Banno, N., Hasegawa, T., Terabe, K., Aono, M.: Appl. Phys. Lett. **91**, 092110 (2007)

5. Terabe, K., Tsuruoka, T., Nayak, A., Ohno, T., Nakayama, T., Aono, M.: Resistive Switching: From Fundamentals of Nanoionic Redox Processes to Memristive Device Applications, p. 515. Wiley-VCH, Berlin (2016)
6. Hasegawa, T., Terabe, K., Sakamoto, T., Aono, M.: MRS Bull. **34**, 929 (2009)
7. Hasegawa, T., Terabe, K., Tsuruoka, T., Aono, M.: Adv. Mater. **24**, 252 (2012)
8. Ohno, T., Hasegawa, T., Tsuruoka, T., Terabe, K., Gimzewski, J.K., Aono, M.: Nat. Mater. **10**, 591 (2011)
9. Tsuruoka, T., Hasegawa, T., Terabe, K., Aono, M.: Nanotechnology. **23**, 435705 (2012)
10. Tsuruoka, T., Terabe, K., Hasegawa, T., Valov, I., Waser, R., Aono, M.: Adv. Funct. Mater. **22**, 70 (2012)
11. Tsuruoka, T., Valov, I., Tappertzhofen, S., van den Hurk, J., Hasegawa, T., Waser, R., Aono, M.: Adv. Funct. Mater. **25**, 6374 (2015)
12. Tsuchiya, T., Terabe, K., Aono, M.: Appl. Phys. Lett. **103**, 073110 (2013)
13. Tsuchiya, T., Terabe, K., Aono, M.: Adv. Mater. **26**, 1087 (2014)
14. Tsuchiya, T., Terabe, K., Aono, M.: Appl. Phys. Lett. **105**, 183101 (2014)
15. Tsuchiya, T., Tsuruoka, T., Terabe, K., Aono, M.: ACS Nano. **9**, 2102 (2015)
16. Tsuchiya, T., Ochi, M., Higuchi, T., Terabe, K., Aono, M.: ACS Appl. Mater. Interfaces. **7**, 12254 (2015)
17. Tsuchiya, T., Moriyama, S., Terabe, K., Aono, M.: Appl. Phys. Lett. **107**, 013104 (2015)
18. Tsuchiya, T., Terabe, K., Ochi, M., Higuchi, T., Osada, M., Yamashita, Y., Ueda, S., Aono, M.: ACS Nano. **10**, 1655 (2016)
19. Tsuchiya, T., Ochi, M., Higuchi, T., Terabe, K.: Jpn. J. Appl. Phys. **55**, 06GJ03 (2016)
20. Terabe, K., Tsuchiya, T., Yang, R., Aono, M.: Nanoscale. **8**, 13873 (2016)
21. Yang, R., Terabe, K., Liu, G., Tsuruoka, T., Hasegawa, T., Gimzewski, J.K., Aono, M.: ACS Nano. **6**, 9515 (2012)
22. Yang, R., Terabe, K., Liu, G., Tsuruoka, T., Hasegawa, T., Aono, M.: Appl. Phys. Lett. **100**, 231603 (2012)
23. Yang, R., Terabe, K., Yao, Y., Tsuruoka, T., Hasegawa, T., Gimzewski, J.K., Aono, M.: Nanotechnology. **24**, 384003 (2013)
24. Aono, M., Bando, Y., Ariga, K.: Adv. Mater. **24**, 150 (2012)
25. Kittel, C.: Introduction to Solid State Physics, 7th edn. Wiley, New York (1996)
26. Sze, S.M.: Physics of Semiconductor Devices, 2nd edn. Wiley, New York (1981)
27. Ahn, C.H., Gariglio, S., Paruch, P., Tybell, T., Antognazza, L., Triscone, J.-M.: Science. **284**, 1152 (1999)
28. Takahashi, K.S., Gabay, M., Jaccard, D., Shibuya, K., Ohnishi, T., Lippmaa, M., Triscon, J.-M.: Nature. **441**, 195 (2006)
29. Bhattacharya, A., Eblen-Zayas, M., Staley, N.E., Huber, W.H., Goldman, A.M.: Appl. Phys. Lett. **85**, 997 (1994)
30. Parendo, K.A., Sarwa B. Tan, K.H., Bhattacharya, A., Eblen-Zayas, M., Staley, N.E., Goldman, A.M.: Phys. Rev. Lett. **94**, 197004 (2005)
31. Funke, K.: Sci. Technol. Adv. Mater. **14**, 043502 (2013)
32. Yamamoto, O.: Sci. Technol. Adv. Mater. **18**, 504–527 (2017)
33. Park, S., Ruoff, S.R.S.: Nat. Nanotechnol. **4**, 217 (2009)
34. Eda, G., Fanchini, G., Chhowalla, M.: Nat. Nanotechnol. **3**, 270 (2008)
35. Eda, G., Lin, Y., Mattevi, C., Yamaguchi, H., Chen, H., Chen, I., Chen, C., Chhowalla, M.: Adv. Mater. **22**, 505 (2010)
36. Eng, A.Y.S., Poh, H.L., Sanek, F., Marysko, M., Matejkova, S., Sofer, Z., Pumera, M.: ACS Nano. **7**, 5930 (2013)
37. Ambrosi, A., Pumera, M.: Chem. Eur. J. **19**, 4748 (2013)
38. Miyoshi, S., Akao, Y., Kuwata, N., Kawamura, J., Oyama, Y., Yagi, T., Yamaguchi, S.: Chem. Mater. **26**, 5194 (2014)
39. Miyoshi, S., Akao, Y., Kuwata, N., Kawamura, J., Oyama, Y., Yagi, T., Yamaguchi, S.: Solid State Ionics. **207**, 21 (2012)

40. Eda, G., Mattevi, C., Yamaguchi, H., Kim, H., Chhowalla, M.: J. Phys. Chem. C. **113**, 15768 (2009)
41. Jung, J.H., Cheon, D.S., Liu, F., Lee, K.B., Seo, T.S.: Angew. Chem. Int. Ed. **49**, 5708 (2010)
42. Zhao, X.-H., Kong, R.-M., Zhang, X.-B., Meng, H.-M., Liu, W.-N., Tan, W., Shen, G.-L., Yu, R.-Q.: Anal. Chem. **83**, 5062 (2011)
43. Jeon, S.-J., Kwak, S.-Y., Yim, D., Ju, J.-M., Kim, J.-H.: J. Am. Chem. Soc. **136**, 10842 (2014)
44. Li, M., Cushing, S.K., Zhou, X., Guo, S., Wu, N.: J. Mater. Chem. **22**, 23374 (2012)
45. Wong, J.J.I., Swartz, A.G., Zheng, R., Han, W., Kawakami, R.K.: Phys. Rev. B. **86**, 060409 (2012)
46. Thackey, M.M., David, W.I.F., Goodenough, J.B.: Mater. Res. Bull. **17**, 785 (1982)
47. Fontcuberta, J., Rodriguez, J., Pernet, M., Longworth, G., Goodenough, J.B.: J. Appl. Phys. **59**, 1918 (1986)
48. Neto, J.M., Nunez, E., Domingues, P.H.: J. Mater. Sci. Lett. **16**, 231 (1997)
49. Sivakumar, V., Kumar, S., Ross, C.A., Shao-Horn, Y.: ECS Trans. **2**, 1 (2007)
50. Yamada, T., Morita, K., Kume, K., Yoshikawa, H., Awaga, K.: J. Mater. Chem. C. **2**, 5183 (2014)
51. Inoue, J., Maekawa, S.: Phys. Rev. B. **53**, R11927 (1996)
52. Drozdov, A.P., Eremets, M.I., Troyan, I.A., Ksenofontov, V., Shylin, S.I.: Nature. **525**, 73–76 (2015)
53. Bardeen, J., Cooper, L.N., Schrieffer, J.R.: Phys. Rev. **108**, 1175 (1957)
54. Glover III, R.E., Sherrill, M.D.: Phys. Rev. Lett. **5**, 248 (1960)
55. Choi, J., Pradheesh, R., Kim, H., Im, H., Chong, Y., Chae, D.-H.: Appl. Phys. Lett. **105**, 012601 (2014)
56. Tsuchiya, T., Miyoshi, S., Yamashita, Y., Yoshikawa, H., Terabe, K., Kobayashi, K., Yamaguchi, S.: Solid State Ionics. **253**, 110 (2013)
57. Tsuchiya, T., Miyoshi, S., Yamashita, Y., Yoshikawa, H., Terabe, K., Kobayashi, K., Yamaguchi, S.: Sci. Technol. Adv. Mater. **14**, 45001 (2013)
58. Kim, S.J., Tsuruoka, T., Hasegawa, T., Aono, M., Terabe, K., Aono, M.: AIMS Mater. Sci. **3**, 245 (2016)

# Artificial Synapses Realized by Atomic Switch Technology

**Tohru Tsuruoka, Takeo Ohno, Alpana Nayak, Rui Yang, Tsuyoshi Hasegawa, Kazuya Terabe, James K. Gimzewski, and Masakazu Aono**

**Abstract** The atomic switch technology can be used to emulate the synaptic plasticity underlying short-term and long-term memory functions in the human brain. These functions are realized by transport of metal ions or oxygen ions in sulfide or oxide matrices. The change in conductance of these devices is considered analogous to the change in strength of biological synapses that varies depending on the strength, frequency, and number of stimulating input pulses. The devices also exhibit sensitivity to the moisture in and temperature of the ambient environment, and configurable multifunction including rectification and synaptic plasticity under different electroforming conditions. These observations indicate that the atomic switch technology has great potential for use as an essential building block for neural computing systems.

T. Tsuruoka (✉) · K. Terabe · M. Aono
International Center for Materials Nanoarchitectonics (MANA), National Institute for Materials Science (NIMS), Tsukuba, Ibaraki, Japan
e-mail: TSURUOKA.Tohru@nims.go.jp

T. Ohno
Faculty of Science and Technology, Graduate School of Engineering, Oita University, Oita, Japan

A. Nayak
Department of Physics, Indian Institute of Technology Patna, Bihta, Patna, Bihar, India

R. Yang
School of Materials Science and Engineering, Huazhong University of Science and Technology, Wuhan, China

T. Hasegawa
Department of Applied Physics, Waseda University, Shinjuku, Tokyo, Japan

J. K. Gimzewski
Department of Chemistry & Biochemistry, University of California Los Angeles, East Los Angeles, CA, USA



M. Aono (ed.), *Atomic Switch*, Advances in Atom and Single Molecule Machines,
https://doi.org/10.1007/978-3-030-34875-5_10

## 1 Introduction

Memory in the human brain is believed to occur via two types of synaptic plasticity: short-term plasticity (STP) and long-term potentiation (LTP). Synaptic plasticity refers to changes that occur in the organization of the brain as a result of information input and human experience. Providing stimuli over longer intervals forms STP through the temporal enhancement of synaptic weight (or synaptic connections), which then quickly decays to its initial state. By contrast, frequent stimulation permanently changes the synaptic weight required to achieve LTP; shorter repetition intervals enable more efficient LTP formation using fewer stimuli. These synaptic behaviors are achieved in a biological system by the action potential from a neuron. Endeavors continue toward identifying appropriate materials and architectures that will bring electronics closer to the complexity of the human brain. Even slight variations in operation based on materials can be rewarding when they are put together to design brain-type elements. It is noteworthy that, in the human brain, various metal ions (Fe, Zn, Cu) are known to influence the process of neural signal transmission, yet each of them has a different role; for instance, the flow of metal ions in the brain has a recognized role in learning and memory [1].

In 2011, we discovered novel artificial synaptic behaviors in an $Ag_2S$-based gap-type atomic switch [2]. The change in conductance of the atomic switch is considered to be analogous to the change in strength of the biological synaptic weight that underlies synaptic plasticity, as illustrated schematically in Fig. 1. Therefore, this type of atomic switch is referred to as an "inorganic synapse". It was subsequently found that a $Cu_2S$-based gap-type atomic switch exhibits similar characteristics and have a strong dependence on air (or moisture) and temperature, showing their ability to perceive environmental factors [3]. STP and LTP behaviors were also observed in a simple metal/insulator/metal (MIM)-structured device, in which metal ions or oxygen ions migrate in a thin oxide film [4, 5]. In this chapter, we describe the synaptic abilities observed in various atomic switch structures, which abilities are controlled by the unique behavior of solid-state electrochemical reactions when confined to the nanometer scale.

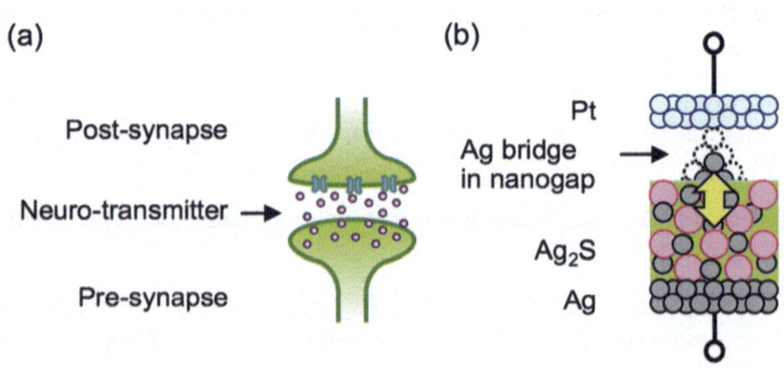


**Fig. 1** (**a**) Biological synapse and (**b**) $Ag_2S$-based gap-type atomic switch (inorganic synapse)

## 2 Gap-Type Atomic Switch Synaptic Behavior

From the earliest stages of the research into the atomic switch, it has shown learning abilities in which the conductance increases gradually under repeated bias sweeping [6]. Subsequently, we found synaptic behavior by applying consecutive voltage pulses [7]. In this section, we describe synaptic behaviors of gap-type atomic switches using sulfides. Utilizing the mixed conductor properties of sulfide materials, we demonstrated not only STP/LTP behaviors but also sensitivity to the environmental conditions. Note that most of Sect. 2.2 is reproduced with permission from Ref. [3]. Copyright 2012 John Wiley & Sons.

### 2.1 $Ag_2S$-Based Switch

Figure 2 shows the change in conductance of a $Ag_2S$-based atomic switch as a function of input electrical pulse repetition time. This artificial inorganic synapse shows two types of conductance behavior, which are observed by controlling the conditions under which the bias voltage causes a $Ag \rightleftarrows Ag^{+}{}_{(Ag2S)} + e^{-}$ electrochemical reaction. In the first behavior, inputting stimuli at a lower repetition rate leads to a temporary increase in conductance, which is followed by a spontaneous decay over time. This behavior corresponds to the STP mechanism. When input pulses were applied at a lower repetition rate, at intervals of 20 s, the atomic switch did not maintain a conductance state higher than 77.5 μS, and decreased steeply with time to its initial low conductance value after each voltage pulse (Fig. 2a). Here, a conductance value of 77.5 μS corresponds to the formation of a single atomic point contact ($1G_0$). Importantly, this decay phenomenon occurred without the application of a

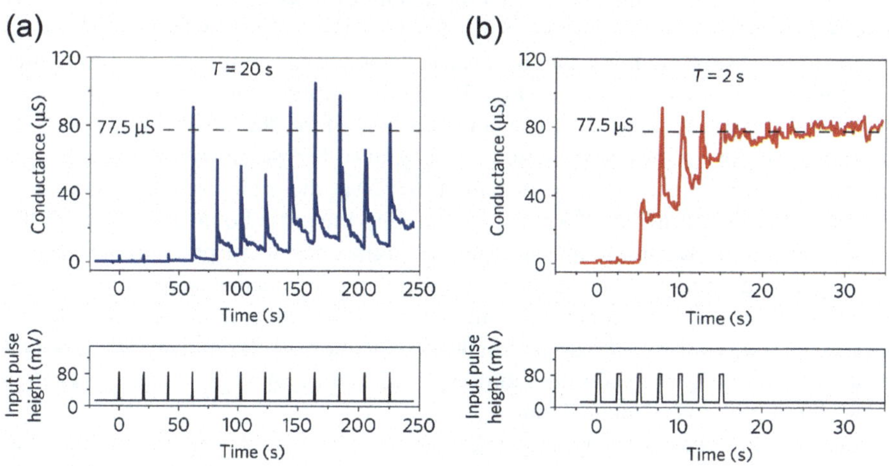


**Fig. 2** (**a**) STP and (**b**) LTP behaviors of a $Ag_2S$-based atomic switch, depending on input-pulse repetition time. Copyright 2011, NPG

voltage, in contrast to the behavior of conventional ReRAMs and memristors [8], for which an applied signal is required to cause a change in conductance. In the second behavior, a persistent enhancement of the conductance can be produced by frequent repetition of stimuli. As the intervals are decreased to 2 s, the atomic switch showed a transition to a higher conductance state, which persisted for a long time even after the input pulse stimuli were stopped (Fig. 2b). This permanent transition to a higher conductance state resembles the persistent increase in the synaptic weight that results from increasing the repetition rate of stimuli by action potentials in a biological nervous system, corresponding to the LTP mechanism.

The STP and LTP synaptic behaviors in the atomic switch can be explained in terms of the formation of a metal bridge and its stability in a nanogap. In the STP mode, the precipitated Ag atoms form an incomplete bridge, and approximately one quantized channel could not be maintained after each input pulse. That is, a large Ag protrusion formed temporarily in the nanogap, which then slowly shrank by dissolving back into the $Ag_2S$ and/or by diffusing away on the $Ag_2S$ surface, gradually decreasing the conductance. On the other hand, in the LTP mode, a complete and robust Ag atomic bridge formed, and a conductance higher than 77.5 μS was maintained after the last input pulse. The key point of this achievement is the use of small input pulses, where each pulse does not turn on the $Ag_2S$-based gap-type atomic switch.

Another type of synaptic behavior has been obtained using a $Ag_2S$-based atomic switch made of small grain $Ag_2S$ [6]. Figure 3 shows the change in resistance of the atomic switch with small $Ag_2S$ crystal grains. The initial resistance before inputting the voltage pulse was set to 1 MΩ. Inputting the voltage pulses resulted in the resistance of the switch remaining constant until the third input signal was applied. However, applying a fourth input pulse caused the resistance to abruptly decrease from 1 MΩ to 10 kΩ. This turn-on switching process shows that the atomic switch stored the information during the first three pulse inputs and turned on at the fourth input pulse (Fig. 3a). The same switching operation was observed for the turn-off process, for which no change in the resistance of the atomic switch was observed during the first several inputs, but the resistance increased significantly to 1 MΩ after a certain number of signal inputs (Fig. 3b). In this experiment, a limited amount of Ag atoms can precipitate from a $Ag_2S$ crystal; therefore, the small $Ag_2S$ grain size must gather Ag cations from neighboring grains for precipitation to occur, as illustrated in Fig. 3c. The energy potential barrier at the grain boundary of a $Ag_2S$ crystal limits the diffusion of Ag cations crossing the grain boundary, although the Ag cations can move freely inside the $Ag_2S$ grains. Thus, Ag precipitation from a small grain requires that the grain be prepared (by increasing the number of Ag cations) using a finite number of input pulses. This behavior is analogous to the experience required to activate a synaptic connection in the human brain, which contrasts with devices exhibiting multilevel and memristive operations that change their output with every input signal.

The atomic switch not only mimics the control of synaptic weight in biology, but may also be useful for implementing a model of memory in the human brain. In psychology, human memory is believed to be created by the dynamic change of

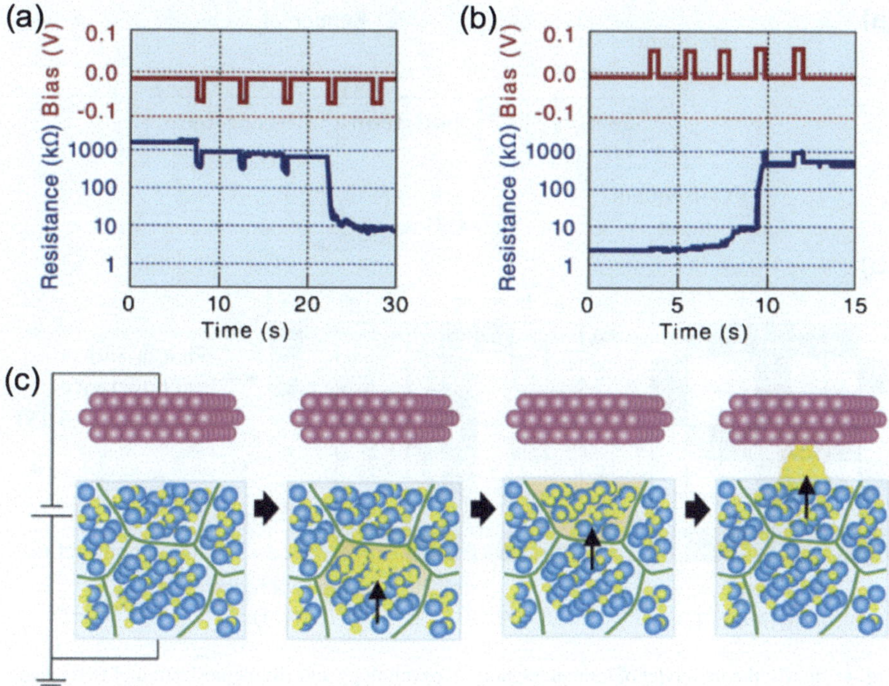


**Fig. 3** Leaning ability of the $Ag_2S$-based atomic switch showing turn-on (**a**) and turn-off (**b**) processes under consecutive voltage pulse application. (**c**) Schematic model of the learning mechanism. Copyright 2010, WILEY-VCH

neural circuits via synaptic plasticity. Although several models of human memory have been proposed, the multistore model developed by Atkinson and Shiffrin in 1968 is one of the most popular psychological memorization models [9]. Figure 4a shows the three types of memory in this model: sensory memory (SM), short-term memory (STM), and long-term memory (LTM). New information from the external environment is stored in the sensory register as an SM for a very short period of time, and then selected information is transferred from a temporary STM in short-term store to a permanent LTM in long-term store. Note that Atkinson and Shiffrin assumed that STM can change to LTM through a rehearsal process, and that frequent rehearsals increase the probability of changing from STM to LTM.

We were successful in introducing the concept of the psychological multistore model into neuromorphic engineering [2]. $Ag_2S$-based atomic switches were used to conduct image memorization to demonstrate psychological behavior. In this demonstration, two images of the numerals '1' and '2' were stored in a $7 \times 7$ $Ag_2S$-based atomic switch array (Fig. 4b). In this array, the change in conductance at each pixel corresponded to the result for a single atomic switch, where a dark color indicates a high conductance value. Here, an image of the numeral '2' was stored using ten inputs with longer intervals ($T = 20$ s), and an image of the numeral '1' was

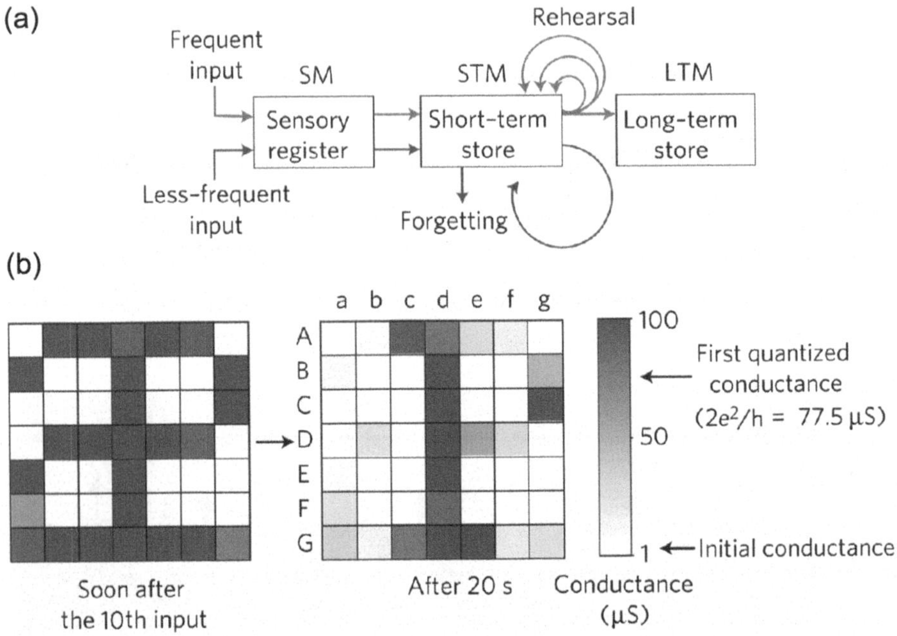


**Fig. 4** (a) Multistore model of human memory in psychology and (**b**) memorization of two images of the numerals '1' and '2' in a 7 × 7 $Ag_2S$-based atomic switch array. Copyright 2011, NPG

simultaneously stored using ten inputs with the same amplitude and width, but with shorter intervals ($T = 2$ s). Soon after the last (10th) input image, the numerals '1' and '2' appeared with higher conductance, which made it difficult to distinguish them from one another. However, the numeral '1' was observed to persist for 20 s after the last input, due to the forgetting of the numeral '2', demonstrating that only the number '1' was transferred to the LTM mode. If a conventional switching device array is used, both images are expected to be stored at the same conductance level due to the same total number of input images, i.e., the same total input power. However, the result indicates that a multistore model best describes the observed behavior. Therefore, this experimental demonstration using a $Ag_2S$-based atomic switch array clearly indicates the implementation of the psychological multistore model, including forgetting behavior.

Forgetting in the human brain is likely one of the most important processes for memorization, because the brain has a limited memorization capacity. Following the first approach to forgetting by Ebbinghaus in 1885 [10], forgetting curves (or retention curves) have been derived in psychology, and repetition rehearsal based on active recall is clearly an appropriate method for increasing the strength of memory. The forgetting curve has been reproduced experimentally using the $Ag_2S$-based atomic switch [2]. One of the forgetting curves can be expressed as a power function $y = b \times t^{-m}$, where $y$ is the memory retention, $b$ is the fit constant for scaling, $t$ is the elapsed time after the $n$-th rehearsal, and $m$ is the rate of the power

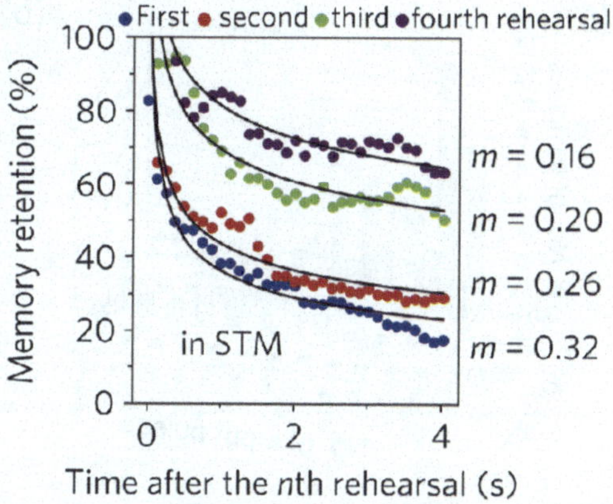


**Fig. 5** Forgetting curves demonstrated by a $Ag_2S$-based atomic switch. Copyright 2011, NPG

function [11]. Figure 5 shows the experimental decay curve for the conductance of a $Ag_2S$-based atomic switch in STM mode: a power function was used to fit the conductance curve to analyze the psychological behavior. As the number of rehearsals increased, the decay speeds, *m*, became smaller and the memory retention ratio increased. This result for the STM mode is reminiscent of the psychological forgetting curve.

As shown in Fig. 2, we can see the formation of SM prior to the STM formation, in which the conductance of the atomic switch increased little during the pulse application, followed by a sharp decay back to its initial value immediately after the pulse. In this operation, more than one input pulse is required to facilitate the transition from SM to STM when a small bias voltage is used, and the number of input pulses required for the first STM formation strongly depends on the amplitude and width of the input voltage pulse. This is shown in Fig. 6 [12]. Increasing the pulse amplitude decreases the number of input pulses required for the first STM formation (Fig. 6a). In addition, the cumulative probability of STM formation was found to depend on the amplitude of the input. For example, six input pulses with $V = 80$ mV brought 91% of switches into STM, whereas only 13% of switches were brought into STM by the same number (six) of pulses with $V = 70$ mV. Input pulses with a broader width also facilitate the first STM formation with a smaller number of input pulses (Fig. 6b). These results resemble the memorization process in the human brain.

## 2.2 $Cu_2S$-Based Atomic Switch

Similar to the $Ag_2S$ system presented in the previous subsection, the $Cu_2S$-based switch also exhibits three different memory behaviors analogous to the SM, STM and LTM modes of the human brain. While $Ag_2S$ is an *n*-type material, $Cu_2S$ is a *p*-type material. In $Cu_2S$, Cu vacancies act as electron acceptors, giving rise to free holes for

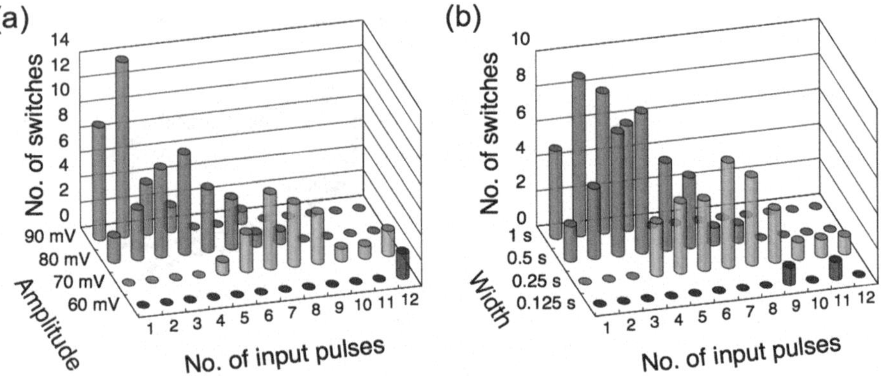


**Fig. 6** First STM formation plotted as a function of the number of input pulses for different (**a**) pulse amplitudes and (**b**) pulse widths. Copyright 2011, AIP

conductivity. Hence, the resistance of $Cu_2S$ increases with an increase of Cu concentration [13]. However, such change in resistance of $Cu_2S$, due to local $Cu^+$ enrichment at the subsurface, can be negligible compared to the resistance of the nanogap of a $Cu_2S$ gap-type atomic switch, where a $Cu_2S$ mixed conductor on a Cu electrode and a counter Pt electrode are opposed in the scanning tunneling microscope. The conductance measurements were carried out in vacuum conditions ($10^{-7}$ Pa) and in air, at different temperatures controlled by indirect resistive heating. Details of the sample preparation and experimental conditions are presented in Ref. [3].

The $Cu_2S$ synapse operation obtained in vacuum for input voltage pulses with varying amplitude ($V = 100$, 150 mV), width ($W = 50$, 500 ms) and intervals ($T = 1$, 10 s) are shown in Fig. 7. For $T = 10$ s, the conductance almost reached $\sim 1G_0$ for all the input pulses of 150 mV and 500 ms, but decayed spontaneously in the interval between the pulses, indicating the STM mode (Fig. 7a). When $T$ was decreased to 1 s, a long-lived transition to the higher conductance state ($\geq G_0$ for at least 20 s) was achieved after ten inputs of the same pulse, which corresponds to the LTM mode (Fig. 7b). When the pulse amplitude was decreased to 100 mV keeping interval at 10 s (Fig. 7c), the SM mode was observed at the 6th pulse prior to observation of the first STM state at the 8th pulse. For the same pulse, when $T$ was decreased to 1 s (Fig. 7e) the first STM mode was achieved after 5 input pulses and the LTM mode was achieved after 11 input pulses. Further, if the width was decreased to 50 ms for the input as in Fig. 7a, the first STM state was achieved after 10 input pulses and a LTM state was not achieved even after 30 input pulses. Figure 7f shows the magnified view of Fig. 7c as indicated by the dashed rectangular box, in which the conductance curves of the successive STM modes are fitted with the exponential decay function $y = y_0 + A\ e^{-t/\tau}$, where $y$ is the conductance, $y_0$ is the conductance offset, $A$ is the fit constant, $\tau$ is the time constant, and $t$ is the time after each input pulse application. Similar analysis has also been presented in the previous subsection in the context of $Ag_2S$ system. Such exponential functions are most commonly used for the quantitative description of retention in human memory [12]. Similar to the

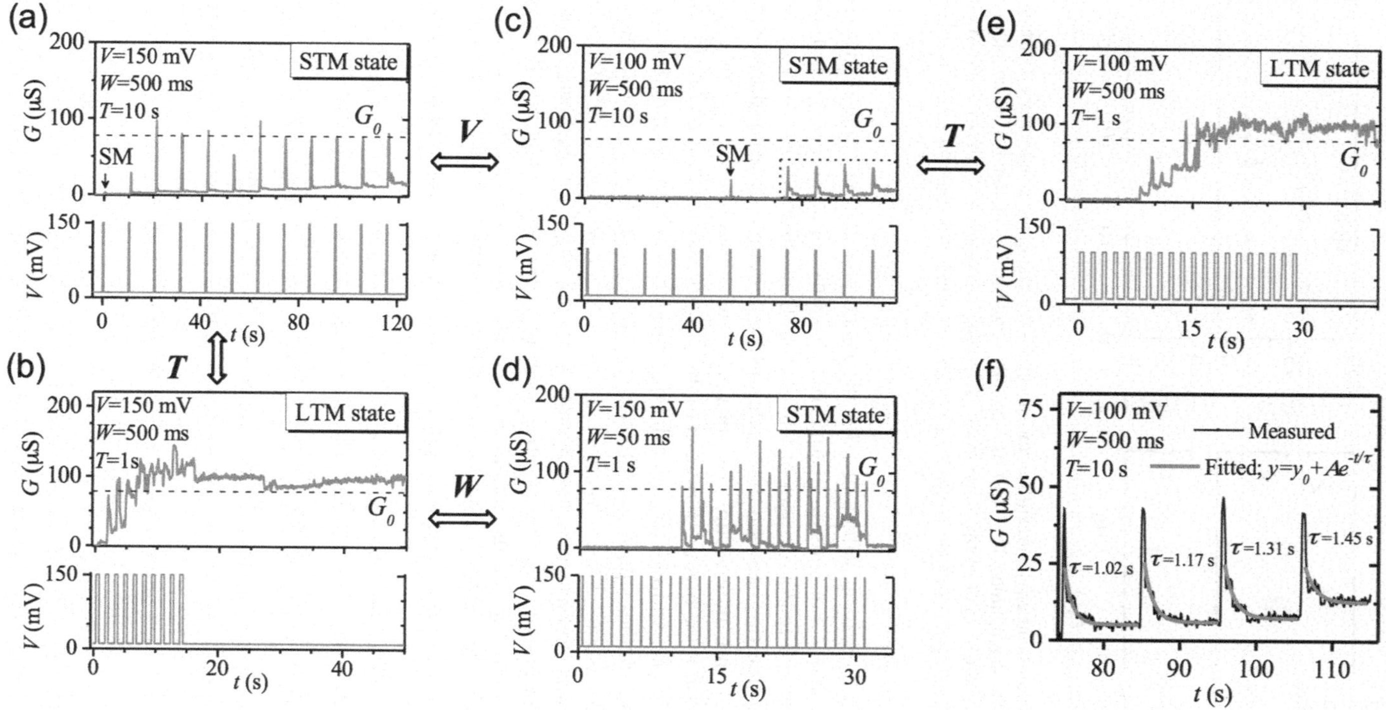


**Fig. 7** Changes in the conductance ($G$) of a $Cu_2S$-based synapse, in vacuum at room temperature, depending on the interval $T$, amplitude $V$ and width $W$ of the input voltage pulse stimulation. (**a**) $V = 150$ mV, $W = 500$ ms, $T = 10$ s, (**b**) $V = 150$ mV, $W = 500$ ms, $T = 1$ s, (**c**) $V = 100$ mV, $W = 500$ ms, $T = 10$ s, (**d**) $V = 150$ mV, $W = 50$ ms, $T = 1$ s, and (**e**) $V = 100$ mV, $W = 500$ ms, $T = 1$ s. (**f**) The values of time constant $\tau$ extracted from the fits of the conductance decay curves shown in the dashed rectangular box in (**c**). The exponential function, $y = y_0 + Ae^{-t/\tau}$, was used to fit the conductance curves, where y is the conductance, $y_0$ is the conductance offset, $A$ is the fit constant and $t$ is the time. Copyright 2012, WILEY-VCH

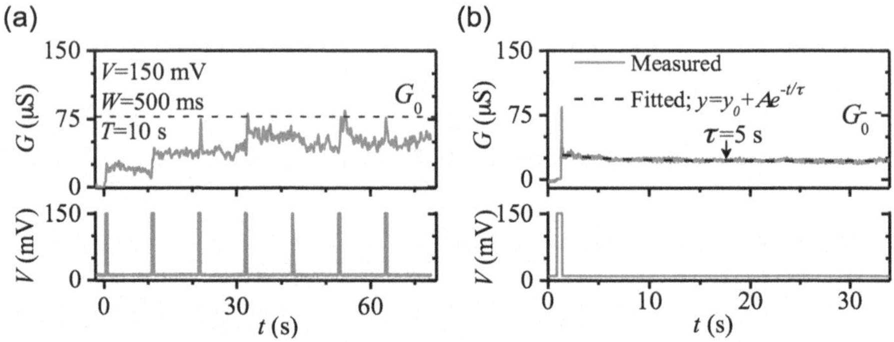


**Fig. 8** Effect of the ambient environment (air/moisture) on the change in conductance $G$ for (**a**) multiple input pulses of amplitude $V = 150$ mV and $W =$ width 500 ms applied at an interval $T = 10$ s, and (**b**) a single input of the same pulse. The time constant $\tau$ is obtained from the fit (dashed line) of the conductance curve using the exponential function. Copyright 2012, WILEY-VCH

$Ag_2S$ system, the increasing values of $\tau$ for the successive curves suggest that the decay of conductance contains a history of previous inputs arising from the non-stoichiometry and rearrangement of $Cu^+$ ions at the subsurface of $Cu_2S$.

Despite the similarities between the $Ag_2S$ and $Cu_2S$ systems, certain differences in their operations are noteworthy. For instance, the operation voltage of $Cu_2S$ synapses was found to be higher compared to the $Ag_2S$ synapses. A possible reason might be the higher activation energy for the migration of $Cu^+$ ions than that of $Ag^+$ ions in sulphides. Another significant difference was found in their operations under varied ambient conditions. While the $Ag_2S$ synapse exhibited almost similar behavior in both ambient and vacuum, the operation of the $Cu_2S$ synapse was highly influenced by the presence of air (relative humidity of ~50%). Compared to its operation in vacuum (Fig. 7), the retention of conductance values after each input pulse was significantly higher when operated in air (Fig. 8a). For example, a conductance state was observed to be stable over 30 s after a single input pulse in air as shown in Fig. 8b. The fitting of the conductance curve using the expression $y = y_0 + A\,e^{-t/\tau}$ yielded the value of $\tau$ to be 5 s, which is much larger than the values obtained in vacuum (Fig. 7f). Further, the difference of $\tau$ was confirmed by time-dependent scanning tunneling microscopy imaging of Cu protrusions grown on $Cu_2S$ in air and in vacuum. A voltage pulse of $V = 200$ mV and $W = 2$ s was applied to grow the protrusion on $Cu_2S$ surface in air. The feedback loop of scanning tunneling microscope was kept active so that the tip was retracted following the growth. The imaging was performed at a sample bias of –100 mV, with a set point tunneling current of 1 nA and a scan speed of 500 nm/s. The dashed circle in Fig. 9a indicates the region on the $Cu_2S$ surface at which the pulse was applied. Figure 9b clearly shows a Cu protrusion in the same region after the pulse application. Successive imaging revealed that the Cu cluster was quite stable and remained for several tens of minutes (Fig. 9c). The height profiles corresponding to the lines drawn on the images are shown on the right-hand side (Fig. 9d).

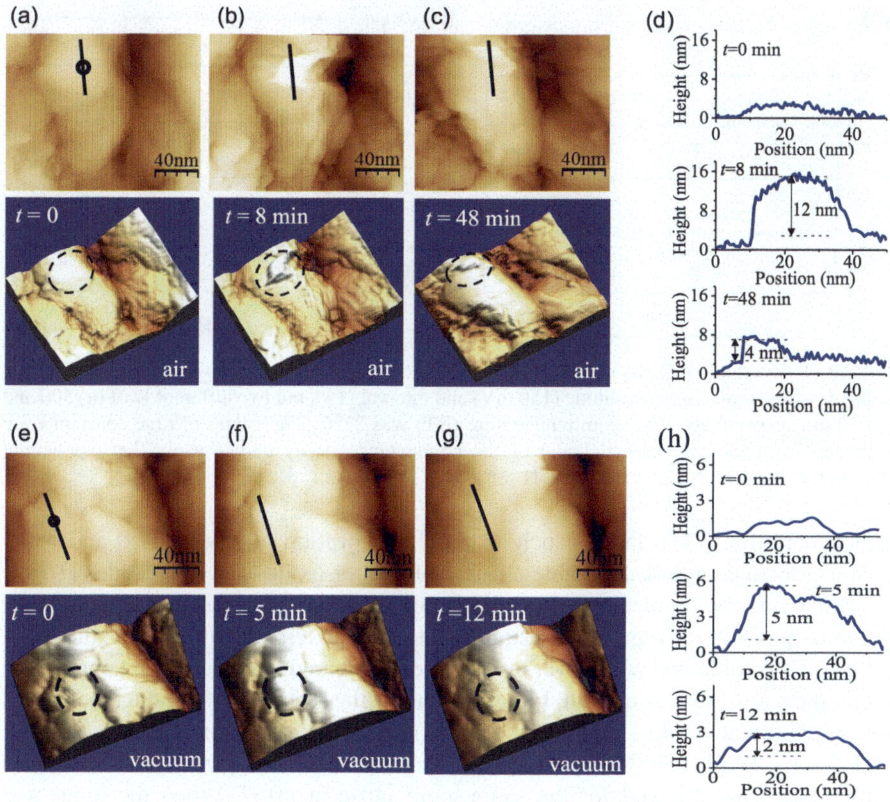


**Fig. 9** (**a–d**) STM images of a Cu protrusion grown in air on a $Cu_2S$ surface upon application of a pulse of 200 mV and width of 2 s. (**a**) Before application of the pulse at the point marked by a circle, (**b**) 8 min and (**c**) 48 min after application of the pulse. (**d**) Respective height profiles corresponding to the lines drawn on the 2D images across the protrusion grown region. (**e–h**) Images of a Cu protrusion grown in vacuum on a $Cu_2S$ surface upon application of the same voltage pulse as in (**a–d**), but for a longer width of 4s. (**e**) Before application of the pulse, (**b**) 5 min and (**c**) 12 min after application of the pulse. (**d**) Respective height profiles. For the sake of clarity, the 2D and 3D image representations are shown one below the other. Copyright 2012, WILEY-VCH

On the contrary, the Cu-protrusions in vacuum shrank rapidly. Since imaging takes time, a voltage pulse of longer width was required to grow a bigger protrusion that would survive for several tens of minutes. Figure 9e–h show successive images of the $Cu_2S$ surface obtained in vacuum before and after growing a Cu protrusion with a voltage pulse of $V = 200$ mV and $W = 4$ s. A cluster of smaller size, compared to that grown in air (Fig. 9b), was imaged 5 min after the pulse application (Fig. 9f). The image of the same region 12 min later shows that the cluster size has further reduced (Fig. 9g). The height profiles are shown on the right-hand side of each image (Fig. 9h). Thus, the time-dependent imaging clearly reveals that the Cu protrusion in air is more stable than in vacuum [3]. In vacuum, the stability of a Cu cluster is limited mainly by the surface diffusion and reincorporation. In air, additional

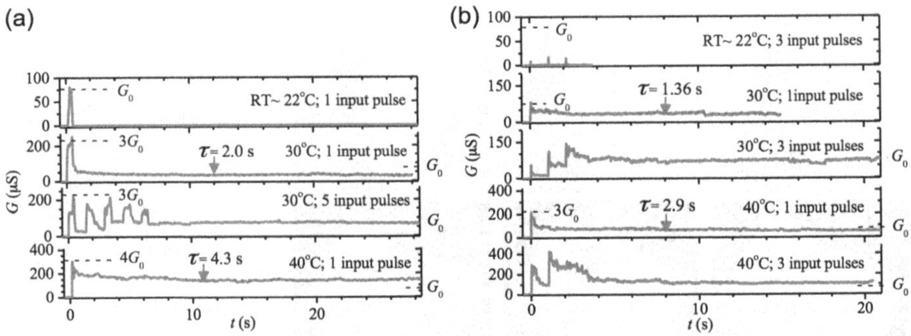


**Fig. 10** Temperature dependence of conductance $G$ of a $Cu_2S$-based synapse in vacuum for input voltage pulses of the same amplitude (150 mV) and interval (1 s), but two different $W$ of (**a**) 500 and (**b**) 50 ms, respectively. The room temperature (RT) was 22°C. The values of time constant $\tau$ are extracted from the fits (dashed lines) of the conductance decay curves using the exponential function. Copyright 2012, WILEY-VCH

atmosphere-dependent factors such as the chemisorption of oxygen [14] and water [15] can form energetically stable clusters of copper oxides and hydroxides [16].

Moreover, the temperature dependence of the $Cu_2S$ synapse indicated a better retention property of the synaptic strength at elevated temperatures. Thus, the LTM state could be obtained with fewer input pulses compared to that at room temperature under the same pulse condition. Figure 10 shows the retention of conductance states for two different widths ($W = 500$ and 50 ms) of the same input pulse of $V = 150$ mV and $T = 1$ s. While the SM state was observed at the room temperature ($\approx$ 22°C), the STM state was achieved for the same input pulse at 30°C. When the input was repeated three times, the LTM state was obtained. However, at 40°C, a single input pulse was enough to achieve the LTM mode (Fig. 10a). Also, the values of conductance and $\tau$ were found to increase with increasing temperature (Fig. 10b), indicating a better retention property of the synaptic strength analogous to a biological synapse at elevated temperatures.

These observations suggest that Cu atoms are precipitated increasingly at elevated temperatures. The rate of precipitation is enhanced due to the exponential dependence on the tunneling current flowing through the inorganic synapse [17]. This temperature behavior of the $Cu_2S$ synapse is consistent with previous observation of the exponentially faster switching of the $Cu_2S$ gap-type atomic switch at higher temperatures [18]. In biology, temperature change is known to have a major impact on the function of the central nervous system [19]. An increase in the amplitudes of facilitation and augmentation and a change in the amplitude of depression at higher temperatures are known to improve the retention time of synaptic strength [20]. Although it is difficult to correlate directly with biological systems at this stage, the $Cu_2S$ synapse shows interesting behavior of facilitating faster LTM formation with fewer or shorter stimulations at elevated temperatures.

# 3 Synaptic Behavior of the Gapless-Type Atomic Switch

As described in the previous section, the gap-type atomic switch clearly exhibits synaptic plasticity, depending on the input pulse conditions. For practical applications, it is important to realize this property in MIM structures with thin oxide films, because of their high compatibility with current CMOS device fabrication processes. In this section, we describe the results for two types of MIM structures that show synaptic plasticity, based on the transport of metal ions and oxygen ions, respectively, in a thin oxide film.

## 3.1 *Ag/$Ta_2O_5$-Based Switch*

If a thin oxide film is sandwiched between an electrochemically active metal electrode (for example, Cu or Ag) and an inert metal electrode (such as Pt and Au), such a structure exhibits bi-stable resistive switching under bias voltage sweeping, in a manner similar to the gap-type atomic switch. The observed resistive switching is attributed to the formation of a metal filament between the two electrodes due to inhomogeneous nucleation on the inert electrode, and dissolution of the metal filament formed assisted by thermochemical reactions [21–23]. From the similarity of its operating mechanism to that of a "gap-type atomic switch", this MIM structure, with an electrochemically active metal electrode, can be referred to as a "gapless-type atomic switch" [24]. We have investigated resistive switching and novel functions of gapless-type, oxide-based atomic switches using $Ta_2O_5$ as a model matrix, and have revealed the redox reactions at the metal/oxide interfaces [25, 26], the effect of moisture absorption by the oxide matrix [27–29], fast-speed switching dynamics [30], quantized conductance, and synaptic behaviors. $Ta_2O_5$ films are usually prepared by RF sputtering or EB deposition. The composition and film density of the oxide matrix was found to affect the switching characteristics, which depend strongly on the ambient humidity conditions [28, 31].

$Ta_2O_5$-based atomic switches show learning abilities that are dependent on the bias conditions, such as sweep cycles and sweep rates. Figure 11a, b represent *I-V* curves of a Ag/$Ta_2O_5$/Pt device with a $Ta_2O_5$ thickness of 15 nm, measured with increased sweep cycles and different sweep rates, respectively [32]. No current compliance was applied in these measurements. The device exhibits bipolar switching with small turn-off voltages, which is a typical characteristic of Ag-$Ta_2O_5$-based atomic switches [23]. With increasing sweep cycles, the on current increased gradually, as seen in Fig. 11a. This is because the diameter of the formed metal filament becomes larger as the voltage sweep was repeated, due to the increased number of Ag ions in the $Ta_2O_5$ matrix. On the other hand, as the sweep rate was decreased, the on current increased while the turn-on voltage slightly decreased (Fig. 11b). This behavior can be explained by a higher concentration of Ag ions in the vicinity of the Pt electrode at lower sweep rates. Once inhomogeneous

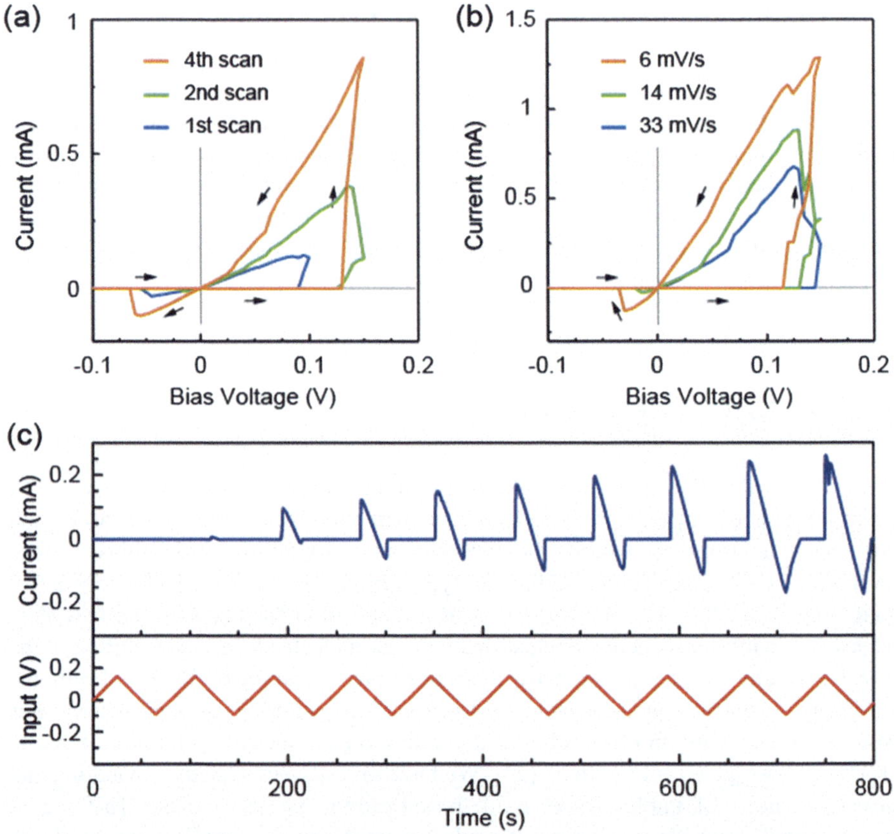


**Fig. 11** Learning abilities of a $Ag/Ta_2O_5/Pt$ atomic switch. (**a**) Sweep cycle dependence, (**b**) sweep rate dependence, and (**c**) time evolution of the device current under repeated voltage sweep

nucleation takes place, the Ag nucleus grows by incorporating Ag ions from the surroundings. As a result, a larger filament is formed at lower sweep rates, thereby increasing the on current. Figure 11c represents the time evolution of the device current under repeated voltage sweeps between 0.18 and –0.15 V. The current appeared from the second sweep cycle and increased gradually with sweep cycles, which corresponds to thickening of the diameter of the formed metal filament. The observed characteristic is similar to that of the model calculation for a proposed [33] and demonstrated [34] by $TiO_2$-based memristor. These results indicate that the history of the input stimulus is stored in the structure, consequently enabling the learning ability of the atomic switch.

When the thickness of the $Ta_2O_5$ film was reduced to less than 10 nm, quantized conductance clearly appeared under application of a constant small bias voltage (less than 0.1 V) [4]. The device conductance also increased and decreased in a stepwise fashion under application of positive and negative voltage pulses, respectively, where each step height corresponds to $G_0$. The distribution of the device conductance, evaluated after the application of each input pulse, showed that the conductance state has peaks at integer multiples of $G_0$ with small distributions around each integer. This result indicates that an atomic point contact with different integer multiples can be formed in a thin $Ta_2O_5$ film. Because a limited number of $Ag^+$ ions is expected to participate in the formation and dissolution of the atomic contact, we believe that it is appropriate to definition this oxide-based MIM structure as a "gapless-type atomic switch".

The conductance state of the Ag/$Ta_2O_5$/Pt device was found to drastically change depending on the interval of the input pulses. Figure 12a represents the device conductance variation under the application of ten input pulses of $V = 0.4$ V with $W = 20$ ms and $T = 2$ s. The device exhibited an enhanced increase in conductance up to ~2$G_0$ with the application of each input pulse. However, the conductance immediately decayed after the input pulses. Each voltage pulse induces the formation of a metal filament, but because of the small input voltage, the filament is unstable and dissolves immediately after the pulse, and the Ag ions diffuse away from the filament position in an ion concentration gradient, as illustrated in the inset of Fig. 12a. If $T$ was decreased to 0.2 s (Fig. 12b), the device initially exhibited a temporary increase for the first few inputs, similar to Fig. 12a. However, with increased input pulses the conductance gradually increased, with accompanying temporary increases, to finally reach ~1$G_0$. This conductance state was maintained for more than 1 min. after the tenth input pulse. The long-lived high conductance state for a shorter time interval can be explained by the formation of a larger metal filament before the Ag ions diffuse completely in a succession of input pulses, resulting in the formation of a stable atomic contact, as illustrated in the inset of Fig. 12b.

The observed conductance behavior for different interval times of input pulses is very similar to that of gap-type $Ag_2S$ and $Cu_2S$ atomic switches [2, 3], and is analogous to the LTP behavior of biological synapses, in which the strength of synaptic weight (or conductance) depends upon the time interval between the stimulating pulses [35, 36]. Our experimental observations indicate that the migration and reduction of metal ions in a thin oxide layer could mimic biological synaptic behavior.

It was also observed that higher input pulse amplitudes result in higher integer multiples of the quantized conductance state. Figure 12c–e show the conductance behavior under the application of ten consecutive pulses with different amplitudes. The conductance reached ~2$G_0$, ~3$G_0$, and ~4$G_0$ for pulses with $V = 0.45$, and 0.5, and 0.55 V, respectively. It can be seen that higher input pulse amplitudes result in higher integer multiples of the quantized conductance state. This result indicates that the atomic switch possesses a unique property in which the conductance strength also depends upon the strength of the stimulating pulses, which is similar to biological synapses. These synaptic behaviors of the $Ta_2O_5$-based atomic switch are difficult to achieve by on/off bi-stable devices based on conventional CMOS transistors.

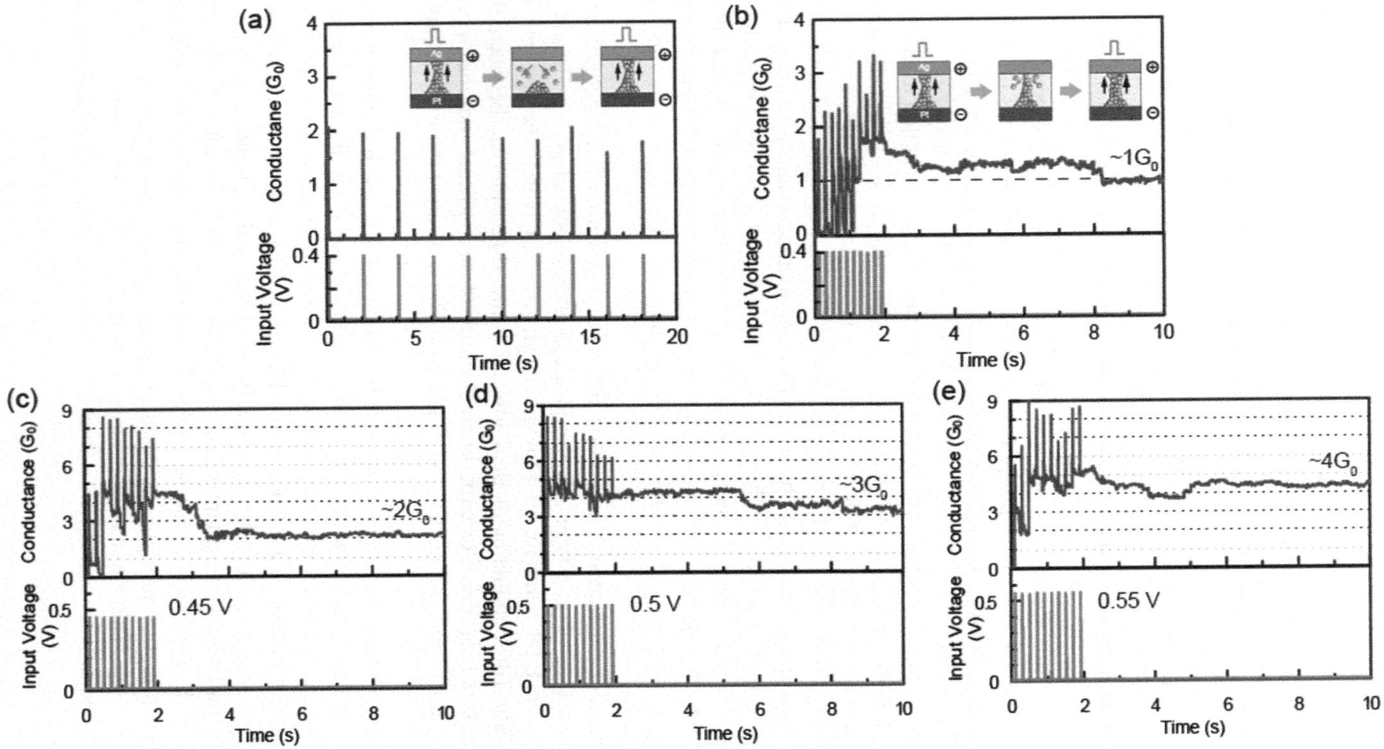


**Fig. 12** Changes in the conductance of a $Ag/Ta_2O_5/Pt$ atomic switch under the application of ten consecutive input pulses of $V = 0.4$ V with W of 20 ms at $T$ of 2 (**a**) and 0.2 s (**b**). The insets schematically illustrate the formation of an atomic contact under consecutive input pulses at respective interval times. Also shown are the results for $T$ of 0.2 s, but different $V$ of 0.45 (**c**), 0.5 (**d**), and 0.55 V (**e**). Copyright 2012, IOP

## 3.2 Pt/$WO_{3-x}$-Based Switch

As mentioned in the preceding subsections, versatile synaptic neuromorphic functions have been achieved in various atomic switch devices by precisely controlling the local metal-ion migration and redox reactions on the nano- and atomic-scales. In these atomic switch systems, an electrochemically active electrode, made from material such as Ag or Cu, is used in a device. Resistive switching in the device occurs through the formation and annihilation of a metallic atom bridge resulting from the migration of highly mobile cations such as $Ag^+$ and $Cu^+$ ions. Actually, resistive switching behavior can also be observed in a device without an electrochemically active electrode. In such cases, the migration of anions, usually oxygen ions play a crucial role. Local oxygen ion migration, which is better described as the migration of an oxygen vacancy ($V_O^{\cdot\cdot}$) and a resultant change in the electronic barrier at the interface, is associated with the resistive switching behavior. Similar synaptic functions, described in the preceding subsections, are also demonstrated in these anion migration devices. In this subsection, we will introduce neuromorphic and electrical multifunction, realized in anion-migration based on a $WO_{3-x}$ layer [37].

The devices have an MIM structure of Pt/$WO_{3-x}$/Pt fabricated on a glass or $SiO_2$/Si substrate. The thin $WO_{3-x}$ film was prepared by RF sputtering from a polycrystalline $WO_3$ target in a gas mixture containing Ar and $O_2$. Details of the device fabrication and characterization methods are presented in Ref. [37]. Both volatile and nonvolatile reconfigurable rectification and resistive switching were realized by adjusting the electric stimuli input parameters, as shown in Fig. 13. The resistance of the pristine device decreased nonlinearly after voltage seeping in both the positive and negative voltage regions (Fig. 13a). However, the current boost induced by voltage sweeping quickly fades over time, which means this resistive switching is volatile. In contrast, after a forming process (Fig. 13c), the device exhibited normal bipolar resistive switching, in which the resistance of the device decreased/increased by applying positive/negative sweeping, as shown in Fig. 13b. In this case, the state with relatively high conductance, obtained after positive voltage sweeping, automatically faded over time but did not return to the original state even over the measurement period, which indicates nonvolatile, or at least partially nonvolatile resistive switching. The positive (negative) forming process is triggered by applying positive (negative) voltage sweeping or voltage pulses on the top Pt electrode to induce a soft-breakdown, with a current compliance to prevent the device from hard-breakdown, see Fig. 13c. Note that the device possesses stable resistive switching only in an oxygen-rich atmosphere (air or $O_2$ gas) after the positive electroforming process [38].

The *I-V* curves of both volatile and nonvolatile resistive switching in Fig. 13 are smooth, without abrupt change, indicating that the device performed analog resistive switching. As shown in the bottom-right panel of Fig. 13a, the volatile resistance of the device is gradually changed by applying consecutive voltage sweeping. The nonvolatile resistance states can also be precisely modified by controlling the

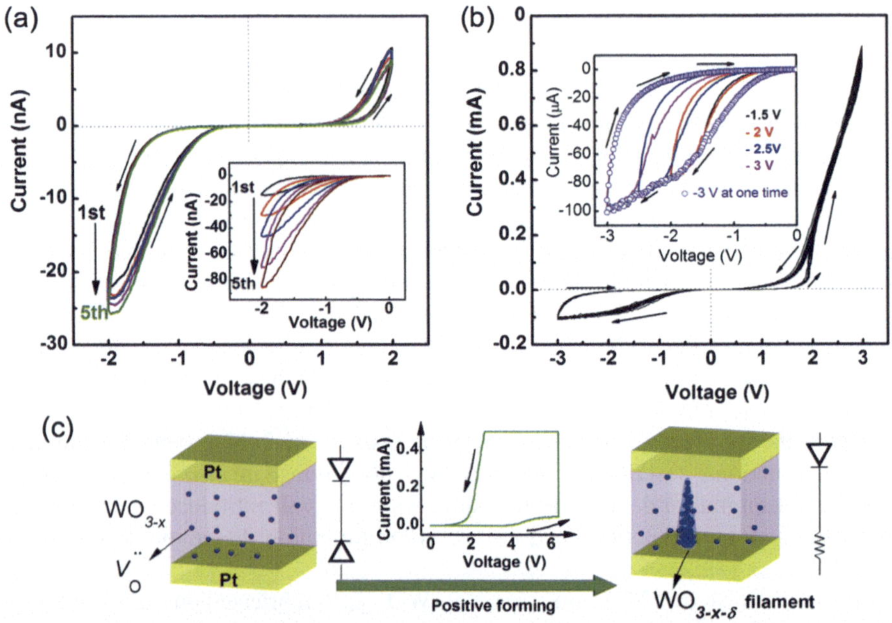


**Fig. 13** Volatile and nonvolatile resistive switching behavior and memorization observed in a Pt/$WO_{3-x}$/Pt device. (**a**) Typical *I-V* curves of volatile resistive switching measured for the as-prepared device. The inset shows five consecutive negative sweep ranges from 0 to 2 V. (**b**) Typical *I-V* characteristics of partial non-volatile resistive behavior. The inset shows the *I-V* curves observed as the sweeping voltage was incrementally increased from –1.5 to –3 V in steps of 0.5 V (solid curves), and that obtained by directly sweeping to –3 V (open circles). (**c**) Schematic illustration and equivalent circuit of the device before and after the positive forming process. The positive forming process is also shown

amplitude of the sweeping voltages, as shown in the inset of Fig. 13b. This analog resistive switching is of particularly concern for neuromorphic devices, since the gradually change of the resistance closely resembles the adaptive learning process of a synapse, and mimics the potentiation/inhabitation of the synaptic weight.

Volatile and nonvolatile rectifications occur in conjunction with volatile and nonvolatile resistive switching behaviors, as shown in Fig. 14. The $WO_{3-x}$ layer is an *n*-type semiconductor with oxygen vacancies as donors. Thus, a Schottky-like barrier forms at the Pt/$WO_{3-x}$ due to the high work function of the Pt electrode. Owing to two Schottky-like barriers forming at the two interfaces and the head-to-head connect in series (Fig. 13c), the device has very low conductance in the pristine state. However, the current in the positive region instantaneously increased while the negative region displayed no distinct increase after application of positive voltage pulses, which lead to the resulting rectification characteristics. The reverse rectification was readily achieved by the application of negative voltage pulses, as shown in Fig. 14a. These rectification characteristics are volatile due to the increased current, which fades quickly and returns to its original state. In contrast, after performing an

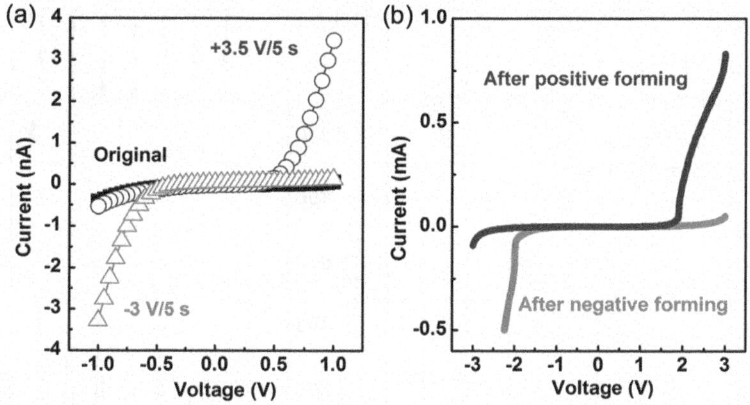


**Fig. 14** Volatile and nonvolatile reconfigurable rectification of a Pt/$WO_{3-x}$/Pt device. (**a**) *I-V* characteristics in the original state without rectification and after applying positive and negative electric pulses with reconfigurable volatile rectifications. (**b**) *I-V* characteristics of the device after the positive and negative electroforming processes, showing nonvolatile rectification

electroforming process, the device shows nonvolatile rectification features. The rectification direction depends on the polarity of the forming process, as shown in Fig. 14b. This reconfigurable rectification and resistive switching, over a wide range of time scales, enables the device to electrically multifunction.

In addition to the electrical multifunction, the device also shows neuromorphic functions; versatile synaptic plasticity. similar to its biological counterparts, our device exhibits analog and incremental properties ranging from volatile to permanent, with inherent learning abilities, as shown in Fig. 15. The value of current peak induced by voltage pulses to the pristine device gradually increases with the number of the pulses, and then quickly decays to its original state during the interval of voltage pulses of $V = 2.8$ V with $W = 0.5$ s and $T = 50$ s, clearly indicating the volatile characteristics of the switching behavior, as seen in Fig. 15a. The instantaneous current peak observed after the seventh pulse is almost five times higher than that observed after the first pulse, even though it seemingly leaks to its phenomenologically original state after each pulse. This means that the volatile resistive switching magnitude depends on the history of the applied electric field. The magnitude of instantaneous current peaks can be precisely controlled by adjusting not only the number, but also the frequency and amplitude of the applied pulses, which is very similar to the STP observed in biology [5].

Neuromimetic transition from STP to LTP can be realized by applying a strong stimulus with high frequency ($V = 3.5$ V with $W = 0.5$ s and $T = 0.5$ s), as shown in Fig. 15b. The first seven pulses induce volatile resistive switching, and a soft breakdown occurs when the eighth pulse is applied, which is similar to what happens in the forming process. The current of the device then remains high, and the behavior of the device changes to nonvolatile bipolar resistive switching. This transition process, from volatile to nonvolatile resistive switching, is very similar to that

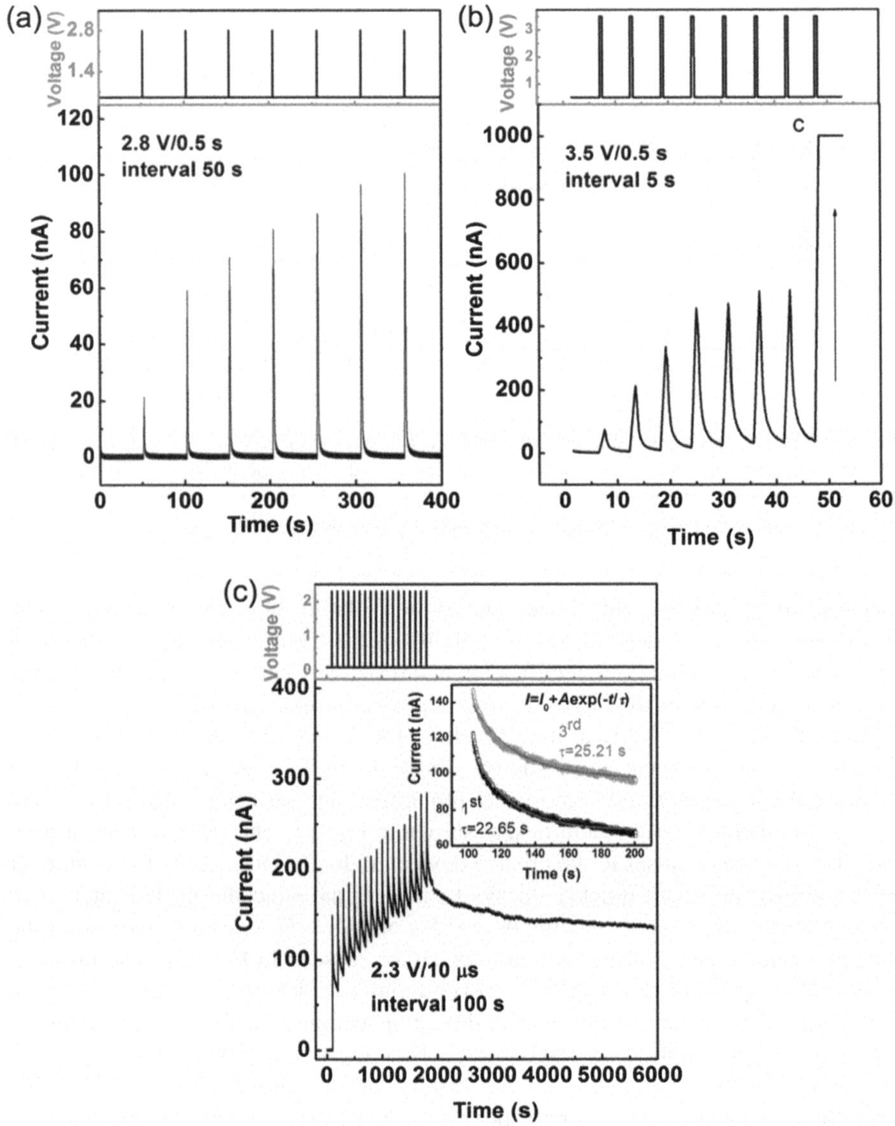


**Fig. 15** Transition from volatile to nonvolatile resistive switching observed in a Pt/$WO_{3-x}$/Pt device. (**a**) Volatile resistive switching obtained by applying seven consecutive pulses of $V = 2.8$ V with $W = 0.5$ s and $T = 50$ s. (**b**) Transition from volatile to nonvolatile switching by successive application of electric pulses with a high amplitude of 3.5 V and a short interval of 5 s. (**c**) Nonvolatile resistive switching and subsequent current state retention characteristics for application of electric pulses of $V = 2.3$ V with $W = 10$ μs. Inset shows current decay curves after the first and third pulses fitted by the exponential function. Copyright 2013, IOP

observed in biological systems, in which the transition from STP to LTP can be achieved by repeatedly stimulating a synapse using a strong potential with high frequency [5].

The LTP in a biological system can remain for long periods of time (several minutes, days, or even years), but it still exponentially fades over time, due to the forgetting effect, at a much slower pace than does STP. This forgetting effect can also be emulated by our device, as shown in Fig. 15c. The current of the device gradually increases by successively applying voltage pulses of $V = 2.3$ V with $W = 10$ μs and $T = 100$ s. After these stimuli, the current of the device decays over time, but does not return to its initial state even after ~4000 s, clearly indicating the nonvolatile characteristics of the switching behavior. The current decay processes can be fitted by an exponential function, as shown in the inset of Fig. 15c. Moreover, the decay time constant increases as the successive current decay process proceeds. These current decay features are strikingly similar to the biological forgetting effect, which endows our device with capabilities that make it an ideal synaptic emulator.

The mechanisms of volatile and nonvolatile resistive switching can be explained as follows. A certain amount of $V_O^{\cdot\cdot}$ is induced during the device fabrication process, as has been verified by X-ray photoelectron spectroscopy (XPS) [37]. For devices without forming processes, a Schottky-like barrier forms at both the top and bottom Pt/$WO_{3-x}$ interfaces, see Fig. 13c. When a positive pulse is applied to the top interface, the external electric field drops mainly at the bottom interface and induces $V_O^{\cdot\cdot}$ migration and accumulation at the bottom interface. This results in the lower portion of the Schottky-barrier to be located at the bottom interface, whereas the Schottky-like barrier at the top interface increase slightly, or at least remains unchanged. Thus, forward rectification appears. After the electric stimuli, the subsequent relaxing of the $V_O^{\cdot\cdot}$ driven by the electrochemical potential gradient results in the volatile characteristics of this rectification and the subsequent resistive switching behaviors. This oxygen migration process has been verified by hard XPS under bias operation measurements [37].

An electroforming process with soft-breakdown is required to trigger the nonvolatile rectification and subsequent resistive switching behaviors. During the forming process, conducting filaments with high oxygen deficient $WO_{3-x-\delta}$ is expected to form in the $WO_{3-x}$ matrix. From combined optical microscopy and cross-sectional TEM observations, we found that the filament is composed of conductive Magnéli phases, with more oxygen vacancies than the mother phases [39]. The observed switching originates from a nano-gap between the top electrode and this conductive filament. On the other hand, the electroforming process exhibited high temperature dependence, suggesting that the Joule heating effect plays a key role in the electroforming process [40]. It was also found that the oxygen migration process during resistive switching occurs mainly between the Pt/$WO_{3-x}$ interface and Pt electrode. The high catalytic activity and oxygen absorbability of Pt electrode are indispensable for realizing stable bipolar resistive switching behavior in a Pt/$WO_{3-x}$ interface. The partial current decay process is attributed to the spontaneous back diffusion of oxygen vacancies after switching.

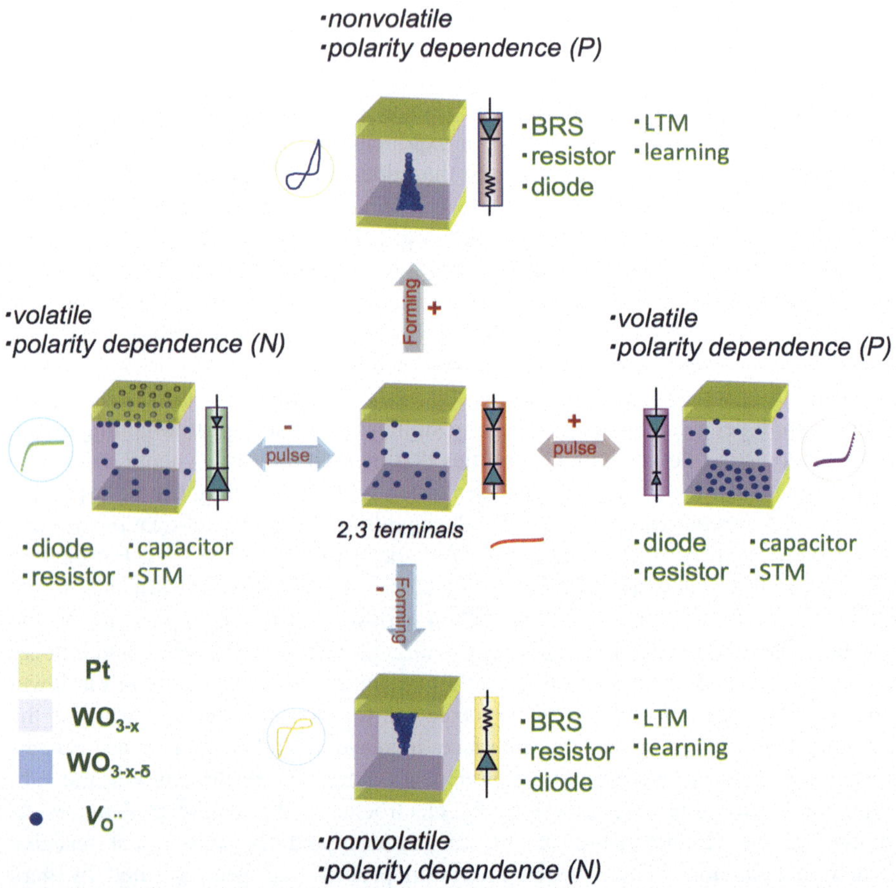


**Fig. 16** Schematic depiction of the configurable electrical and neuromorphic multifunction realized in the $WO_{3-x}$ based devices by using the principle of atomic switch operation. *P* and *N* indicate the dependence of positive and negative polarity on the external field of volatile and nonvolatile electric and neuromorphic multifunction, respectively. Copyright 2012, ACS

Based on the above results, we demonstrated an on-demand nanodevice with configurable electrical and neuromorphic functions in a Pt/$WO_{3-x}$/Pt device, as illustrated schematically in Fig. 16. In the as-prepared original state, there is a certain number of oxygen vacancies in the $WO_{3-x}$ layer. A Schottky-like barrier forms at both the top and bottom interfaces, and the height or width of the Schottky-like barrier can be modulated by applying small bias, which in turn induces volatile resistive switching, rectification, and STP behavior. The oxygen migration direction directly depends on the polarity of the external electric field, hence, the volatile rectification direction can be reversely controlled, as shown in the right- and left-side illustrations in Fig. 16. In these cases, the device can be used as a diode, a capacitor, a resistor, and an artificial synapse with STP. When an electroforming process is triggered by applying a large bias, conductive filaments with highly oxygen deficient Magnéli phases ($WO_{3-x-\delta}$) are formed. The residual Schottky-barrier is located at either the top or bottom interface, depending on the polarity of the forming voltage. Thus, the devices exhibit nonvolatile rectification and partially nonvolatile bipolar resistive switching with configurable polarity, as shown in the upper and lower illustrations in Fig. 16. In these cases, the device can be used as a diode, a resistor, a switching element, and an artificial synapse with learning functions in long-term dynamics.

## 4 Summary

In this chapter, we described the synaptic plasticity realized by various atomic switch structures. Gap-type and gapless-type atomic switches using metal cations can mimic the key features of the learning and memorization abilities in the human brain, such as SM, STM, and LTM modes in the psychological multistore model. The synaptic plasticity underlying these modes can be controlled by input voltage pulse stimulations, which is similar to what occurs with biological synapses. Depending on the matrix electrolyte material, sensitivity to the moisture in and temperature of the ambient environment was also demonstrated, which suggests the possibility of achieving advanced artificial synaptic elements with the ability to sense environmental conditions. This result is a step toward mimicking the uniqueness of the human brain in its ability to perceive the environment. We also demonstrate an on-demand multiple electrical and neuromorphic functions using oxygen anion migration in a $WO_{3-x}$ thin film. Configurable multifunction, including volatile and nonvolatile resistive switching, rectification and synaptic plasticity in both short-term and long-term dynamics, were attained. It should be emphasized that all the functions described herein can be achieved in a single device, without the need of external programming, which is difficult to achieve by on/off bi-stable switching devices based on conventional CMOS transistors. Thus, our devices, based on atomic switch technology, have great potential for use as essential components for configurable circuits, analog memories, and digital-neural fused network systems.

## References

1. Arnesano, F.: The role of copper ion and the ubiquitin system in neurodegenerative disorders. In: Pignataro, B. (ed) Ideas in Chemistry and Molecular Sciences: Where Chemistry Meets Life. WILEY-VCH, Weinheim (2010)
2. Ohno, T., Hasegawa, T., Tsuruoka, T., Terabe, K., Gimzewski, J.K., Aono, M.: Short-term plasticity and long-term potentiation mimicked in single inorganic synapses. Nat. Mater. **10**, 591 (2011)

3. Nayak, A., Ohno, T., Tsuruoka, T., Terabe, K., Hasegawa, T., Gimzewski, J.K., Aono, M.: Controlling the synaptic plasticity of a $Cu_2S$ gap-type atomic switch. Adv. Funct. Mater. **22**, 3606 (2012)
4. Tsuruoka, T., Hasegawa, T., Terabe, K., Aono, M.: Conductance quantization and synaptic behavior of a $Ta_2O_5$-based atomic switch. Nanotechnology. **23**, 435705 (2012)
5. Yang, R., Terabe, K., Yao, Y., Tsuruoka, T., Hasegawa, T., Gimzewski, J.K., Aono, M.: Synaptic plasticity and memory functions achieved in a $WO_{3-x}$-based nanoionics device by using the principle of atomic switch operation. Nanotechnology. **24**, 384003 (2013)
6. Hasegawa, T., Ohno, T., Terabe, K., Tsuruoka, T., Nakayama, T., Gimzewski, J.K., Aono, M.: Learning abilities achieved by a single solid-state atomic switch. Adv. Mater. **22**, 1831 (2010)
7. Hasegawa, T., Nayak, A., Ohno, T., Terabe, K., Tsuruoka, T., Gimzewski, J.K., Aono, M.: Memristive operations demonstrated by gap-type atomic switches. Appl. Phys. A. **102**, 811 (2011)
8. Jo, S.H., Chang, T., Ebong, I., Bhadviya, B.B., Mazumder, P., Lu, W.: Nanoscale memristor device as synapse in neuromorphic systems. Nano Lett. **10**, 1297 (2010)
9. Atkinson, R.C., Shiffrin, R.M.: Human memory: A proposed system and its control processes. In: Spence, K.W., Spence, J.T. (eds.) The Psychology of Learning and Motivation: Advances in Research and Theory, vol. 2, pp. 89–195. Academic, New York (1968)
10. Ebbinghaus, H.: In: Ruger, H.A., Bussenius, C.E. (eds. trans.) Memory: A Contribution to Experimental Psychology. Teachers College, Columbia University, New York (1913)
11. Rubin, D.C., Wenzel, A.E.: One hundred years of forgetting: A quantitative description of retention. Psychol. Rev. **103**, 734 (1996)
12. Ohno, T., Hasegawa, T., Nayak, A., Tsuruoka, T., Gimzewski, J.K., Aono, M.: Sensory and short-term memory formations observed in a $Ag_2S$ gap-type atomic switch. Appl. Phys. Lett. **99**, 203108 (2011)
13. Kundu, M., Hasegawa, T., Terabe, K., Aono, M.: Effect of sulfurization conditions on structural and electrical properties of copper sulfide films. J. Appl. Phys. **103**, 073523 (2008)
14. Urban, J., Sack-Kongehl, H., Weiss, K.: HREM studies of the structure and the oxidation process of copper clusters created by inert gas aggregation. Z. Phys. D. **36**, 73 (1996)
15. Ruuska, H., Tapani, A.P.: MP2 study on water adsorption on cluster models of Cu(111). J. Phys. Chem. B. **108**, 2614 (2004)
16. ter Maat, H., Hogendoorn, J.A., Versteeg, G.F.: The removal of hydrogen sulfide from gas streams using an aqueous metal sulfate absorbent: Part II. The regeneration of copper sulfide to copper oxide-an experimental study. Sep. Purif. Technol. **43**, 199 (2005)
17. Nayak, A., Tsuruoka, T., Terabe, K., Hasegawa, T., Aono, M.: Theoretical investigation of kinetics of a $Cu_2S$-based gap-type atomic switch. Appl. Phys. Lett. **98**, 233501 (2011)
18. Nayak, A., Tsuruoka, T., Terabe, K., Hasegawa, T., Aono, M.: Switching kinetics of a $Cu_2S$-based gap-type atomic switch. Nanotechnology. **22**, 235201 (2011)
19. Schiff, S.J., Somjen, G.G.: The effects of temperature on synaptic transmission in hippocampal tissue slices. Brain Res. **345**, 279 (1985)
20. Klyachko, V.A., Stevens, C.F.: Temperature-dependent shift of balance among the components of short-term plasticity in hippocampal synapses. J. Neurosci. **26**, 6945 (2006)
21. Tsuruoka, T., Terabe, K., Hasegawa, T., Aono, M.: Forming and switching mechanisms of a cation-migration-based oxide resistive memory. Nanotechnology. **21**, 425205 (2010)
22. Tsuruoka, T., Terabe, K., Hasegawa, T., Aono, M.: Temperature effects on the switching kinetics of a Cu-$Ta_2O_5$-based atomic switch. Nanotechnology. **22**, 254013 (2011)
23. Tsuruoka, T., Hasegawa, T., Terabe, K., Aono, M.: Operating mechanism and resistive switching characteristics of two- and three-terminal atomic switches using a thin metal oxide layer. J. Electroceram. **39**, 143 (2017). https://doi.org/10.1007/s10832-016-0063-9
24. Hasegawa, T., Terabe, K., Sakamoto, K., Aono, M.: Nanoionics switching devices: "Atomic switches". MRS Bull. **34**, 929 (2009)

25. Tappertzhofen, S., Valov, I., Tsuruoka, T., Hasegawa, T., Waser, R., Aono, M.: Generic relevance of counter charges for cation-based nanoscale resistive switching memory. ACS Nano. **7**, 6396 (2013)
26. Tsuruoka, T., Valov, I., Tappertzhofen, S., van den Hulk, J., Hasegawa, T., Waver, R., Aono, M.: Redox reactions at Cu(Ag)/$Ta_2O_5$ interfaces and the effects of $Ta_2O_5$ film density on the forming process in atomic switch structures. Adv. Funct. Mater. **25**, 6374 (2015)
27. Tsuruoka, T., Terabe, K., Hasegawa, T., Valov, I., Waser, R., Aono, M.: Effects of moisture on the switching characteristics of oxide-based, gapless-type atomic switches. Adv. Funct. Mater. **22**, 70 (2012)
28. Mannequin, C., Tsuruoka, T., Hasegawa, T., Aono, M.: Composition of thin $Ta_2O_5$ films deposited by different methods and the effects of humidity on their resistive switching behavior. Jpn. J. Appl. Phys. **55**, 06GG08 (2016)
29. Tsuruoka, T., Valov, I., Mannequin, C., Hasegawa, T., Waser, R., Aono, M.: Humidity effects on the redox reactions and ionic transport in a Cu/$Ta_2O_5$ atomic switch structure. Jpn. J. Appl. Phys. **55**, 06GJ09 (2016)
30. Tsuruoka, T., Hasegawa, T., Valov, I., Waser, R., Aono, M.: Rate-limiting processes in the fast SET operation of a gapless-type Cu-$Ta_2O_5$ atomic switch. AIP Adv. **3**, 032114 (2013)
31. Mannequin, C., Tsuruoka, T., Hasegawa, T., Aono, M.: Identification and roles of nonstoichiometric oxygen in amorphous $Ta_2O_5$ thin films deposited by electron beam and sputtering processes. Appl. Surf. Sci. **385**, 426 (2016)
32. Tsuruoka, T., Hasegawa, T., Aono, M.: Synaptic plasticity and memristive behavior operated by atomic switches. https://doi.org/10.1109/CNN.2014.6888615 (2014)
33. Strukov, D.B., Snider, G.S., Stewart, D.R., Williams, R.S.: The missing memristor found. Nature. **453**, 80 (2008)
34. Yang, J.J., Pickett, M.D., Li, X., Ohlberg, D.A.A., Stewart, D.R., Williams, R.S.: Memristive switching mechanism for metal/oxide/metal nanodevices. Nat. Nanotechnol. **3**, 429 (2008)
35. Magleby, K.L.: The effect of repetitive stimulation on facilitation of transmitter release at the frog neuromuscular junction. J. Physiol. **234**, 327 (1973)
36. Atluri, P.P., Regehr, W.G.: Determinants of the time course of facilitation at the granule cell to Purkinje cell synapse. J. Neurosci. **16**, 5661 (1996)
37. Yang, R., Terabe, K., Liu, G., Tsuruoka, T., Hasegawa, T., Gimzewski, J.K., Aono, M.: On-demand nanodevice with electrical and neuromorphic multifunction realized by local ion migration. ACS Nano. **6**, 9515 (2012)
38. Yang, R., Terabe, K., Tsuruoka, T., Hasegawa, T., Aono, M.: Oxygen migration process in the interfaces during bipolar resistance switching behavior of $WO_{3-x}$-based nanoionics devices. Appl. Phys. Lett. **100**, 231603 (2012)
39. Tan, Z.H., Yang, R., Terabe, K., Yin, X.B., Zhang, X.D., Guo, X.: Synaptic metaplasticity realized in oxide memristive devices. Adv. Mater. **28**, 377 (2016)
40. Yang, R., Terabe, K., Guo, X.: The Joule heating effect on the electroforming process for the resistive switching behavior in Pt/$WO_{3-x}$/Pt memristive devices, oral presentation. In: PRiME2016, Honolulu, HI, 2016.10.05

# Atomic Switch Networks for Neuroarchitectonics: Past, Present, Future

**R. Aguilera, K. Scharnhorst, S. L. Lilak, C. S. Dunham, M. Aono, A. Z. Stieg, and J. K. Gimzewski**

**Abstract** Artificial realizations of the mammalian brain alongside their integration into electronic components are explored through neuromorphic architectures, neuroarchitectectonics, on CMOS compatible platforms. Exploration of neuromorphic technologies continue to develop as an alternative computational paradigm as both capacity and capability reach their fundamental limits with the end of the transistor-driven industrial phenomenon of Moore's law. Here, we consider the electronic landscape within neuromorphic technologies and the role of the atomic switch as a model device. We report the fabrication of an atomic switch network (ASN) showing critical dynamics and harness criticality to perform benchmark signal classification and Boolean logic tasks. Observed evidence of biomimetic behavior such as synaptic plasticity and fading memory enable the ASN to attain a cognitive capability within the context of artificial neural networks.

---

R. Aguilera · K. Scharnhorst · S. L. Lilak · C. S. Dunham
Department of Chemistry and Biochemistry, UCLA, Los Angeles, CA, USA

M. Aono
International Center for Materials Nanoarchitectonics (MANA), National Institute for Materials Science (NIMS), Tsukuba, Ibaraki, Japan

A. Z. Stieg
WPI Center for Materials Nanoarchitectonics (MANA), National Institute for Materials Science (NIMS), Tsukuba, Ibaraki, Japan

California NanoSystems Institute (CNSI), UCLA, Los Angeles, CA, USA

J. K. Gimzewski (✉)
Department of Chemistry and Biochemistry, UCLA, Los Angeles, CA, USA

WPI Center for Materials Nanoarchitectonics (MANA), National Institute for Materials Science (NIMS), Tsukuba, Ibaraki, Japan

California NanoSystems Institute (CNSI), UCLA, Los Angeles, CA, USA
e-mail: gimzewski@cnsi.ucla.edu

 
M. Aono (ed.), *Atomic Switch*, Advances in Atom and Single Molecule Machines,
https://doi.org/10.1007/978-3-030-34875-5_11

## 1 Introduction

In 1965 Gordon Moore, co-founder and chairman of Intel, published a paper on the number of transistors per integrated circuit (microchip) doubling every 2 years. This trend became known as Moore's Law and meant processing speed doubled every 2 years. Recently, Moore's law shows signs of slowing down due to physical and economic constraints. In the next few decades individual elements will approach the scale of a few atoms and in turn the fundamental limits of miniaturization [1]. Feature sizes are constrained by the optical diffraction limit which defines minimum feature size by wavelength divided by two times the numerical aperture of the microscope [2]. Even before physical boundaries are approached, the economical limitations of continuing miniaturization and massive integration will be reached. With extremely small features fabrication costs increase dramatically, making them unsuitable for cost-effective mass production. Thus, as the end of Moore's Law is approached [3] processing power will no longer increase unless alternative individual elements or architectures of integrated circuits are explored.

Modern computers ubiquitously use the Von Neumann architecture first described by John von Neumann in 1945. This architecture separates memory and processing components within integrated circuits [4]. Partitioning data and instruction storage from arithmetic/logic processing leads to a limitation of information transfer known as the Von Neumann bottleneck [5]. When calculating more complex problems requiring mass amounts of data or instructions from memory storage, processors sit idle. In order to maximize information processing and ultimately overcome the Von Neumann bottleneck, significant changes to computer architectures must be made.

Carver Mead, a scientist at Cal Tech that coined the term "Moore's Law" became known for his bio-inspired work in the mid 60s. Particularly his attempts to emulate neural functionality directly into analog hardware implementations and pioneering the "Neuromorphic" computation field. His neuromorphic research was specifically based on analog metal oxide semiconductor (MOS) processors. By 1995 Mead and collaborators demonstrated a single silicon transistor 'synapse' capable of analog learning. However, the ensuing rapid development in digital microprocessors using VLSI superseded early analog computing approaches. In turn, digital neural network software led to today's deep learning and machine learning algorithms.

Concurrently, the growth of the internet-of-things, cloud computing and the rapid explosion of unstructured data has placed new demands on computers in our increasingly interconnected world. Examples of interconnected data are those from satellites, sensors, economic markets, commerce, global climate patterns, social media and consumer habits. This combinatorial complexity challenges the inherently serial Von Neumann architecture of computers which, at their core, scale poorly into supercomputers requiring massive increases in hardware and energy consumption. Currently, China's Tianhe-2 supercomputer uses 18 MW at full power [6] and is still very far from performing many tasks that a human brain can achieve using merely 20 W. In terms of dealing with complex tasks, the scalability of today's computers cannot keep up in a realistic manner with the world in which we live.

In response, we find ourselves in a new era of challenges promulgating research with the aim to develop science and technology that emulate how the brain works and to utilize that knowledge for creating a paradigm shift in computation towards cognition. New technologies such as functional magnetic resonance imaging (fMRI) and advanced image processing have also made major inroads into neuroscience, although the theory of understanding how our brain works is at an early stage. Nevertheless, fundamental scientific experiments to emulate and create brain-like behavioral characteristics are underway.

In this chapter we discuss one such approach using devices, based on the atomic switch's synaptic-like properties, connected in a network called the "Atomic Switch Network" (ASN). The device is inspired, not only by its synaptic function, but also by the brain's inherent network characteristics in the neocortex and its distributed memory.

## 2 Frameworks for Neuromorphic and Bio-inspired Computing

In order to advance the field of computing, new computational hardware paradigms are required including artificial neural networks, deep learning, reservoir computation, and neuromorphic computing. Additionally, identifying ways to apply these systems for use with established algorithms such as deep learning will play a key role in ushering in the next generation of computing [7].

### *2.1 Neuromorphic Computing and Artificial Neural Networks*

Beautifully designed by the natural world, the human brain is capable of complex multisensory tasks and decision-making using architecture distinct from the hyper-engineered grids of von Neumann computers. Modern neurobiology describes neurons using the formulations of the Hodgkin-Huxley model [8], emulating single neurons as a circuit of capacitors, nonlinear resistors, and a current source. Individual neurons connect to one another through synapses, creating a connectome architecture [9] or network allowing cascading ions to transmit information. Network connectivity determines the efficiency of interneuron communication and has been heuristically observed to follow a small-world network topology [10, 11]. Found throughout nature from galaxy formation to sociological trends, small-world dynamics form the core of understanding complex systems nonlinearizable by current methods. Amazingly, cognitive behavior and memory association practiced by the brain utilize these salient features to be able to perform both computation and information storage within a single synapse [12]. Collectives of interacting neurons found in the brain act as compartmentalized networks ascribed to specific brain functions, but are capable of growing through heuristic learning. Computing and cognitive capabilities developed through an evolutionary scheme selectively prune weakly correlated neurons and enforce strongly correlated neurons as described in Hebbian learning [13]. These features drastically increase the power of the human brain to complete complicated tasks too computationally expensive for current CMOS technology.

Mead drew upon the natural computing capability of neuron networks to develop the concept of neuromorphic hardware in the mid 1980s. Neuromorphic engineers attempt to emulate neuron functionalities and brain architectures to be capable of performing similar multisensory complex tasks such as associative-dissociative memorization of unstructured data, pattern recognition, and chaotic series prediction, to name a few. Initial attempts to harness brain dynamics for computation relied upon contemporary CMOS technology to construct purpose-built field programmable gate arrays (FPGAs), and supercomputer assisted software within the field of machine learning. Artificial neural networks first developed as the perceptron [14] by Frank Rosenblatt adopted the same connectionist theory as biological neurons. Implemented using camera photocells while contemporary variations are realized in software, the perceptron consisted of a collection of nodes representing neurons with each node owning a number of weighted connections $w_i$ transmitting information $x_i$. Information propagates from external sensors towards individual nodes. Information traveling through the connections and converge at their respective nodes, activating the node depending on a transfer function or learning rule $f_j$.

$$y_j(t) = f_j\left(\sum w_i \cdot x_i(t)\right)$$

The overall task is then computed using the sum of all node operations while programmable control is achieved through modification of the weights in a learning procedure [15].

Conceptually, Mead's neuromorphic computing described a system possessing analog circuit elements capable of emulating biological features of the brain. The fundamental element of neuromorphic computing is the spiking neuron, and functions similarly to the logic gates of traditional von Neumann architectures. While traditional logic gates evaluate data as binary states of either 0 or 1, spiking neurons transmit a series of one or more spikes within a fixed period of time where the number of spikes represent the data as discrete continuous values, 0, 1, 2, etc. Thus, neuromorphic computing more closely resembles power-efficient analog systems [16]. A neuromorphic system contains many of these spiking neurons connected in a complex network, which must then be conditioned or trained, using spike-timing-dependent plasticity in order to perform specific tasks such as pattern recognition.

The key advantage of a neuromorphic system is its highly interconnected and parallel architecture providing remarkably reduced power consumption relative to traditional von Neumann architectures. Additionally, neuromorphic systems by design remove the significant bottleneck between memory and processing. This combination of improved bandwidth between memory and processor, along with parallel processing will theoretically allow neuromorphic computing systems to perform complex operations much faster and more efficiently than conventional systems [17].

Artificial neural networks, or ANNs, are a biologically inspired computing architecture incorporating functional elements designed to mimic the brain's neural networks. The networks form by a collection of artificial neurons. These neurons are interconnected via structures functionally resembling synapses, such that each neuron can transmit a signal it generates or receives to another neuron it is connected with [17]. Each neuron and synapse may possess a different "weight", dependent on plasticity and history of past signals, which modifies the strength of the transmitted/received signal. The neurons themselves are organized into three layers—input, hidden, and output—and signals traverse these layers one or more times as the network works to solve a given task. Key advantages for this type of architectural framework include fault tolerance characteristics and reliability within hardware [17].

The most basic model of an artificial neural network, the feedforward neural network, only allows signals to propagate in one direction, forward, towards the output layer [17]. An alternative operational mode to the feedforward neural network exists in the form of recurrent neural networks (RNN). Here the connections between units of the network form a feedback loop, which allows the internal memory of the RNN to process arbitrary input sequences [17]. This characteristic widens the RNN's application range to include tasks such as speech recognition. Both feedforward and recurrent artificial neural networks are discussed further in Sect. 3.3.

## 2.2 *Artificial Neural Networks*

Artificial neural networks (ANNs) have been developed to have brain-like functions using three layers of neurons, input, hidden and output. Hidden layers are unobserved neurons which alter the flow of information between inputs and outputs achieving a level of complexity that prevents outputs being simple functions of inputs. Connections between each layer act as synapses with tunable (weighted) properties. The presence of a feed-back loop also allows for dynamical complexity to dominate the system.

ANNs require training cycles for the tuning of synaptic weights fitted to a desired algorithm. In altering the hysteresis of each synapse, desired outputs can be retrieved from task specific inputs. ANN training techniques can be broken into three primary classes: supervised, unsupervised and reinforcement learning. Supervised learning simultaneously introduces an input vector paired with the desired outputs and adjusts its weights through an implemented learning algorithm (i.e. Manhattan update rule) [18]. Unsupervised learning, also commonly referred to as self-organization, outputs are trained to respond to a cluster of inputs. This results in a system designed to respond to specific input stimuli, and fabricate its own representations of the transmitted information. Reinforcement learning can be thought of as a trial-and-error system akin to repeating a maze periodically until the optimal route has been determined. The ANN is not given a specific path to take, and must be tuned through numerous training cycles [19].

Recurrent neural networks (RNNs) are a sequence based model of ANNs. RNN's can be separated into two primary classes, the RNN will either be presented with a constant, single input and stabilize to a desired state or presented with variable inputs fluctuating over time with the intention of yielding time-dependent outputs [20]. RNN's are capable of using their internal memory to process input sequences allowing for complex applications including speech and handwriting recognition [20, 21].

Convolutional neural networks (CNNs) function similar to a general ANN, however neurons are now arranged in three dimensions and characterized by height, width and depth. This manipulates the manner in which patterns are constrained and scales more efficiently than a typical ANN for larger pattern recognition systems. CNNs utilize three core layers: the convolutional layer, pooling layer and fully-connected layer (this is present in a typical ANN). The convolutional layer computes output neurons in relation to a dot product between the input region and its corresponding weights. After the convolutional layer, an element based activation function is applied followed by the pooling layer which down samples the width and height dimensions while leaving the CNN volume intact; both of these act as fixed pre-determined functions. Finally, the fully-connected layer will compute class scores which relate the pattern to its desired category. Training through backpropagation is only necessary for the convolutional and fully-connected layers. CNNs core design achieves much more efficient pattern classification than a regular ANN and makes it readily implementable in deep neural networks for image and speech recognition.

Recursive neural networks (rNNs) are another model applicable to deep learning that function by recursively applying an identical set of weights to a given system. This achieves structured predictions based on variable inputs; this architecture also allows for the flow of data in any given direction. While CNNs are good deep learning models for pattern recognition, rNNs offer the ability to predict variations in hierarchical systems, demonstrated in natural language processing, and can be regarded as a linear modification of RNNs though are unable to parse tree-like hierarchies [20].

Multi-layer perceptron's (MLPs) are a class of ANNs which implement feed forward loops. All hidden and output layers in the MLP act as a neuron responding to stimuli (inputs) which rely on a non-linear, typically sigmoidal or hyperbolic tangent, activation function. The activation function acts as a weighted tolerance value governing the output function. MLPs typically utilize backpropagation algorithms for training and the weights can be updated iteratively after each testing phase (often results in chaotic alterations) or in unison after all weights have been analyzed (batch learning, often yields more stable alterations) [22]. By defining classes to analyze, MLPs can be applied to many different tasks including 3D image recognition and handwritten image recognition.

## 2.3 Deep Learning

Modern neuromorphic system design strives to build the same adaptive learning capacity found in biological neural systems into an electronic device. One way to incorporate this system design is through the employment of machine learning algorithms. Machine learning algorithms attempt to emulate neuronal design in software. The learning process can be accomplished in a supervised or unsupervised manner, with the key distinction being whether the data fed into the algorithm is labeled (supervised) or unlabeled (unsupervised) [23]. Presently, supervised learning is the most common form of machine learning, regardless of the algorithm employed.

Historically, machine learning algorithms and techniques struggled to process natural data sets in raw form and only through extensive work by engineers could machine learning be tailored to perform tasks such as speech recognition [23]. This has changed with the development of new methods, feature learning or representation learning, which are capable of identifying the minimum features that represents each class of object for detection and/or classification within a data set. Those methods led to the creation of the well-publicized machine learning technique known as deep-learning that simulates a multi-layered neuron architecture. Its purpose is for organizing unstructured data and applying this learned information to various tasks, such as computer vision, speech recognition and bioinformatics. In the case of supervised learning, the process starts by amassing a large data set where each element of the data set is labeled with its corresponding category. Next, the machine is trained by supplying it with data representative of each category, for which the machine then produces an output in the form of a vector of scores for every category. The goal is to train the machine so that the vector of scores for a particular piece of data has the highest possible score for the category to which it belongs. This is achieved by computing a function that measures the error between the output scores and a desired score pattern, after which the machine modifies its own parameters, or weights, via a gradient vector—a function that adjusts weights depending on whether the error would increase or decrease based on that adjustment—in order to reduce the error. Consequently, training can be an arduous process for which the best results typically require incredibly extensive data dates and hundreds of millions of weights [23].

More advanced machine learning algorithms attempt to completely emulate neuronal design in software, simulating multi-layered neurons similar to the perceptron in an architecture known as 'deep learning' [24]. Organized in a hierarchal tree, the output of one perceptron is fed forward to a connected perceptron to compute complex tasks. The interaction between layers increases the difficulty in the learning procedure due to the nonseparability of individual weights from the internal recurrent response. Researchers have developed methods of mitigating the difficulties through feedback learning and error backpropagation schemes that predict network behavior [24, 25].

Contemporary industries alleviate big data demands by relying on these machine learning algorithms distributed across hundreds of servers and graphics processing units (GPUs), which are discussed in a later section, in an attempt to emulate a brain's capability of organizing unstructured data. However, CMOS-based neural networks are trending towards costly power consumption as transfer speeds between buffer memory and logic components as well as miniaturization capabilities are approaching fundamental limits.

## 2.4 *Reservoir Computing*

Reservoir computing (RC) is an emerging paradigm that promotes computing using the intrinsic nonlinear dynamics of an excited system called a reservoir. Maass et al. initially proposed a version of RC called the Liquid State Machine (LSM) as a model for cortical microcircuits. Independently, Jaeger introduced a variation of RC called the Echo State Machine (ESM) as an alternative RNN approach for control tasks. Variations of both LSM and ESM have been proposed for many different machine learning and system control tasks. Büsing et al. conducted a comprehensive study of reservoir performance using different metrics as a function of the node connectivity K, the logarithm of the number of states per node m, and the variance of the weights in the reservoir [26].

At its core, the RC paradigm utilizes a reservoir's capability to project an input's information into a mathematical higher representation space, similar to a Fourier transform. A variety of spatially distributed mathematical operations occur throughout the system according to system properties and dimensionality. Computation occurs as a recursive learning algorithm that inscribes a filter on the system such that the projection spans the correct mathematical operations and the desired process is achieved. The reservoir essentially outputs a series of nonlinear transformations of the input which are then trained at the output layer by synaptic weights using linear regression. A system with sufficiently rich dynamics can remember perturbations by an external input over time which compared to other approaches, has many key advantages of using RC such as:

1. Computationally inexpensive training (low programming overhead).
2. Flexibility in the physical reservoir implementation (cost-effective fabrication).
3. A high tolerance to material variation, defects, and faults (robustness).

These factors make RC particularly suitable for emerging unconventional computing paradigms, such as computing using physical phenomena [27] and self-assembled electronic architectures [28]. The high nonlinearity, and thus the high dimensionality, ensures convergence of this algorithm in practical time. Reservoir computing differs from earlier attempts at computing with random assemblies of nano-cells or switches, e.g., Tour et al. [29]. Such systems lacked a formal framework and required complex and time-consuming optimization steps in order to produce a desired output function. Another problem of these early attempts was the lack of scalability, and thus the difficulty to obtain more complex functions.

# 3 Hardware Paradigms for Neuromorphic Computing

The previous section described a few framework paradigms for next-generation computing. This section will discuss several examples of the physical implementation of these paradigms, including hardware configurations utilized for deep learning and neuromorphic computing. Throughout the design of a new architecture, its desired neuromorphic properties need to be considered. An array of different metrics can benchmark devices for the implementation of a specific neural network. Considerable factors include efficiency, reconfigurability, and scalability.

## *3.1 Neuromorphic Chips*

Neuromorphic refers to any artificial neural system whose core design and architectures are based upon the biological central nervous systems. Traditional computing architecture efficiency is governed by how many floating-point operations per second (FLOPS) per watt can be performed. In developing neuromorphic systems, spiking neurons are fabricated into the hardware; these can be benchmarked by their synaptic operations per second (SOPS) per watt [30]. Floating-point operations are inefficient and slow; even the most powerful super-computers are not capable of obtaining real-time performance on detailed large-scale simulations of neural systems [30]. Synaptic operations offer an advantage in that operations are achieved directly through the hardware; this allows for real-time operation independent of synapse density and/or coupling. This advantage is achieved through circumventing the von Neumann bottleneck; a single component of hardware running specific tasks alleviates the overhead from multiple components communicating with each other achieving real-time analysis.

Silicon neurons are a promising avenue for mimicking biological synaptic/neuronal interactions. Unlike the previously mentioned hardware, a silicon neuron can be broken into computational blocks which can be functionalized for task specific neuronal activity. The synaptic block is capable of carrying out both linear and non-linear input spikes with short and long-term plasticity mechanisms available. Some blocks are a group of sub-blocks designed to computationally represent the theoretical model in which they are based upon. Finally, dendrite and axon circuit blocks account for the spatial structure and interconnectivity of the overall system allowing for an intricate and fully connected array of neurons [31]. Sub-segments of each element can be manipulated for the desired implementation of a specific task or model.

The clear desire for neuromorphic architectures has led to further investigation and developments of different synthetic synapse models. In recent years, atomic switch systems have garnered much interest and are now being modeled and investigated as synthetic synapses with the hopes of scaling them into a connected network yielding synaptic densities and topographies similar to that of the human brain.

**ASICs**

Carver Mead's work in pioneering analog VLSI implementations for neural systems set the foundation for early neurocomputing architectures, or neuroarchitectures. His work clearly outlined the need for neurons and synapses (weighted connections between neurons) for real-time processing on a single device capable of parallel processing. Mead's investigation of neural networks denoted that various frameworks could be implemented for general purpose neurocomputers, however, the neuroarchitecture will need to be uniquely designed while accommodating the desired neural networks [7]. This presents the potential to develop rigid and highly efficient task specific neurocomputers or, by contrast, versatile and tunable neurocomputers utilizing a hierarchy of neural networks.

Early implementations of neurocomputers were achieved using application-specific integrated circuits (ASICs) fabricated using CMOS VLSI technology. As their name suggests, they were developed for the execution of specific tasks and cannot be reconfigured at a later time by the end user. Each ASIC works as either as a master or slave node interconnected in a ring or bus network which broadcast signals based on their topology. The master node functions to control neurocomputation while the slave nodes enable parallel processing, the backbone of ASIC's efficiency for neuroarchitectures. This system utilizes external memory for the storage of neuron outputs and synaptic weights [32]. Efficiency and speed is ultimately governed by the number of neurons on a chip. An issue with ASIC neuroarchitectures is that they are developed for a specific neural network and are not reconfigurable devices [33] despite this they are notably proficient at specific tasks and highly power efficient.

## *3.2 FPGAs*

Field-programmable gate arrays (FPGAs) while worse in raw performance than ASICs, offer the added benefit of reconfigurability. FPGAs are designed to be manipulated by the end user allowing for a great degree of flexibility in designing FPGA based neuroarchitectures capable of functioning in an array of neural networks for prospective optimization in ASIC implementations. Despite their lower performance, FPGA's re-configurability and capacity to perform parallel processing has overshadowed ASICs in the development of neurocomputers [33].

As FPGA architectures continue to garner interest, programming frameworks have been developed to simplify their manipulation. The Open Computing

Language (OpenCL) platform has been designed to allowing programming through a heterogeneous array of hardware components including FPGAs. Intel has demonstrated a convolutional neural network (CNN) implementation using this platform which currently outperforms traditional CPU based CNNs with a significant increase in efficiency [34]. Advances in software continue to expand the use of FPGAs for a wide array of applications.

Continued developments in FPGA architectures have demonstrated their proficiency as accelerators in deep convolutional neural networks [35], Qiao et al. have developed an FPGA accelerator compatible with Caffe software, which is currently implemented in a wide array of CNNs. A comparison of their FPGA (Xilinx Zynq) accelerator to a traditional CPU (Intel Xeon X5675) and GPU (Nvidia Tesla K20) based CNNs shows FPGAs are superior in versatility and power efficiency to CPUs [34]. While GPUs still offer the capability of much higher performance (486 Gflops to the Zynq's 77.8 Gflops), the large disparity in power consumption (235 W to the Zynq's 14.4 W) implicates FPGAs as the most power efficient device to be implemented in the acceleration of CNNs. Future advances in GPU and FPGA technology may alter this trend.

## *3.3 Graphics Processing Units (GPUs)*

The term graphics processing unit, or GPU, was promulgated by NVIDIA in 1999 to mark the release of the world's first such device, the GeForce 256 [36]. This GPU was touted as an incredible advancement in the world of computer hardware, possessing approximately 23 million transistors (NVIDIA boasted at the time that this was twice as complex as a Pentium III processor) and possessed 50 gigaflops of floating-point calculation capability [36]. Nearly two decades later, GPUs have made enormous strides in power and capability. The most powerful GPU as of 2017 is NVIDIA's Tesla V100, a device possessing 21 billion transistors, over 5000 cores providing 120 teraflops of performance for deep learning applications while drawing only 300 W of power [37]. Thus, these units have increased in processing power by over four orders of magnitude over the last 18 years.

As GPU power increased over time, interest in them from individuals and organizations outside the computer gaming community grew significantly. In the quest to expand upon the capabilities of existing computing architectures and platforms, academic research groups began evaluating GPU's viability for performing certain tasks previously delegated to supercomputers. In 2012, researchers at the University of Toronto demonstrated that GPUs could be used to drastically improve tasks related to computer vision and deep learning, such as image reconstruction, by using the GPUs to run deep neural networks [38]. These GPUs displayed significant performance gains relative to traditional computer processor-based neural networks thanks to their massively parallel architecture involving thousands of individual processor units and exceptional processor-to-memory bandwidth.

Since that time, companies, particularly NVIDIA, have helped lead the way toward an artificial intelligence revolution with the use of GPUs [39]. GPUs are now utilized extensively by companies and universities involved in the fields of big data, machine learning, and genomics, among others [39–41]. GPUs are also seeing increased utilization in areas of academic research as more coding languages and libraries, such as C and FORTRAN, are updated to take advantage of the parallel processing power of this architecture.

## *3.4 Purpose-Built Chips and History*

With an exponentially growing interest in neuromorphic architectures, many researchers began further pursuing their development. In 2005, Fast Analog Computing with Emergent Transient States (FACETS) launched a research initiative funded by the UK to implement brain-like hardware architecture for neuromorphic computing. The project concluded in 2010 with a VLSI implementation of a traditional CMOS fabricated device containing 400 neurons and 100,000 synapses [42]. In 2011, brain-inspired multiscale computation in neuromorphic hybrid systems (BrainScaleS) intended to expand upon FACETS work and concluded in 2015 with an architecture containing 1.6 million neurons and 400 million synapses. In the middle of the BrainScaleS project further interest in neuromorphic architectures arose.

In 2013, the Information and Communication Technologies (ICT) of the EU launched the Human Brain Project. This comprehensive research initiative aimed to advance our understanding of the human brain through numerous fields including neuroscience and computation. In the field of neuromorphic architectures, this flagship intends to refine and expand upon the work of FACETS and BrainScaleS. The same year, the United States launched the Brain Initiative under the Obama administration with similar intent and currently funds numerous agencies including DARPA, NIH and NSF. Through this funding neuromorphic based projects continue to progress.

As these architectures advance, hardware has begun to be designed for task specific applications. IBM has developed TrueNorth, a brain inspired device suitable for complex applications which utilize neural networks. TrueNorth is a 5.4 billion transistor chip containing 4096 neurosynaptic cores interconnected to an intrachip that utilizes 1 million programmable spiking neurons and 256 million configurable synapses; it consumes a mere 70 mW and is able to process 46 billion SOPS, per watt. This greatly exceeds the limit of energy efficient super computers which are only capable of processing 4.5 billion FLOPS per watt. This device has demonstrated high fidelity multi-object recognition in real-time capable of discerning objects based on different classes (i.e. person vs. cyclist). TrueNorth highly excels in task specific applications which do not require reconfigurability similar to ASICs.

The Spiking Neural Network Architecture (SpiNNaker) is on ongoing project which aims to have 1 million ARMs processors in parallel. Unlike TrueNorth which is optimized for specific tasks, SpiNNaker has implemented a versatile, tunable neuromorphic architecture in the same vein as FPGAs at the cost of power efficiency. Both architectures have been designed for scalability and can accommodate cascades of devices allowing for easier implementations of larger systems.

Qualcomm Technologies Zeroth Machine Intelligence platform has also developed a deep learning software development kit (SDK) which has brought the power of deep learning to mobile devices. This breakthrough has removed the need to connect to a cloud server to utilize the benefits of deep learning and gives mobile phones the innate capability to perform numerous complex tasks like facial recognition, object tracking and natural language processing [43].

## *3.5 ASNs for Computing*

ASN devices exhibit various properties associated with atomic switches and other memristive systems [44], the latter defined as a system with its internal resistance based on electric flux [45]. Such properties include but are not limited to a requisite forming step and distributed, frequency-dependent hysteretic switching among a collection of dynamically interacting elements. In addition, the functional topology of ASNs has been shown to produce a diversity of complex behaviors, ranging from distributed memory function to emergent critical dynamics similar to those found in both fMRI/EEG of biological brains and multi-electrode array (MEA) studies of neuronal populations [15].

Observations of power-law scaling in various device dynamics and a 'fading memory' property of learned states have implicated as an essential component for applications of reservoir computing (RC) using critical states. Initial progress in the use of ASNs as nonlinear reservoirs capable of task performance in the RC paradigm has shown through simulation and experimental implementation of a benchmark task known as waveform generation.

Based on extensive studies of the dynamical response of ASNs (Fig. 4), these devices have been identified as an ideal platform for hardware-based reservoir computation [46]. ASN devices have shown, through both experiment and simulation [47–49], to be a viable platform for hardware-based RC toward applications in pattern recognition, prediction and logic. Based on their ability to integrate, segregate, store and respond to external stimulus, the utility of ASNs as nonlinear reservoirs has been demonstrated through implementation of multiple benchmark tasks including: (1) waveform generation [48] and (2) various logic operations (AND, OR, XOR) (see Sect. 3.2). The speed, density, and [50] scalability of the ASN serve to overcome major hurdles in the RC paradigm.

## 4 Building the Atomic Switch Network

Designing a system capable of neuromorphic computation requires adequate functionality in terms of non-linearity, persistent activity, and recurrent structures. In addition to these metrics, a chip must be power efficient in order to be a viable option for industrial applications. Keeping these requirements in mind our group sought to harness the inherent properties of individual atomic switches (non-linearity, quantized conductance, and memory) and use them as artificial neurons in this device. Through these means, a device was fabricated using both top-down and bottom-up methods. The atomic switch network (ASN) consists of highly recurrent structures that produce dynamic activity through space and time. These networks produce non-linear responses and their emergent behavior is much more complex than that of its individual junctions. This dynamic system's emergent distributed behavior also provides a diverse array of output signals. Configured as neuromorphic chips, ASN devices are capable of performing alternative types of computing.

**Atomic Switches as Synthetic Synapses**

Establishing specific connections between patterns of electrical activity and brain function is a difficult task that requires studying general features of neuronal structure in order to determine the essential properties required to construct a device capable of learning in a physical sense. These features are believed to include synaptic plasticity, allowing physical reconfiguration of the network to enable functional differentiation and the development of hierarchical structures, which all possess correlated memory distributed throughout the dynamically coupled synapses. Therefore, it can be inferred that learning capacity is connected to dynamic activity within the brain. Specifically, a near-critical or "edge of chaos" operational regime [51] has been associated with the fast, correlated response to stimulation necessary for computation and learning. Though extremely attractive as a construct for developing computational machinery whose operation results from intrinsic critical dynamics, the production of such a device in hardware has proven a daunting task, with ASN devices being one of the few successful demonstrations in the scholarly literature.

The first experiment to measure the transition from an electron quantum tunneling to single point contact regime was reported in 1987 using a scanning tunneling microscope (STM) in ultra-high-vacuum (UHV) on a silver surface [52]. Current-distance characteristics showed that, at sufficiently small tip-surface gaps, an abrupt increase in conductance, G, of $\sim \frac{2e^2}{h} \approx \frac{1}{13} k\Omega$ which is the quantized unit of conductance. Subsequent theoretical analysis verified that at small gap distance the effective tunnel barrier collapses prior to point contact via ballistic electron injection [53]. Later work demonstrated further jumps of $\sim n\frac{2e^2}{h}$, where n $=$ 1, 2, 3... in the conductance occur as the contact area is increased. Such observations were not limited to STM experiments; even two macroscopic wires brought in contact also displayed this effect, albeit in a less controlled manner. Quantized conduction, also introduces Landauer's concept of transmission where the term t is the transmission [54].

$$G = \frac{2e^2}{h} \sum_n t_n$$

In 2002, experiments by Terabe et. al found that Ag atoms could be transported through an STM tip made of silver coated with silver sulfide and deposited on a surface in a controlled manner [55]. The characteristics of this process also occurred via quantized conduction. However, the mechanism involved ion migration under the influence of an electric field, a process called 'electroionics' meaning that in addition to electron motion, ion motion also occurs simultaneously. Normally ionic diffusion processes on the macro-scale are considered to be slow, but when they are induced on the nanometer scale, they are actually quite fast and can occur on a (sub-) nanosecond time scale depending on the geometry and dimensions of the junction. In 2005, using junctions fabricated using conventional microelectronics, Terabe et al demonstrated atomic switching in silver sulfide junctions with discrete and reversible quantized jumps from n = 1 to 10. This was the birth of the "atomic switch". Since that date, a number of researchers have observed quantized conduction in a wide range of materials including sulfide junctions of copper, tungsten sub-oxides as well as various metal-doped polymers.

Aside from the fundamental science of their quantization, interesting electronic features of atomic switches are pinched hysteresis, large ON/OFF conduction ratio, MHz switching speeds and volatility characteristics as well as CMOS compatibility because of their potential in digital electronic memory applications. Indeed, NEC recently have incorporated atomic switch technology into field programmable gate arrays (FPGAs) where a reduction in device foot print, speed, and energy consumption was achieved by replacing certain memory tasks, normally using transistors, into the circuitry [56].

Additional atomic switch functionality was reported in 2011 when studying switching near-threshold conditions [57, 58]. It was found that atomic switches have an on-off memorization property of past switching events. For instance if switching is performed infrequently, the switches remain in the on-state only briefly whereas if frequent switching events are made in rapid succession then the on-state persists for a longer time. A series of careful experiments were able to relate these physical observations to a psychological model of learning call the Akinson-Schriffin multi-store model. The essence of the model involves sensory memory (SM), short-term memory (STM) and long-term memory (LTM). New information arrives to the brain as sensory memory and that information is passed to short-term memory. In the absence of similar stimulation, information is forgotten. However, if the process is repeated many times the information is moved into long-term memory. Think of learning to play a tune on a piano by diligently practicing repeatedly. In terms of bio-inspiration the operational characteristics of the atomic switch under threshold switching also related to characteristics of biological synapses. The atomic switch therefore has also been called a synthetic synapse where memory is represented by conduction state.

The next step in creating a 'brain inspired' device is the fabrication of networks of synthetic synapses (Atomic Switches). Taking the neocortex as a biologically inspiration, self-assembly was used to incorporate atomic switches into a dense dendritic tangle of silver nanowires resulting in a density of ~$10^8$ connections/cm$^2$. In response to electrical inputs which inject energy into the network, these networks exhibit self-organization, critical power law dynamics and spatio-temporal non-linear outputs at a multiple electrodes. The device is called an Atomic Switch Network (ASN) and is described in detail in the next section.

## 4.1 Network Fabrication

Several routes to fabricate functionally complex recurrent networks have been experimentally explored, including: seed free networks, random seed networks, and patterned seed networks. The seeds are small areas of deposited copper that react in solution to generate silver wires through electroless deposition. The patterned seed networks proved the most versatile, and utilized a combination of top-down with bottom-up fabrication, a powerful general fabrication approach known as nanoarchitectonics. Initial approaches on implementing atomic switches into a network topology consisted of pipetting 150 μL of an isopropanol suspension (149.8 mg Ag/L) of monodisperse silver nanowires (120–150 nm × 20–50 μm, Aldrich) onto a substrate and allowing to air dry. These devices were then activated using the technique described in Sect. 4.2 and non-linear IV curves were subsequently observed. However, the non-uniformity in the dispersion of nanowires initially caused concern in the area of spatially distributed activity.

With a density controlled network in mind, electrochemistry was used to grow a recurrent silver network via copper seeds. Following the galvanic reaction below, network growth occurs through an electroless deposition (ELD) reaction via individual atom displacement reactions between $Ag^+$ and $Cu^0$ based on respective electric potentials. A spontaneous ELD reaction is preferred over an electrically induced reaction to minimize artifacts and maintain the delicate nature of the electrochemical reactions. Here, silver atoms are oxidized while copper is reduced during the galvanic displacement reaction.

$$Cu(s) + 2\,Ag^{+}(aq) \rightarrow 2\,Ag(s) + Cu^{2+}(s) \quad \Delta E = -1.26\ V$$

Using this ELD reaction, a random seed network was fabricated by pipetting a 1 mL aliquot of copper microspheres 1–10 μm (99.995% purity, Alfa-Aesar) which was then air dried. Silver nitrate (50 mM) was pipetted (20 μL) onto the center of the device. Following the ELD described above, silver dendritic structures

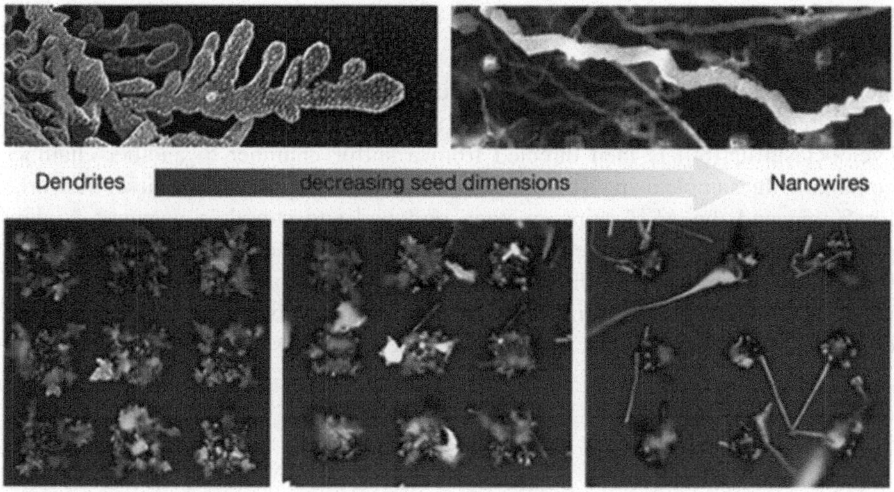


**Fig. 1** Above are SEM images depicting the morphological transition seen from changing seed size. Branching dendritic crystals occur above 10 μm posts, while below 3 μm nanowire growth along the (111) lattice is observed. Intermediary post sizes yield a mixture of dendrites and nanowires (Avizienis Crystal Growth & Design)

spontaneously formed a complex network. Again, network density exhibited non-uniformity due to the stochasticity associated with drop casting metal suspensions.

Successful implementations of the ELD reaction above allowed us to design a technique using highly patterned top-down photolithography combined with complex spontaneous and self-organized growth. The patterned seed networks consist of a 2 μm layer of AZ nLOF 2020 (a negative photoresist), a soft bake, followed by UV photolithography, and a post-exposure bake. This resist is developed in MF26A, rinsed with isopropanol, and a 300 nm layer of copper is then deposited and lifted off overnight in acetone. At the end of this process, a patterned grid of copper posts 300 nm high is left. The size and pitch of these posts were refined over time to give the most desirable silver crystal growth [59].

When first designing a purpose-built device to emulate mammalian brain activity, dendritic silver structures were desired. Experimentally we find that by changing the size of the copper posts, a morphological transition occurred and a seed site of $1 \times 1$ μm up to $3 \times 3$ μm leads to fine long rhizome-like nanowires. Seeds between $3 \times 3$ μm and $10 \times 10$ μm yield a mixture of nanowires with branched dendritic structures, while posts larger than $10 \times 10$ μm produce predominantly dendrites [59] (Fig. 1).

## 4.2 Network Functionalization

Devices are functionalized through the deposition of an insulator onto metallic nanowires. In the case of Ag|$Ag_2S$|Ag, a temperature controlled sublimation of cyclooctasulfur ($S_8$) is then directed from a sulfur chamber to another chamber containing the sample using a carrier gas. Sulfur gas is exposed to the silver networks for 5 min and the ASN chips are then removed and stored in vacuum. A slow diffusion reaction governs the movement of sulfur into the silver lattice, which can be assessed by monitoring the electrical this time resistance of the device and if necessary consecutive sulfurizations are performed to obtain the optimum sulfide coverage. The surface chemical reaction can be written as:

$$S_8(g) + 16\,Ag(s) \rightarrow 8Ag_2S(s)$$

Once the desired resistance is reached, the network is initialized in a process called electroforming by sweeping a triangular voltage waveform across the network via the contact electrodes. The triangular sweeps induce a current flow across the device through complex pathways, and with adequate voltage, the device is able to form tiny filaments. A multitude of junctions form creating a conductive pathway to ground, and thus increased current. This increasing current flow through the device can be seen in Fig. 5 and depicts this electroformation of numerous nanofilamentary pathways throughout the atomic switch network.

## 4.3 Device Fabrication

Over the years there have been multiple generations of ASN devices, starting with simple two electrode devices all the way up to 128 electrodes chips. Devices with 4, 16 and 128 electrodes are shown in Fig. 2. Patterned seed networks consist of a 2 μm layer of AZ nLOF 2020 (a negative photoresist), a soft bake, followed by UV photolithography, and a post-exposure bake. This resist is developed in MF26A, rinsed with isopropanol, and a 300 nm layer of copper is then deposited and lifted off overnight in acetone. Copper lift off occurs leaving a patterned grid of copper posts ~300 nm thick. These patterned seeds enable a level of network density control not previously realized in earlier experiments. The process provides a level of reproducible control over the spontaneous growth of materials that are CMOS compatible and most importantly for alternative computing: structurally complex.

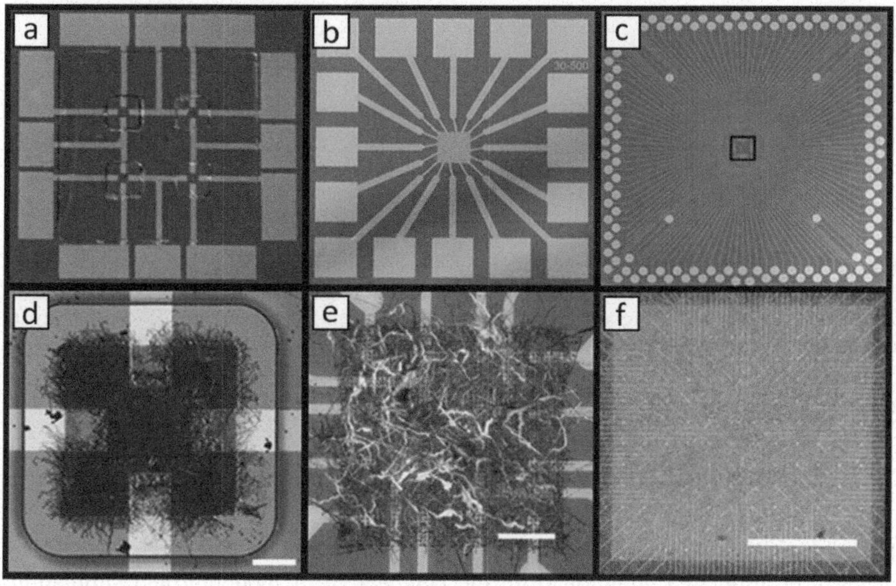


**Fig. 2** Different generations of ASN devices are shown above. Chips include (**a**) 4 electrode device, (**b**) 16 electrode, and (**c**) 128 electrode ASN. Inner networks of each device are shown on the bottom row (**d–f**, scale bars = 100 μm, 1 mm)

## *4.4 Measurement Platform*

The ASN measurement platform was custom designed in CAD and 3D printed (MakerBot Replicator 2.0). ASN chips sit on the bottom of the platform and are attached with gold spring-loaded pins. These pins connect the measurement hardware to the outer electrodes of the ASN chip. Measurement hardware consists of a source measure unit ((National Instruments Model 4141), two data acquisition cards (National Instruments Model 6368), and a switching module (National Instruments Model 2532). These units allow for the ability to define each electrode as an input or output (I/O) signal, specific signals to source, and concurrently record these input signals as well as record spatially-distributed voltage traces that percolate through the network. All measurements on the multielectrode array are performed and recorded simultaneously. Control software for the ASN device was coded in LabView 2012 (National Instruments). Post analysis work was done in MatLab 2010b (MathWorks).

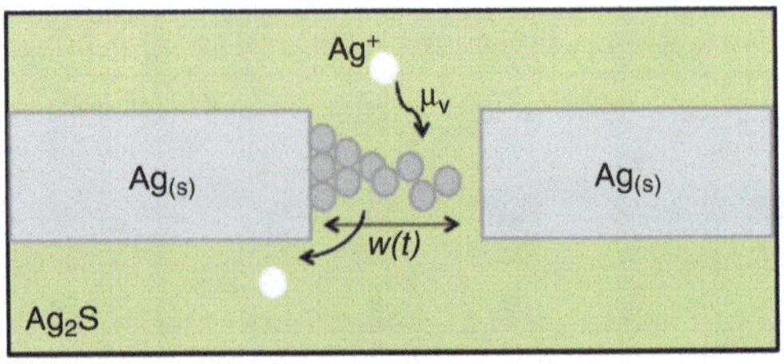


**Fig. 3** Atomic switches are comprised of an Ag|$Ag_2S$|Ag junction. Applied electrical bias causes Ag cation migration to the cathode where it is reduced, forming a stable metallic filament, resulting in resistance change. This migration is modeled by the filament length w(t), Ag cation mobility $\mu_v$ and additional stochastic terms (Sillin Nanotechnology 2013)

# 5 Results: Atomic Switch Network Dynamics

## *5.1 Operational Characteristics of the Atomic Switch*

Experimental studies into the operational characteristics of atomic switches has recently promulgated attention from the perspective of modeling and simulation. Atomic switches are known to operate through two mechanisms: (1) formation/dissolution of conductive filaments, and (2) a phase transition between monoclinic acanthite (α) and body centered cubic argentite (β) within $Ag_2S$. Application of a bias voltage across the junction has been shown using transmission electron microscopy to induce the formation of nanoscale conducting channels across the $Ag_2S$ interface through a bias-catalyzed phase transition, converting the surrounding α-$Ag_2S$ matrix to the conductive β-$Ag_2S$ phase which exhibits high super ionic mobility (Fig. 3).

**Voltage Pulsed STM/LTM**

In the absence of continued applied bias, the conductive channels eventually return to their stoichiometric, thermodynamically favored equilibrium state, reverting the atomic switch to its initial high resistance. This transition gives rise to a weakly memristive behavior prior to the formation of Ag filaments across the interface. Continued application of bias voltage results in a concurrent increase in current through the device, which then further drives migration of silver cations toward the cathode. At the cathode mobile silver cations are subsequently reduced to metallic $Ag^0$, forming a highly conductive Ag nanofilamentary wire. The completion of this filament results in a strong transition to an ON state defined by a significant increase in conductivity with a typical conductance ON/OFF ratio of ~$10^5$ [59]. Removal of the applied bias results in filament dissolution as the device again returns its thermodynamic equilibrium state. The completion and dissolution of this filamentary structure characterizes strongly memristive behavior. Continuous application of a bias voltage serves to increase filament thickness as additional silver cations are reduced, causing thickening of the metallic filament. This dynamic process has been shown to alter the dissolution time constant, and can be externally controlled by changing the input bias pattern (e.g. pulse frequency). Such changes in volatility can be interpreted as long-term or short-term memory (LTM and STM) (Fig. 4).

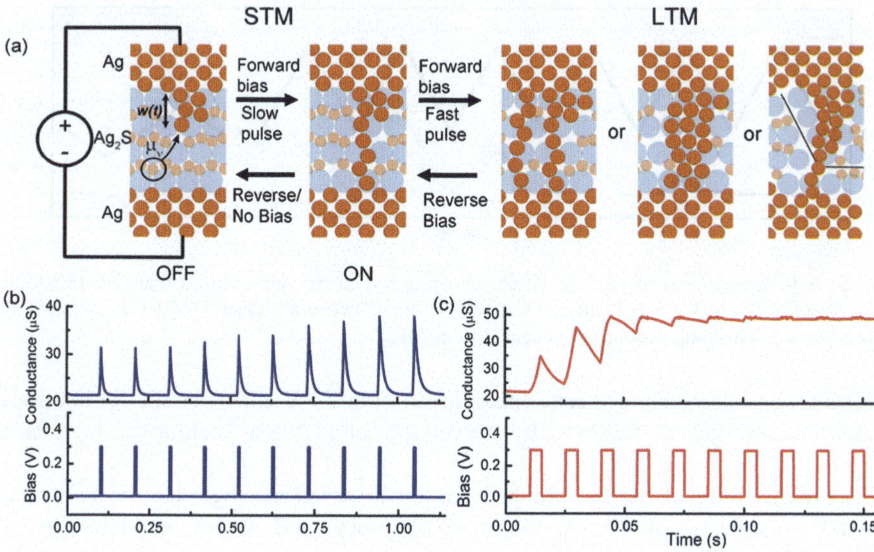


**Fig. 4** Resistive switching and long/short term memory effects in at an Ag-$Ag_2S$-Ag junction arise from (**a**) increased $Ag^+$ mobility in the presence of an externally applied electric field. (**b**) Short pulses reduce $Ag^+$ to form a conductive Ag filament which will quickly re-dissolve in the absence of an applied bias acting as short term memory. (**c**) Longer pulses of the same amplitude are capable of generating long lasting filaments acting as longer term memory. This likely arises from a combination of multiple filament formations, thicker filaments formed and $Ag^+$ ions which have irreversibly crossed a grain boundary until the external bias is removed (AZ Stieg Memristor Networks 2014)

## 5.2 Device Activation and Switching

The ionic resistive atomic switch has been shown to exhibit fascinating electrical properties analogous to short- and long-term plasticity at single synaptic junctions while operating as a two-terminal device controlled through formation/annihilation of a metal filament within a Metal-Insulator-Metal (MIM) interface. However, the behavior of a collection of atomic switches directly coupled both spatially and electrically is as of yet unknown. In common with the current understanding of switching mechanisms in devices based on $Ag_2S$, $TiO_2$, and $Ta_2O_5$, our network required an initial forming step to create a short-lived high conductivity 'ON' state. As measured by current-voltage (I-V) spectroscopy, ASN devices demonstrated non-linear I-V characteristics comprised of a sequential decrease in network resistance with consecutive bias sweeps followed by an abrupt transition to the activated 'ON' state. This cascade-type activation required a higher switching voltage (~7 V); however, they exhibit similar time constants (μs) to single atom switches. The switching behavior in the fractal network is attributed to an increased spreading resistance of many individual switching interfaces. In contrast, un-sulfurized control devices comprised of a purely metallic network have substantially lower resistance

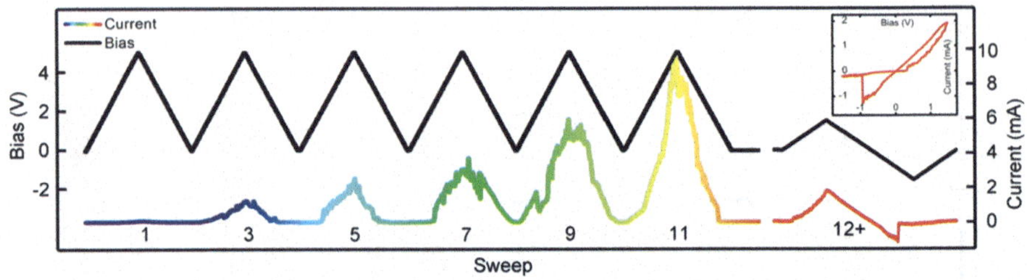


**Fig. 5** Activation sweep of an ASN device showing the electrically induced filament formation step. A signal of +/− 3 V was input in one corner of the device at a frequency of 10 Hz and current was collected across the network (Avizienis PLoS 2012)

(<100 Ω) and demonstrated repeatable linear, ohmic I-V characteristics at intermediate voltages (±3 V) followed by irreversible breakdown (melting) at high bias (Fig. 5).

Reproducibility of the switching behavior observed in single atomic switches, i.e. I-V hysteresis, short- and long-term memory, and device activation, were validated using the ASN simulation. In addition, the simulations faithfully reproduced the various emergent properties specific to the ASN architecture. These efforts allow detailed investigation of internal dynamics of the network where it would otherwise have been experimentally impractical. Similar to the electroforming step observed in individual memristive elements, ASNs must undergo an activation process before they display memristive and emergent behaviors [44, 60]. Freshly fabricated ASNs contain $Ag_2S$ interfaces in their low-temperature insulating phase and function as quasi-ohmic resistors. Bias voltage sweeps of the virgin-state network devices exhibited weak memristive/soft switching behavior as silver cations initially migrate into the junctions and are characterized by pinched hysteresis current-voltage curves with a small RON/ROFF ratio and a smooth transition between the two states (Fig. 6a). Continued application of a bias voltage produced an abrupt, nearly discontinuous jump to a state of higher conductance (Fig. 6b). Repeated stimulation with bipolar bias voltage sweeps produced strong memristive/hard switching behavior, typified by abrupt switching between two distinct resistance states (Fig. 6c). While parameters such as threshold voltage and the RON/ROFF are to an extent device specific, the qualitative transition from weak to strong memristive behavior is a general property of the ASNs.

This observed phase transition has been theoretically predicted in simulations of memristor networks [61] and was reproduced in ASN simulation [49]. The transition from soft to hard switching results from the emergence of distinct spatial patterns corresponding to individual hard and soft switching elements (Fig. 6a′). The initial weakly memristive state was characterized by a large fraction of soft switching junctions. As net flux through the network increased, connections became increasingly polarized and conductive. Continued stimulation eventually yields the formation of a percolative pathway comprised of conductive, hard switching elements across the simulated network (Fig. 6b′). Completion of this pathway results in a

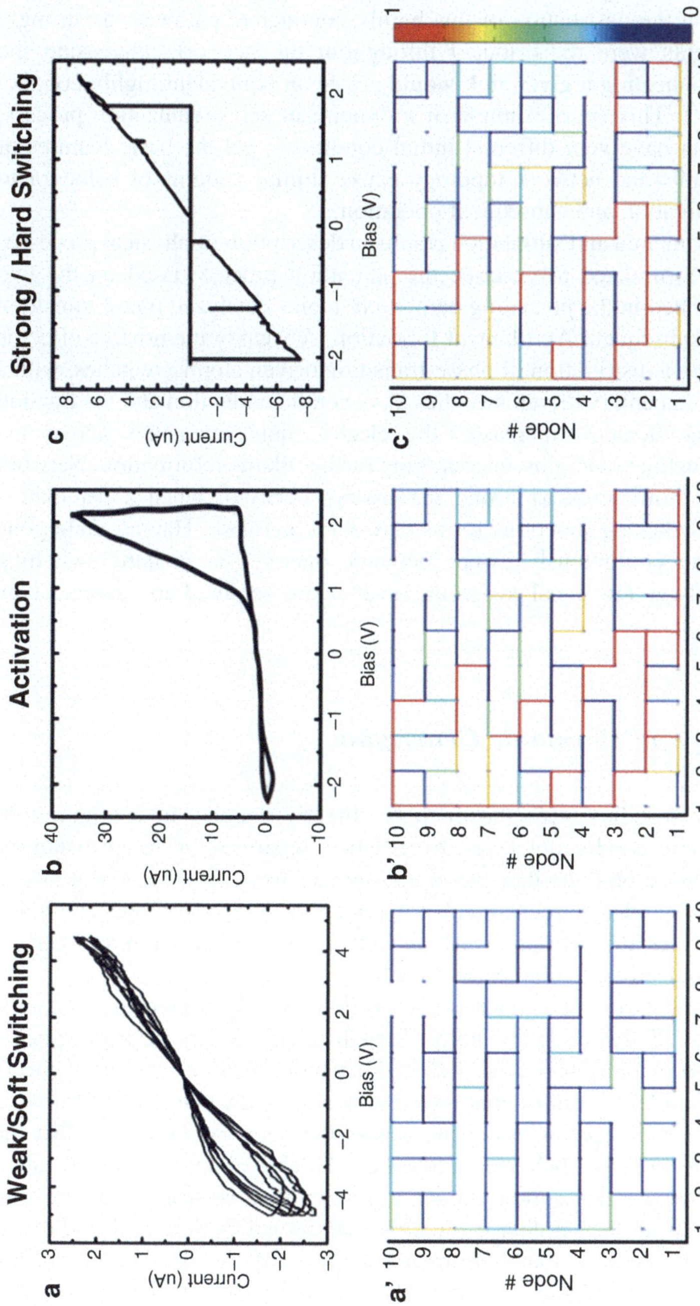


**Fig. 6** Simulations of device activation confirmed experimental results where (**a**) an initial weakly memristive switching (α = 10%) repeats indefinitely until (**b**) a transition in behavior from weak (blue, α = 10%) to strong (red, α = 0%) switching occurs. Strongly memristive switching (**c**) persists indefinitely with α = 0%. Internal network connectivity maps (**a′–c′**) reveal conductive pathways in the soft switching, transitional, and hard switching state (Sillin Nanotechnology 2013)

dramatic change in conductance associated with the activated state and a concurrent shift from weak to strong memristive behavior. Subsequent hard switching was observed following the destruction of this highly conductive pathway, as strongly memristive elements were redistributed throughout the network, increasing the probability that connecting a given link would create an equivalent highly conductive path (Fig. 6c′). This is an example of a dynamical self-organization process: different ASNs can have very different initial conditions, yet the basic features of their functional units and network topology cause similar patterns of behavior to emerge during activation and subsequent operation.

Based on experimental and simulation results, a description of physical processes in the ASN was formulated to describe the activation process based on the two mechanisms described in the preceding paragraph: a bias-catalyzed phase transition of $Ag_2S$, and the subsequent Ag filament formation. A weakly memristive effect is caused primarily by a distribution of phase-transition driven atomic switches, with a small fraction of filamentary driven switches. As overall conduction and the fraction of hard switching elements increases, the electric field intensifies across the remaining soft switching junctions, encouraging further filament formation. Network response changes from weak to strong memristive behavior when a percolative pathway of hard switching junctions forms across the network. Having undergone this transition, the continuously swept network operates as a hard switching memristor, as only a few local switching events are required to reconnect an equivalent path.

## 5.3 *Coupling and Harmonic Generation*

Individual atomic switches were shown to be directly coupled in configurations using a shared ionic conducting layer, even when separated by large distances. Spatially distributed atomic switch junctions interact through local variations in ionic concentration and electrochemical potential that depend on the combined electrical resistance of the entire network as well as the configuration, or state, of all other electro-ionically interconnected switches.

While ‘weak’ and ‘strong’ behavior can be exhibited by single elements, the most interesting features of this complex atomic switch device are its network-specific properties. Infrared imaging was used track Joule heating from current flow during DC bias sweeps in order to confirm distributed network conductance. The IR images (Fig. 7) show power dissipation occurring across the network, indicating that the phase change in network I-V behavior was not attributable to percolation but rather due to the sum of parallel current flows, meaning that network structure and connectivity actively influence device function. Additional evidence for the distribution of switch function stemmed from the analysis of the device’s frequency response. Theoretical simulations indicate that second harmonic generation will occur under an applied sinusoidal voltage in networks whose percentage of hard switching junctions exceeds the percolation threshold. Furthermore, the relative

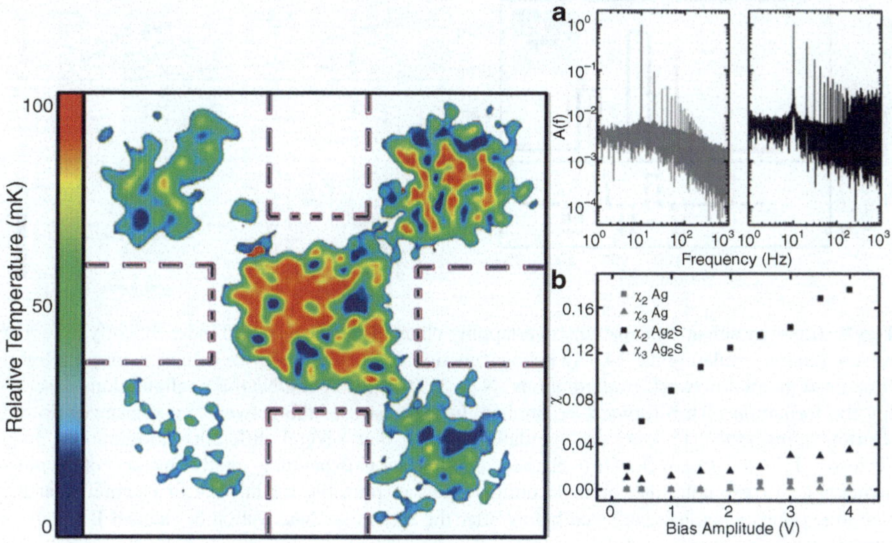


**Fig. 7** Network-specific behaviors. (Left) Representative IR image (sensitivity <20 mK) of Joule heating in atomic switch network during bias sweeps indicating current flow distributed throughout the device. Electrode positions are indicated by dashed lines. The image was taken from data integrated during <1 min of bias. (Right) Frequency Response (**a**) Fourier transforms of a functional device's current response (black) to a 2 V, 10 Hz sinusoidal input signal shows enhanced overtones of the input signal with respect to a control device (gray). (**b**) Plot of normalized amplitudes ($\chi$) of 2nd and 3rd harmonic generation for varying sinusoidal signal voltages in both functional (black) and control (gray) networks. Sulfurized networks generate higher harmonics, as was theoretically predicted for random memristor/resistor networks with 80% or more strongly memristive (switching) elements

magnitude of higher harmonics increases with the relative number of hard switching junctions. Device response to a 10 Hz sinusoidal voltage signal revealed a large increase in higher frequency components after functionalization (Fig. 7). The proportion of higher harmonics generated increases with signal amplitude, indicating that the network contains a distribution of switching voltage thresholds. As the bias voltage increases, so does the number of memristive junctions operating in the hard switching regime, producing a larger degree of higher harmonic generation. This confirms the IR observation of distributed activity throughout the network, with different regions activating at different voltages.

## 5.4 *Memory and Plasticity*

A general objective in designing a functional device platform included a direct interface the between memory/logic elements embedded in ASN architecture and

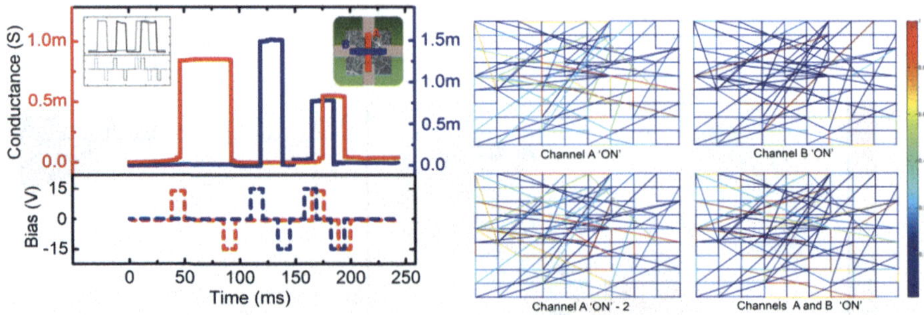


**Fig. 8** (Left) Simulation of spatially overlapping channels being modified independently by write/rewrite pulses, emulating the 2-bit switching functionality of actual device behavior (inset). (Right) Simulated internal network configurations (N = 219) at different ON/OFF configurations describing the formation of feedforward assemblies. In ON states of the network, conductances do not distribute uniformly. In fact, the simulation shows that several different configurations may correspond to the same ON/OFF channel configuration depending on the history of channel switching. For example, the internal configurations responsible for the ON of channel A at the two time points when it is activated before/after the activation/deactivation of channel B (blue), is shown

externally controlled input/output contact electrodes allowing for data processing. In the current device configuration, ASNs can only be electrically probed using the macroscopic interface electrodes. It was therefore essential to confirm that these electrodes could be effectively coupled to local ensembles of atomic switches within a particular spatial region of the network. Experimental observations of network plasticity [44] as a mechanism for the formation of feedforward pathways within ASNs was addressed through simulation [49] as seen in Fig. 8. Monitoring the conductance of all electrode combinations throughout the stimulation regimen revealed dynamic patterns of activity in regions free from intentional manipulation. The coexistence of localized changes in network connectivity alongside complex system-wide correlations suggest a capability for autonomous, higher-dimensional information processing through formation of specialized functional regions.

## 5.5 *Fluctuations, Correlations and Power Laws*

Our group examined the ASN device for emergent properties considered fundamental to brain function, which are not observed for individual atomic switches operating in simpler geometries, namely recurrent dynamics and the activation of feedforward subnetworks [44]. The presence of recurrent loops and dynamics within the ASN devices were demonstrated by applying a constant DC bias (Fig. 9a) across a particular region of the network. This produced persistent, bidirectional fluctuations—both increases and decreases—in network conductivity.

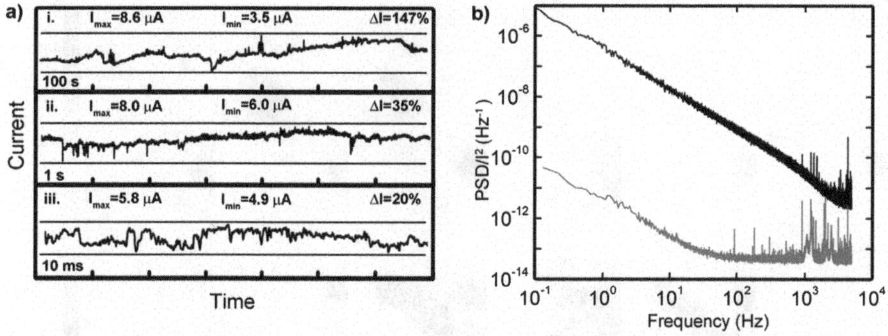


**Fig. 9** DC response (**a**) time traces of current response to 2 V DC bias show current increases and decreases at all time scales around a mean of 5.81 μA (standard deviation 0.88 μA), behavior specific to recurrent AS networks. (**b**) Fourier transforms of DC bias response for Ag control (grey) and functionalized Ag-$Ag_2S$ (black) networks. The power spectrum of the functionalized network displays 1/f power law scaling, indicating a high level of temporal correlation and memory (Avizienis PLoS 2012)

In the absence of recurrent structures within the network, conductivity would increase monotonically under constant DC bias, as in the case of a single atomic switch. However, bidirectional fluctuations in the current response persisted for several days under constant applied voltage, demonstrating that the complex network connectivity inherently resists localized positive feedback that would lead to the serial formation of a single, dominant high conductivity pathway between electrodes. Previously unreported current fluctuations of this kind are ascribed to recurrent loops in the network that create complex couplings between switches, resulting in network dynamics that do not converge to a steady state even under constant bias. A single switch turning ON does not simply lead to an increased potential drop across the next junction in a serial chain, but redistributes voltage across many recurrent connections that can ultimately produce a net decrease in network conductivity. These fluctuations are not attributable to noise, as shown by comparing the Fourier transformed current responses (Fig. 9b) of the devices to constant voltage before and after functionalization. The formation of atomic switch junctions expands the degree of correlation in current fluctuations, producing 1/f-like behavior across the entire sampled range. This behavior is distinct from that of control devices (unsulfurized silver network, grey line in Fig. 9b), which flattens to white noise and some high energy, high frequency fluctuations attributed to arcing between neighboring wires.

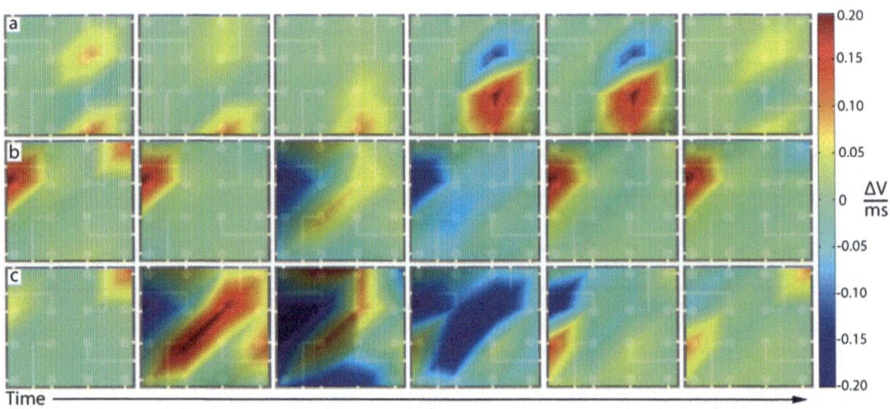


**Fig. 10** Spatial temporal switching activity in the ASN is seen in rows (**a**)–(**c**). Each row represents 3 ms with a 0.5 ms frame rate while a 5 V DC bias was applied at the upper right electrode and grounded at the lower left electrode. Both (**a**, **b**) localized switching, and (**c**) distributed switching affecting the entire network are observed (Demis Nanotech. 2015)

## *5.6 Distributed Switching/Correlations*

Further examples of correlations within the network exist as local and global switching events are monitored in the form of voltage fluctuations at each electrode. With an applied DC bias, the network moves through different patterns of activity. These patterns include local perturbations seen in rows (a) and (b) of Fig. 10, as well as large cascading switching events distributed throughout the whole network (Fig. 10c). Potential changes vary through time from low to high as current flows throughout the complex architecture. Electrodes many orders of magnitude larger than the individual atomic switches capture a general potential map along with the separability of outputs. Thus, the highly recurrent and interconnected coupling of individual switches leads to the emergence of distributed activity, which enables for a wide array of outputs and in turn potential applications in alternative computing paradigms.

## *5.7 Temporal Metastability and Criticality*

To our knowledge, the atomic switch network represents a unique implementation of a purpose-built self-assembled network composed of coupled non-linear elements that clearly demonstrate the essential characteristics of criticality, specifically power-law scaling of: 1/f fluctuations, energetic avalanches, as well as temporal metastability. The emergent complex behaviors observed in their temporally metastability indicate a capacity for memory and learning via persistent critical states with

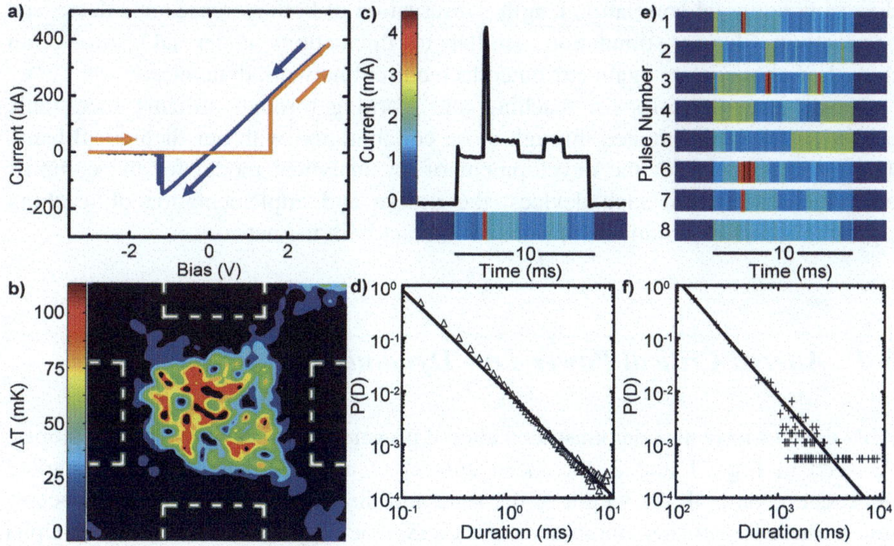


**Fig. 11** Electrical characteristics of complex nanoelectroionic networks. (**a**) Experimental I-V curve demonstrating pinched hysteresis; RON = 8 KΩ, ROFF > 10 MΩ. (**b**) Ultrasensitive IR image of a distributed device conductance under external bias at 300 K. (**c**, **e**) Representative network current response to a 2 V pulse showing switching between discrete, metastable conductance states. (**d**, **f**) Temporal correlation of metastable states observed during pulsed stimulation demonstrated power law scaling for probability, P(D), of duration. Power law scaling existed for residence time both (**d**) within a single 10 ms pulse and (**f**) over 2.5 s during extended periods of pulsed stimulation (Stieg Adv. Mat 2012)

potential utility for the creation of physically intelligent machines capable of evolution and learning. In addition to the exciting behaviors described in Sect. 5.6, complex networks of coupled nonlinear elements such as this commonly manifest non-trivial spatiotemporal evolution through dynamic system reconfigurations. Such reconfigurations, readily described by critical dynamics, enable enhanced maintenance of system correlations and more effective signal propagation. Indicators of criticality typically include power-law scaling of 1/f fluctuations and temporal metastability. Analysis of the power spectral density of network conductivity in the activated state revealed 1/f power law scaling over five orders of magnitude with $f \approx 1.4$ (data not shown). Temporal metastability was observed during sub-threshold pulsed voltage stimulation, analogous to methods employed in neuroscience to probe cortical cultures. Under typical conditions, the current response fluctuated over a wide range of metastable conductance states associated with discrete network configurations (Fig. 11c–f), as classified by residence times in a given state ranging from milliseconds (within a single stimulation pulse) to several seconds (across hundreds of pulses). Comparing the probability of state duration with its time duration indicated a power law distribution with ≈1.8 (Fig. 11e–f), indicating a diverging temporal correlation length. Observation of both increased and decreased conductivity during stimulation, similar to fluctuations under DC stimulation described above, were again attributed to recurrent network dynamics.

Finally, opportunities for teaching and learning through utilizing metastable critical states were initiated through close collaboration with our theoretical team. This process required: the development of an analytical model for the complex behavior observed in such devices, the design and implementation of teaching algorithms, and exploration of means to interact with the network.

## 5.8 Altered Critical Power-Law Dynamics

ASN devices have also demonstrated altered power spectral density, or PSD, slopes. As seen in Figs. 1 and 2, the mean current of a given state showed a marked dependence on both the length of the state and the probability (P(D)) of its occurrence where longer state durations and increased activity, characterized by 1/f alpha power-law scaling of PSD, were observed for intermediate mean current values. In an effort to control these properties, a current-controlled feedback loop was implemented in a similar fashion to that shown in Fig. 3. Real-time maintenance of a defined current set-point between two arbitrarily defined electrodes was readily achieved through application of an applied bias voltage (Fig. 14). While persistent fluctuations in ASN conductance are known to exhibit non-trivial spatiotemporal correlations characterized by power-law scaling of PSD [44, 46], procedures to control such correlations have yet to be reported. Utilization of the current-control approach provides a direct method to tune network dynamics as seen in Fig. 12, where higher current setpoints generated larger, more rapid network reconfigurations in the form of resistance switching as indicated by steeper PSD slopes (α) for both current and local voltage. Reliable transitions between resistance states, and thus

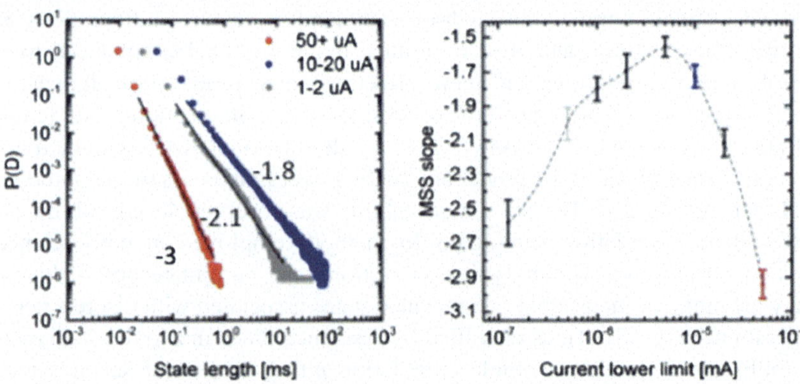


**Fig. 12** Representative probability distribution P(D) of metastable state duration (left) obeyed power law scaling with exponents dependent on the mean current during a given state (right)

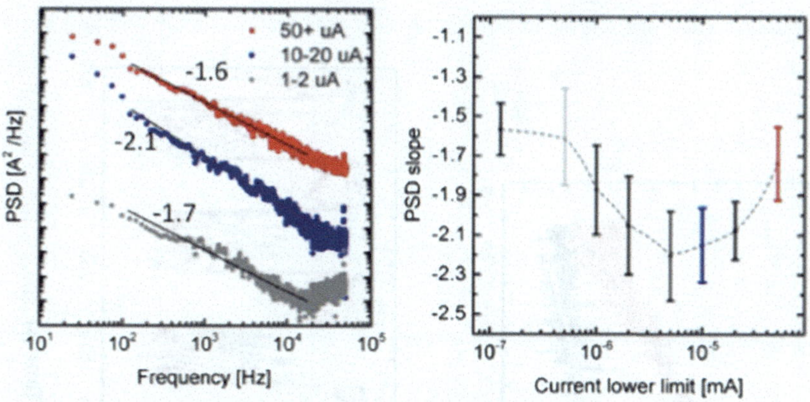


**Fig. 13** Representative example of power law ($1/f^{\alpha}$) scaling (left) of the PSD slope ($\alpha$) which was observed to depend on the mean value of the current output (right)

regimes of operational dynamics, was achieved as seen in Fig. 13. Finally, cross-correlation analysis of spatiotemporal correlations in device activity at various points throughout the network provided further support of the current set-point as a control parameter for network dynamics. During periods of limited activity, observed correlations were attributed to shared background noise. Periods of activity, at higher current set-points, resulted in a diversity of voltage recordings throughout the network characterized by broadening of correlation coefficients.

# 6 Computing with the Atomic Switch Network

## *6.1 Theoretical Constructs*

The ASN is one of a limited number of CMOS compatible platforms capable of performing RC [62]. Within the context of the RC formalism, each atomic switch is a functional node in the reservoir and the connective weights between each node are mediated by the silver nanowires. The multi-electrode array on which the network is grown can be adjusted to control input and readout functionality of the electrodes to measure all nodes in 10–50 μm regions of the network. All necessary criteria such as short-term memory, increased fault tolerance, and an arbitrarily scalable number of higher-dimensional outputs are fulfilled by the ASN. The memristive behavior of individual atomic switches bestows the ASN with a fading memory characteristic. This ensures that previous inputs to the network do not exert considerable influence over the current state. The power-law dynamics (Fig. 12) indicate that the system has a scale-free topology that allows it to operate at the "edge-of-chaos," a dynamical regime providing a balance between memory and instability. The non-linear transformations Fig. 14 are an intrinsic behavior of the system that can be harnessed to

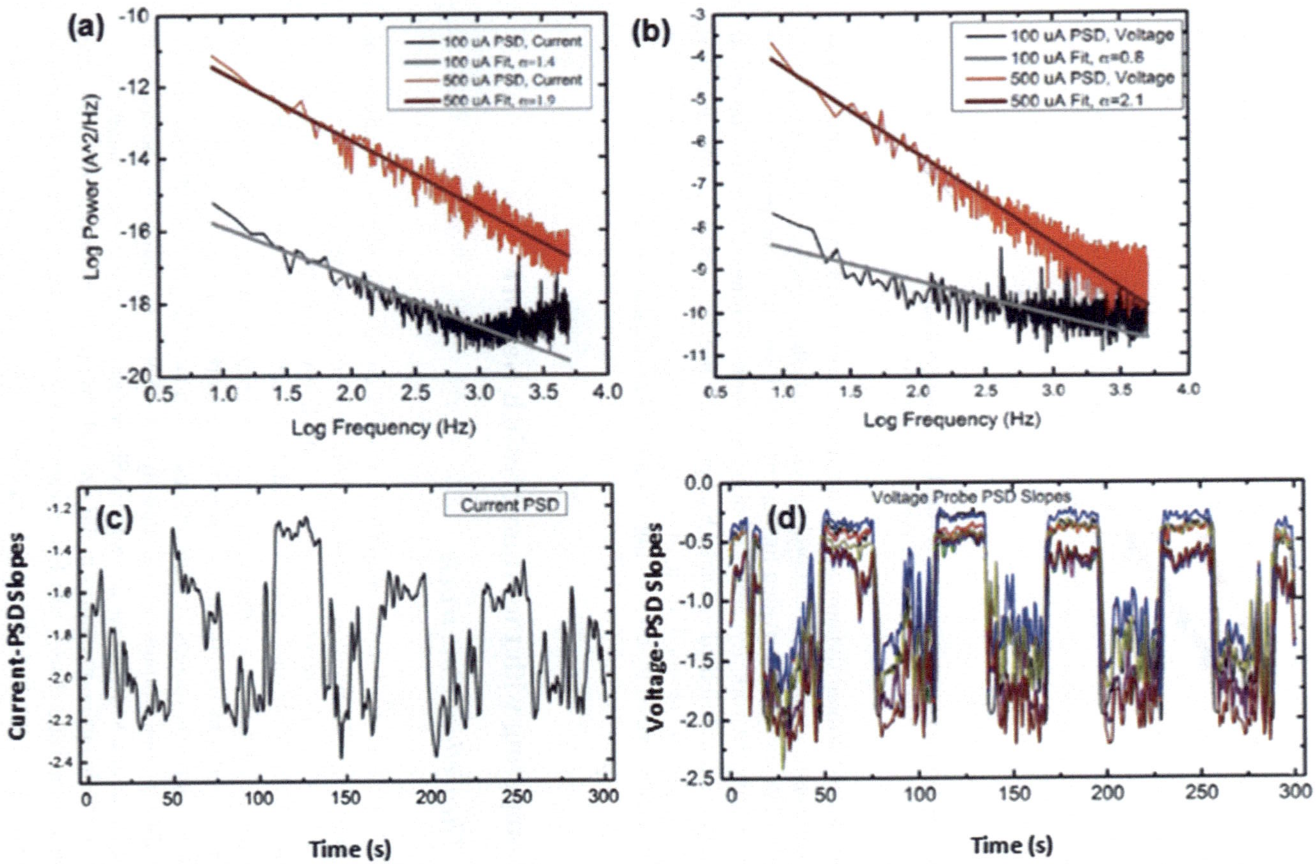


**Fig. 14** Dependence of power law scaling (PSD α value) on current setpoint using a current-controlled feedback loop operating in real-time through control of the applied bias voltage. PSD slope (α) of the current output (left) and simultaneously acquired voltage (right). Repeated cycling of current setpoint demonstrated an ability to tune the PSD slope (α), and hence temporal dynamics, of the current output (top) and local voltage throughout the ASN device (bottom)

increase performance. These transformations manifested as separable unique voltage signals and are a consequence of the input signal entering a higher representational space that is then used in reconstructing the desired target waveform. Unlike current computational models of explicitly programmable algorithms, RC relies on systems operating in a regime where they are able to 'learn' through experience, circumventing the need for intelligent programming.

Generally, RC utilizes a randomly connected network, dubbed the 'reservoir,' composed of interacting elements called neurons with a topology based on mammalian brain neural networks. Information propagates through network connections via electrical or electrochemical signals which are preferentially directed towards neurons with the greatest connective strength. Neurons may be activated by the incoming signal based on a set of learning rules causing them to undergo internal modifications and excitatory transformation of the incoming signal. This stimulated response of the neuron is then communicated through the neuron's outgoing connections, thereby allowing the signal to percolate throughout the network. Computation is achieved by recording outputs from a few neurons assigned during initialization and these neurons are trained to achieve the desired functionality. Training resembles the evolutionary phenomenon observed in biological and natural systems adapting due to environmental changes. External stimulation of the reservoir allows the system to independently evolve into a number of diverse neurons due to signal propagation and local interactions which activate and modify neuron properties. Clusters of neurons are capable of displaying distinct activity and emergent behaviors due to local interactions interplaying with signal percolation. Training selects a number of clusters desirable for computation and reinforces these neurons for specialization into relevant mathematical processors. After the system is properly trained using sample tasks, the neurons are passivated such that the system retains the knowledge and experience for future computation. RC is both a simple and elegant construction that avoids the need for absolute control over programmable elements while capable of producing powerful processors due to evolutionary concepts. Performance is controlled by neuron connectivity and distribution of strong and weak connections. Operational utility is thus dependent on the reservoir's statistical characteristics and global parameters, focusing on emergent qualities of the network instead of individual elements.

The non-linear dynamics exhibited by the ASN device is uniquely positioned to be exploited using the paradigm of reservoir computing [49, 63]. Interactions between atomic switches in the network produce non-linear transformations of input signals not present in Ohmic resistors or single switches. The additional non-linear transformations of the input signal increases the diversity of the output signals and thereby increase output separability and potential for computational capability. In the reservoir computing approach, the input signal undergoes a projection into a higher dimensional representation space, which can be thought of as an expansion of the output into a sum of mathematical elements [64, 65]. Having a rich collection of elements provides a large repertoire of possible transformations of the input via mathematical representation vectors. The separability and utility of multiple transformations enable the construction of mathematical algorithms by integrating desired representations. Within this representation space, reservoir computing is able to identify useful elements of the output function to attempt to solve a specific task. Depending on the degree of non-linearity, which in turn determines the size and resolution of the reservoir, a task may be optimally solved in this higher dimensional representation space. In the ASN device, variations in connectivity, atomic switch density, and junctions contribute to the degree of non-linearity by creating a larger range of functional elements that have different activation and operational voltages.

## *6.2 Implementations*

### 6.2.1 Waveform Regression

Simulations of ASNs have indicated that the system has the fundamental capacity to perform waveform regression [66]. From simulation results, performance depended on the level of higher harmonics produced and the harmonic distortion (Fig. 16) required for the specific task. For example, the cosine task only requires a shift in its periodicity and, therefore, does not require extensive higher harmonic generation. Conversely, the square wave task requires infinitely distributed harmonics to produce a straight line through wave interference. Further, voltage dependent simulations showed that increasing device activation controlled these harmonic generations. Here, the device was expected to perform in a similar way with task difficulty increasing from cosine, triangle, sawtooth, and square due to increasing harmonic requirements. Device initialization and activation to achieve the best performance is described in the previous section.

Experimental performance of various waveform regression tasks using ASN devices are presented in Fig. 15. To implement waveform regression the ASN was stimulated with a bipolar sinusoidal voltage, inducing switching activity and placing the network in an active state. The output potentials measured at each electrode were then combined using the Moore-Penrose linear regression and optimized during a training period [67–69]. Two-second epochs of data were used to evaluate the ASN's computational capability, where 1 s of data was allocated for both training and testing. Performance was measured during a 1 s period after training where the ASN accomplished various tasks (Fig. 15). The performance of the ASN was quantified by calculating the normalized mean squared error between the target and generated waveforms [70]. Here, the difference between error and unity was used to calculate accuracy.

The ASN was capable in achieving up to ~90% accuracy using 62 of the 64 measurement electrodes for each task. Task complexity increased from cosine to square wave due to the increasing mismatch between the sinusoidal input and the target waveform. In the case of cosine generation, the overall waveform of the input is preserved save for a shift in its periodicity. The cosine generation was the simplest task where the ASN performed with the highest accuracy, ~90%. Note that the

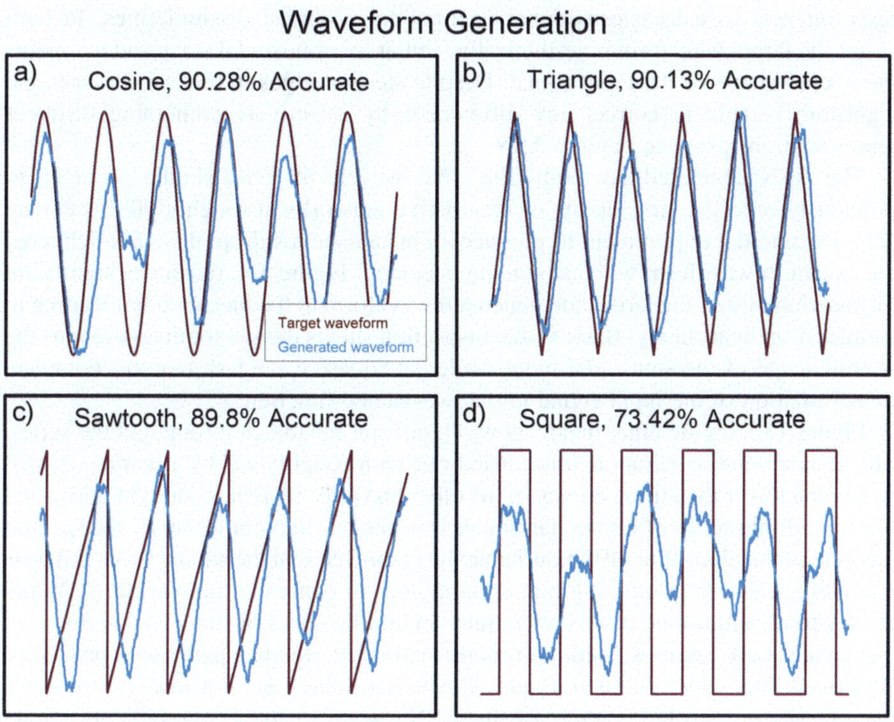


**Fig. 15** Computation of a sinusoidal wave into various waveforms. The above figure shows several waveforms (sawtooth, square, triangle, and cosine) produced using the ASN as a computational device using the setup in Fig. 4. Each plot contains the desired signal (red) and the computed signal (blue) with their accuracy w.r.t. the desired signal shown above the curves. All tasks share an 11 Hz frequency for their waveforms and share the same dataset with only differences in the target task. The dataset was approximately 1 min long, divided into 2 s epochs, and 1 s within each interval was allocated for training and testing. A 1 s excerpt which best represents device behavior during testing are shown above (Sillin Nanotechnology 2013)

cosine regression shown in Fig. 15a would not be possible using a grid of regular resistors due to their intrinsic linear response. Since individual atomic switches have a non-linear memristive response, it is possible to harness that state function into the highly recurrent structure of the ASN. The highly recurrent structure allowed higher levels of coupled interactions that cannot be captured by a single atomic switch, resulting in emergent behaviors. Particularly, the network was capable of producing delayed responses and enabled the network to shift the phase of the input signal by a half-wavelength, producing a cosine.

Figure 15b shows hardly any mismatch in the triangle generation, achieving a similar ~90% accuracy and visually validates the performance metric used throughout our analysis. A similar argument is used to explain the high performance of the triangle wave when compared to the cosine task. The determining factor for reservoir performance is the level of similarity between the target and input signal, where the reservoir acts as a transformational operator to minimize dissimilarities. In both cases, the target waveform is aesthetically similar to a sinusoidal wave and maintains the overall shape of the input signal. Despite steeper edges in the triangle task, the algorithm is able to correct any differences by selectively combining different representations produced by the ASN.

The ASN generated sawtooth (Fig. 15c) waveforms with similar accuracy to previously reported simulations of memristive networks at roughly 90% accuracy [71]. Despite the requirement to produce an instantaneous drop, the ASN delivered the sawtooth waveform with astounding accuracy. Figure 15c illustrates significant mismatch between the target and generated waveform at the turning point leading to a minor drop in accuracy. Basic visual inspection shows the sawtooth task retains the overall shape of the sinusoidal input while the square wave task requires complete transformation of the input signal into a two-valued function.

Figure 15d, on the other hand, shows significant mismatch throughout the series. The square wave generation was carried out with roughly a 78% accuracy, which was much lower than the accuracy of the other tasks. To recreate a straight horizontal line, an infinite series of higher harmonics is necessary in order to satisfy the spectral theorem in the algorithm [49]. Fourier analysis showed that the square wave task was relatively selective in utilizing higher harmonics to construct the waveform. While the sawtooth and square wave both require an infinite series of sinusoidal harmonics, the square wave requires continuous constructive interference patterns to produce a horizontal line, which limits it to odd or even harmonics and drastically diminishes the regression algorithm. In this case, the ASN was only capable of producing a finite number of higher harmonics. However, further post-processing such as setting a threshold on the voltage to binarize the data can be performed to expand the device's response to a square wave input, a necessity for reliable Boolean logic computing [72].

It was found that the ASN was capable of replicating computing performances typical of reservoirs with $10^3$ output signals [73]. Theoretical studies predicts the performance to scale with an increasing number of output signals due to the dependence on the regression algorithm [74]. However, how can a reservoir with much fewer output signals outperform reservoirs with output signals orders of magnitude higher than the ASN? Further inspection of the mathematical formalism [74] show that performance is additionally characterized by the uniqueness of each output signal. Obtaining a set of unique signals allows us to linearly combine the output signals into a number of unique solutions, where the number of unique solutions scales with the number of unique output signals. The larger set of solutions increases the size of the "net" we cast which increases the probability and approximation of producing the correct solution (Fig. 16).

### 6.2.2 Logic

Expanded efforts to assess their performance in Boolean logic operations using non-temporal inputs based on randomized Boolean input streams. Zero and one

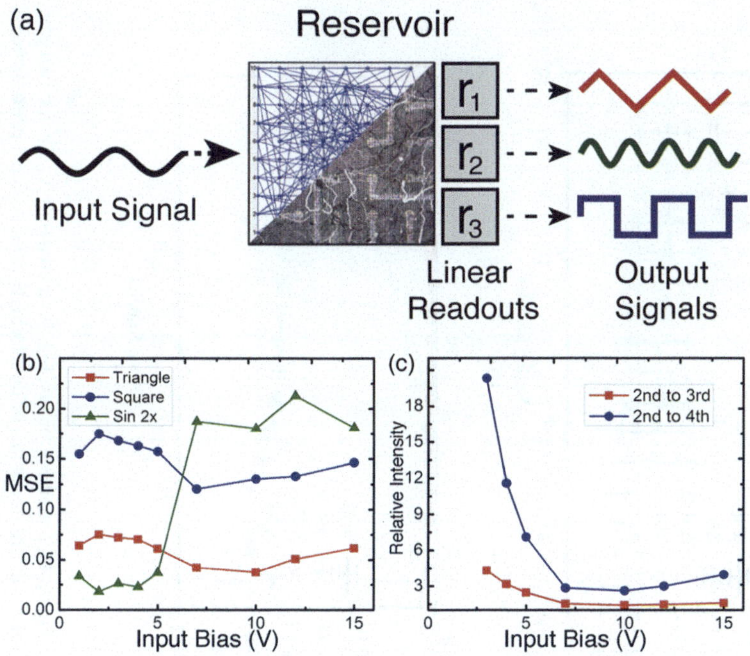


**Fig. 16** (**a**) Schematic of network simulation used in the waveform generation RC task, with specific electrodes chosen as inputs/outputs (16 output electrodes). RC was implemented using a 10 × 10 node network with a 5 V, 10 Hz sinusoidal input signal and tasked to produce 10 Hz triangle/square and 20 Hz sinusoidal waveforms. (**b**) Mean-squared error (MSE) for each task with respect to driving amplitude showed minimal error in triangle/square waveform generation task at 10 V, corresponding to the onset of higher harmonic generation (see red curve of Fig. 6b). Performance in the 20 Hz sinusoidal waveform generation task decreased when (**c**) the relative amplitude of the average 2nd harmonic intensities of the readouts becomes increasingly diminutive. These results correspond to a strong dependence on the 2nd harmonic for 20 Hz sine generation and the need for HHG in triangle/square generation as expected by Fourier analysis (Sillin Nanotechnology 2013)

bits were converted to negative and positive DC voltage pulses, respectively. Next, a linear readout layer was applied to an array of voltage outputs from the device to reconstruct target output signals for the given task. ASNs produced nearly perfect results at low voltages for AND, OR, and NAND with more than 95% confidence. XOR, which requires non-linearity to solve, was able to be partially solved at high voltages with more than 95% confidence owed to stable, non-temporal, non-linear behaviors in the device under optimized operational conditions. As opposed to previous works which have investigated temporal computation in ASNs, this work was the first to demonstrate semi-predictable, non-temporal, non-linear behavior within the device. These results demonstrate that the device connectivity is complete enough to perform complex computations. With a more comprehensive view of ASN behavior, these devices will be capable of performing functions currently implemented in CMOS while occupying less area and processing more inputs simultaneously (Fig. 17).

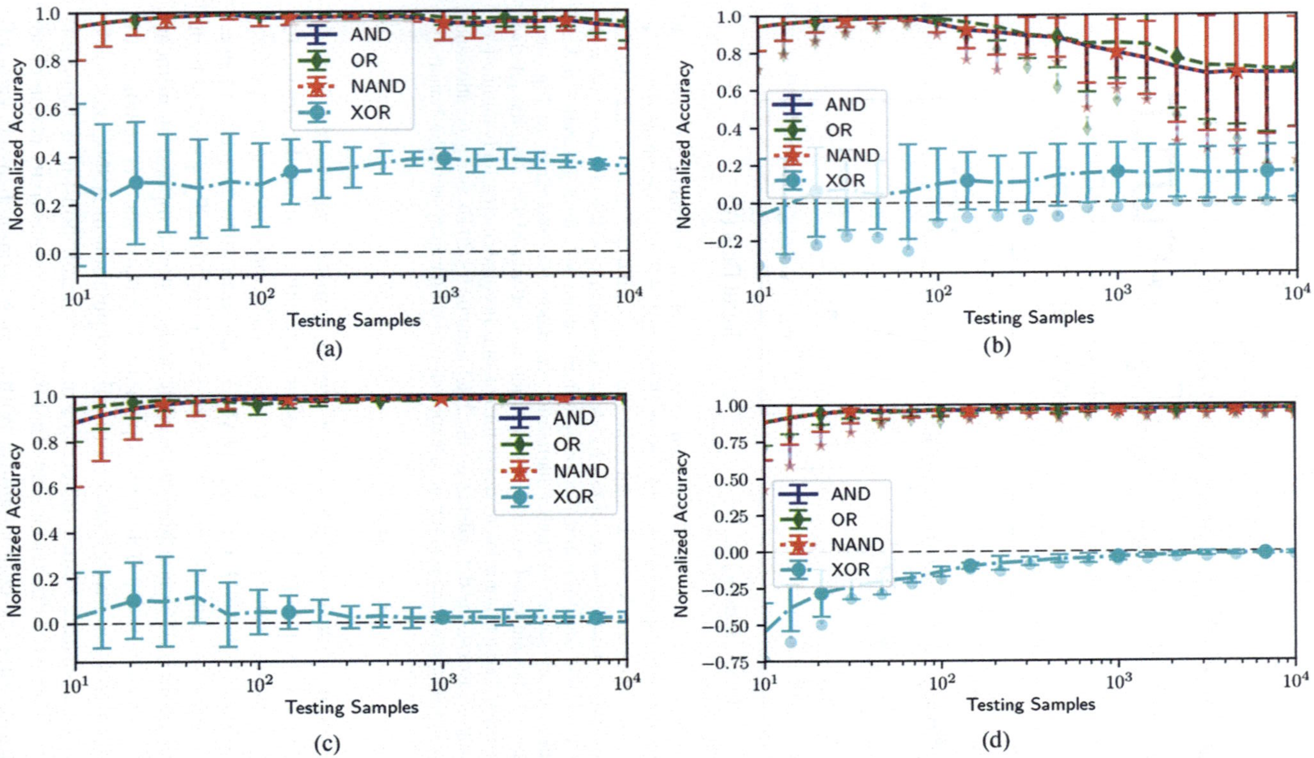


**Fig. 17** Accuracy on all Boolean logic functions learned using 1000 training samples. Figure 4a, b used 3 V inputs; Fig. 4c, d used 0.01 V. The left two plots represent normalized accuracy of a single electrode combination. Error bars represent the standard deviations of the electrode combination's performance at different points in time. The right two plots are the 95% confidence accuracies across all electrode combinations; that is, we are 95% sure that a random electrode combination, at any point in time, will produce a regression that will outperform the mean line shown. Faded markers below lines indicate the worst-performing electrode combination's 95% confidence accuracy

## 7 Outlook

If we consider the future of A.I. certainly we are at a turning point in history where increasingly we see the practically limitless applications in diverse areas such as driverless vehicles, medical diagnosis—healthcare, climate prediction, even social comfort. Our digital computer systems of today are nevertheless being pushed to the ultimate limits of fabrication with the end of Moore's Law in clear sight. In other words, it will become prohibitively expensive to maintain the advances required for A.I. to create a post-human world. The poor scaling of computer hardware and software required as combinatorial complexity increases will not be overcome in future von Neumann machines.

Predictions of increasing computational capacity in comparison to that of the human brain or the "singularity", as coined by Ray Kurtzweil are unlikely using digital approaches. The scenario of computational equivalence to a human was proposed to give rise to a sudden and massive increase in dominantly non-biological intelligence, and we propose it can best be approached by using biological inspiration in making a system that has physical operational characteristics closer to ourselves. The Atom Switch and networks of them have a potential role to play in Hybrid–CMOS morphic systems where methods such as Reservoir Computation in physical analogue hardware can be integrated into a morphic system that utilizes the optimum performances of CMOS digital with the ASN approach. Already Atomic Switches have successfully integrated into FPGA devices by NEC reducing energy consumption, footprint, size and transistor count [56]. The Atomic Switch technologies are more robust in terms of sensitivity to electromagnetic noise and radiation than Flash making them candidates for robotic and space satellite applications. The key differences of the ASN approach to conventional computation are in the elimination of programming and error correcting each step of a calculation with the RC paradigm. Likewise, ASN devices use distributed fading memory not RAM similar to living systems. The ability to handle multiple tasks in parallel is another advantage of such an approach. Although accurate calculations of arithmetic operations will always be superior in digital systems, analog systems such as the ASN excel in decision making, or noisy and error prone data that have no precise solution but rather a range of outcomes with a best guess of the outcome probabilities a bit like Newtonian vs. Quantum mechanics where deterministic solutions are replaced by probabilities. The potential impact of A.I. in society has become quite heated in terms of the dangers it poses to our society and discussions of government regulation (Elon Musk) and possibly imposing taxes on A.I. robots and systems have even been proposed (Bill Gates). However, in many advanced countries there is a future need for such technology as baby-boomers retire and the population of a young work force declines A.I. will be essential in healthcare, welfare, national security, and in many other areas of societal enhancements.

## References


1. Waldrop, M.M.: The chips are down for Moore's law. Nature. **530**, 144–147 (2016)
2. Abbe, E.: Contributions to the Theory of the Microscope and the Microscopic Perception. Springer (1873)
3. International Technology Roadmap for Semiconductors (2015)
4. Neumann, J.V.: First draft of a report on the EDVAC. IEEE Ann. Hist. Comput. **15**, 27–75 (1993)
5. Backus, J.W.: Can programming be liberated from the von Neumann style? A functional style and its algebra of programs. Comm. ACM. **21**, 613–641 (1978)
6. Dongarra, J.: Visit to the National University for Defense Technology Changsha, China. University of Tennessee (2013)
7. Mead, C.: Neuromorphic electronic systems. In: Proceedings of the IEEE. IEEE (1990)
8. Hodgkin, A.L., Huxley, A.F.: A quantitative description of membrane current and its application to conduction and excitation in nerve. J. Physiol. **117**, 500–544 (1952)
9. Haimovici, A., Tagliazucchi, E., Balenzuela, P., Chialvo, D.R.: Brain organization into resting state networks emerges at criticality on a model of the human connectome. Phys. Rev. Lett. **110** (17), 178101 (2013)
10. Wang, X.F., Chen, G.: Complex networks: small-world, scale-free and beyond. IEEE Circ. Syst. Mag. **3**, 6–20 (2003)
11. Sporns, O.: Small-world connectivity, motif composition, and complexity of fractal neuronal connections. Biosystems. **85**, 55–64 (2006)
12. Abbott, L.F., Nelson, S.B.: Synaptic plasticity: taming the beast. Nat. Neurosci. **3**, 1178 (2000)
13. Hebb, D.O.: Organization of Behavior. Wiley, New York (1950)
14. Rosenblatt, F.: The perceptron: a probabilistic model for information storage and organization in the brain. Psychol. Rev. **65**, 386–408 (1958)
15. Hopfield, J.J.: Artificial neural networks. IEEE Circ. Dev. Mag. **4**, 3–10 (1988)
16. Gomes, L.: Neuromorphic chips are destined for deep learning—or obscurity. IEEE Spectrum (2017)
17. Schuman, C.D., Potok, T.E., Patton, R.M., Douglas Birdwell, J., Dean, M.E., Rose, G.S., Plank, J.S.: A survey of neuromorphic computing and neural networks in hardware. arXiv (2017)
18. Christie, P., Stroobandt, D.: The interpretation and application of Rent's rule. IEEE Trans. VLSI Syst. **8**, 639–648 (2000)
19. Abraham, A.: Artificial neural networks. In: Sydenham, P.H., Thorn, R. (eds.) Handbook of Measuring System Design. Wiley, New York (2005)
20. Medsker, L., Jain, L.C.: Recurrent Neural Networks: Design and Applications. CRC, Boca Raton, FL (2001)
21. Graves, A., Mohamed, A., Hinton, G.: Speech recognition with deep recurrent neural networks. ArXiv (2013)
22. Khotanzad, A., Chung, C.: Application of multi-layer perceptron neural networks to vision problems. Neural Comput. Appl. **7**, 249–259 (1998)
23. LeCun, Y., Bengio, Y., Hinton, G.: Deep learning. Nature. **521**, 436–444 (2015)
24. LeCun, Y., Boser, B.E., Denker, J.S., Henderson, D., Howard, R.E., Hubbard, W.E., Jackel, L. D.: Handwritten digit recognition with a back-propagation network. In: Touretsky, D.S. (ed.) Advances in Neural Information Processing Systems. Morgan Kaufmann, San Mateo (1990)
25. Hinton, G.E.: Learning multiple layers of representation. Trends Cogn. Sci. **11**(10), 428–434 (2007)
26. Büsing, L., Schrauwen, B., Legenstein, R.: Connectivity, dynamics, and memory in reservoir computing with binary and analog neurons. Neural Comput. **22**, 1272–1311 (2010)
27. Sojakka, C. F.: Pattern recognition in a bucket. In: Wolfgang Banzhaf, J. Z. (ed.) European Conference on Artificial Life: Advances in Artificial Life (2003)
28. Goudarzi, A., Teuscher, C., Gulbahce, N., Rohlf, T.: Emergent criticality through adaptive information processing in Boolean networks. Phys. Rev. Lett. **108**, 128702 (2012)

29. Tour, J.M., Cheng, L., Nackashi, D.P., Yao, Y., Flatt, A.K., Angelo, S.K.S., Mallouk, T.E., Franzon, P.D.: Nanocell electronic memories. J. Am. Chem. Soc. **125**, 13279–13283 (2003)
30. Merolla, P.A., Arthur, J.V., Alvarez-Icaza, R., Cassidy, A.S., Sawada, J., et al.: A million spiking-neuron integrated circuit with a scalable communication network and interface. Science. 345, 668–673 (2014)
31. Indiveri, G., Linares-Barranco, B., Hamilton, T., van Schaik, A., Etienne-Cummings, R., et al.: Neuromorphic silicon neuron circuits. Front. Neurosci. 5, 1–23 (2011)
32. Shimokawa, Y., Fuwa, Y., Aramaki N. A parallel ASIC VLSI neurocomputer for a large number of neurons and billion connections per second speed. In: IEEE International Joint Conference on Neural Networks. Singapore (1991)
33. Omondi, A.R., Rajapakse, J.C.: FPGA Implementations of Neural Networks. Springer, Dordrecht (2006)
34. Nurvitadhi, E., Sheffield, D., Sim, J., Mishra, A., Venkatesh, G., Marr, D. Accelerating binarized neural networks: comparison of FPGA, CPU, GPU, and ASIC. In: International Conference on Field-Programmable Technology (FPT), IEEE (2016)
35. Qiao, Y., Shen, J., Xiao, T., Yang, Q., Wen, M., Zhang, C.: FPGA-accelerated deep convolutional neural networks for high throughput and energy efficiency. Concurr. Comput. Pract. Exp. **29** (2016)
36. NVIDIA launches the world's first graphics processing unit: GeForce 256 [Online]. http://www.nvidia.com/object/IO_20020111_5424.html (1999)
37. Jager, C.: Nvidia unveils Volta: the most powerful GPU ever [online]. https://www.lifehacker.com.au/2017/05/nvidias-unveils-volta-gv100-the-most-powerful-gpu-ever/ (2017)
38. Krizhevsky, A., Sutskever, I., Hinton, G. E.: Imagenet classification with deep convolutional neural networks. In: Advances in Neural Information Processing Systems (2012)
39. Saxena, A.: Deep learning pioneers boost research at NVIDIA AI labs around the world [online]. https://blogs.nvidia.com/blog/2017/07/11/deep-learning-pioneers-boost-research-at-nvidia-ai-labs-around-the-world/ (2017)
40. Romero, A., et al.: Diet networks: thin parameters for fat genomics. ArXiv (2017)
41. Finn, C., Levine, S.: Deep visual foresight for planning robot motion. ArXiv (2017)
42. Meier, K.: The FACETS project. Available https://facets.kip.uni-heidelberg.de/images/4/48/Public%2D%2DFACETS_15879_Summary-flyer.pdf (2010)
43. Qualcomm helps make your mobile devices smarter with new Snapdragon machine learning software development kit. https://www.qualcomm.com/news/releases/2016/05/02/qualcomm-helps-make-your-mobile-devices-smarter-new-snapdragon-machine (2016)
44. Avizienis, A.V., Sillin, H.O., Martin-Olmos, C., Shieh, H.H., Aono, M., Stieg, A.Z., Gimzewski, J.K.: Neuromorphic atomic switch networks. PLoS One. **7**(8), e42772 (2012)
45. Yang, J.J., Strukov, D.B., Stewart, D.R.: Memristive devices for computing. Nat. Nanotechnol. **8**, 13–24 (2013)
46. Stieg, A.Z., et al.: Self-organization and emergence of dynamical structures in neuromorphic atomic switch networks. In: Adamatzky, A., Chua, L. (eds.) Memristor Networks. Springer, Cham (2014)
47. Demis, E.C., Aguilera, R., Sillin, H.O., Scharnhorst, K., Sandouk, E.J., Aono, M., Stieg, A.Z., Gimzewski, J.K.: Atomic switch networks nanoarchitectonic design of a complex system for natural computing. Nanotechnology. **26**, 204003 (2015)
48. Demis, E.C., Aguilera, R., Scharnhorst, K., Aono, M., Stieg, A.Z., Gimzewski, J.K.: Nanoarchitectonic atomic switch networks for unconventional computing. Jpn. J. Appl. Phys. **55**, 1102B2 (2016)
49. Sillin, H.O., Aguilera, R., Shieh, H.H., Avizienis, A.V., Aono, M., Stieg, A.Z., Gimzewski, J. K.: A theoretical and experimental study of neuromorphic atomic switch networks for reservoir computing. Nanotechnology. **24**, 384004 (2013)
50. Scharnhorst, K.S., Carbajal, J.P., Aguilera, R.C., Sandouk, E.J., Aono, M., Stieg, A.Z., Gimzewski, J.K.: Atomic switch networks as complex adaptive systems. Jpn. J. Appl. Phys. **57**, 03ED02 (2018)

51. Langton, C.G.: Computation at the edge of chaos: phase transitions and emergent computation. Phys. D. **42**, 12–37 (1990)
52. Gimzewski, J.K., Möller, R.: Transition from the tunneling regime to point contact studied using scanning tunneling microscopy. Phys. Rev. B. **36**(2), 1284–1287 (1987)
53. Lang, N.D.: Theory of single-atom imaging in the scanning tunneling microscope. Phys. Rev. Lett. **56**, 1164–1167 (1986)
54. van Houton, H., Beenakker, C.: Quantum point contacts. Phys. Today. **49**(7), 22–27 (1996)
55. Terabe, K., Nakayama, T., Hasegawa, T., Aono, M.: Formation and disappearance of a nanoscale silver cluster realized by solid electrochemical reaction. J. Appl. Phys. **91**, 10110–10114 (2002)
56. NEC. NEC integrates nanobridge in the Cu interconnects of Si LSI. https://phys.org/news/2009-12-nec-nanobridge-cu-interconnects-si.html (2009)
57. Ohno, T., Hasegawa, T., Tsuruoka, T., Terabe, K., Gimzewski, J.K., Aono, M.: Short-term plasticity and long-term potentiation mimicked in single inorganic synapses. Nat. Mater. **10**, 591–595 (2011)
58. Hasegawa, T., Nayak, A., Ohno, T., Terabe, K., Tsuruoka, T., Gimzewski, J.K., Aono, M.: Memristive operations demonstrated by gap-type atomic switches. Appl. Phys. A. **102**, 811–815 (2011)
59. Avizienis, A.V., Martin-Olmos, C., Sillin, H.O., Aono, M., Gimzewski, J.K., Stieg, A.Z.: Morphological transitions from dendrites to nanowires in the electroless deposition of silver. Cryst. Growth Des. **13**(2), 465–469 (2013)
60. Stieg, A.Z., Avizienis, A.V., Sillin, H.O., Martin-Olmos, C., Aono, M., Gimzewski, J.K.: Emergent criticality in complex turing B-type atomic switch networks. Adv. Mater. **24**, 286–293 (2011)
61. Oskoee, E.N., Sahimi, M.: Electric currents in networks of interconnected memristors. Phys. Rev. E. **83**, 031105 (2011)
62. Goudarzi, A., Lakin, M.R., Stefanovic, D., Teuscher, C.: A model for variation-and fault-tolerant digital logic using self-assembled nanowire architectures. In: IEEE/ACM International Symposium on Nanoscale Architectures. ACM, pp. 116–121 (2014)
63. Verstraeten, D.: Reservoir computing: computation with dynamical systems. PhD thesis, Ghent University (2009)
64. Legenstein, R., Maass, W.: What makes a dynamical system computationally powerful? In: Haykin, S., Principe, J.C., Sejnowski, T.J., McWhirter, J. (eds.) New Directions in Statistical Signal Processing: From Systems to Brain. MIT Press, Cambridge, MA (2005)
65. Lukoševičius, M., Jaeger, H.: Reservoir computing approaches to recurrent neural network training. Comput. Sci. Rev. **3**, 127–149 (2009)
66. Wyffels, F., Schrauwen, B.: A comparative study of reservoir computing strategies for monthly time series prediction. Neurocomputing. **73**, 1958–1964 (2010)
67. Castro, L.N.D.: Fundamentals of natural computing: an overview. Phys. Life Rev. **4**, 1–36 (2007)
68. Modha, D.S., Ananthanarayanan, R., Esser, S.K., Ndirango, A., Sherbondy, A., Singh, R.: Cognitive computing. Commun. ACM. **54**, 62–71 (2011)
69. Yu, S., Kuzum, K., Philip Wong, H. S.: Design considerations of synaptic device for neuromorphic computing. In: IEEE International Symposium on Circuits and Systems, Melbourne, VIC. IEEE, pp 1062–1065 (2014)
70. Schrauwen, B., Verstraeten, D., Van Campenhout, J.: An overview of reservoir computing: theory, applications and implementations. In: 15th European Symposium on Artificial Neural Networks, pp. 471–482 (2007)
71. Bürger, J., Goudarzi, A., Stefanovic, D., Teuscher, C.: Hierarchical composition of memristive networks for real-time computing. In: IEEE/ACM International Symposium on Nanoscale Architectures (NANOARCH). IEEE (2015)

72. Gacem, K., Retrouvey, J.M., Chabi, D., Filoramo, A., Zhao, W., Klein, J.O., Derycke, V.: Neuromorphic function learning with carbon nanotube based synapses. Nanotechnology. **24**, 384013 (2013)
73. Snyder, D., Goudarzi, A., Teuscher, C.: Computational capabilities of random automata networks for reservoir computing. Phys. Rev. E. **87**, 042808 (2013)
74. Carbajal, J.P., Dambre, J., Hermans, M., Schrauwen, B.: Memristor models for machine learning. Neural Comput. **27**, 725–747 (2015)

# A List of Papers Related to the Atomic Switch

**Tohru Tsuruoka and Masakazu Aono**

As described in the previous chapters, research on the atomic switch has started to be used as resistive switching devices, but now its applications for the device are expanding to programmable logic gates, transistors, sensors, flexible electronics, and neuromorphic circuits. We have also realized various novel functions, for the control of electrical, optical, and magnetic properties, based on atomic switch technology. Although many of our papers are cited in the preceding chapters, we provide here a complete chronological list of all the peer-reviewed papers relating to the atomic switch, which papers have been published by the authors of this book and their many collaborators. We hope that this list will help our readers to understand the history and diversity of research on the atomic switch.

**2001**

1. "Quantum point contact switch realized by solid electrochemical reaction"
   K. Terabe, T. Hasegawa, T. Nakayama, and M. Aono
   Riken Review **37** (2001) 7–8.
   ISSN:0919-3405

**2002**

2. "Ionic/electronic mixed conductor tip of a scanning tunneling microscope as a metal atom source for nanostructuring"
   K. Terabe, T. Nakayama, T. Hasegawa, and M. Aono
   Applied Physics Letters **80** (2002) 4009–4011.
   DOI: 10.1063/1.1480887

T. Tsuruoka (✉) · M. Aono
International Center for Materials Nanoarchitectonics (MANA), National Institute for Materials Science (NIMS), Tsukuba, Ibaraki, Japan
e-mail: TSURUOKA.Tohru@nims.go.jp; 1944ma@mth.biglobe.ne.jp

 
M. Aono (ed.), *Atomic Switch*, Advances in Atom and Single Molecule Machines,
https://doi.org/10.1007/978-3-030-34875-5_12

3. "Formation and disappearance of a nanoscale silver clusters realized by solid electrochemical reaction"
K. Terabe, T. Nakayama, T. Hasegawa, and M. Aono
Journal of Applied Physics **91** (2002) 10110–10114.
DOI: 10.1063/1.1481775

**2003**

4. "Nanometer-scale switches using copper sulfide"
T. Sakamoto, H. Sunamura, H. Kawaura, T. Hasegawa, T. Nakayama, and M. Aono
Applied Physics Letters **82** (2003) 3032–3034.
DOI: 10.1063/1.1572964

**2004**

5. "A nonvolatile programmable solid electrolyte nanometer switch"
T. Sakamoto, S. Kaeriyama, H. Sunamura, M. Mizuno, H. Kawaura, T. Hasegawa, K. Terabe, T. Nakayama, and M. Aono
2004 IEEE International Solid-State Circuits Conference Digest of Technical Papers 290–291.

**2005**

6. "Quantized conductance atomic switch"
K. Terabe, T. Hasegawa, T. Nakayama, and M. Aono
Nature **433** (2005) 47–50.
DOI: 10.1038/nature03190
7. "Ionic-electronic conductor nanostructures: template-confined growth and nonlinear electrical transport"
C. H. Liang, K. Terabe, T. Hasegawa, R. Negishi, T. Tamura, and M. Aono
Small **10** (2005) 971–975.
DOI: 10.1002/smll.200500155
8. "A nonvolatile programmable solid-electrolyte nanometer switch"
S. Kaeriyama, T. Sakamoto, H. Sunamura, M. Mizuno, H. Kawaura, T. Hasegawa, K. Terabe, and M. Aono
IEEE J. Solid-State Circuits **40** (2005) 168–176.
DOI: 10.1109/JSSC.2004.837244
9. "Three terminal solid-electrolyte nanometer switch"
T. Sakamoto, N. Banno, N. Iguchi, H. Kawaura, S. Kaeriyama, M. Mizuno, K. Terabe, T. Hasegawa, and M. Aono
2005 International Electron Device Meeting, 489–492.
DOI: 10.1109/IEDM.2005.1609383

**2006**

10. "Template synthesis of M/$M_2S$ (M = Ag,Cu) hetero-nanowires by electrochemical technique"
C. H. Liang, K. Terabe, T. Hasegawa, and M. Aono

Solid State Ionics **177** (2006) 2527–2531.
DOI: 10.1016/j.ssi.2006.02.037
11. "Formation and metastable silver nanowire of hexagonal structure and their structure transformation under electron beam irradiation"
C. H. Liang, K. Terabe, T. Hasegawa, and M. Aono
Japanese Journal of Applied Physics **45** (2006) 6046–6048.
DOI: 10.1143/JJAP.45.6046
12. "Fabrication of nanoscale gaps using a combination of self-assemble molecular and electron beam lithographic technique"
R. Negishi, T. Hasegawa, K. Terabe, M. Aono, T. Ebihara, H. Tanaka, and T. Ogawa
Applied Physics Letters **88** (2006) 223112.
DOI: 10.1063/1.2209208
13. "Effects of sulfurization conditions and post deposition annealing treatment on the structural and electrical properties of silver sulfide films"
M. Kundu, K. Terabe, T. Hasegawa, and M. Aono
Journal of Applied Physics **99** (2006) 103501.
DOI: 10.1063/1.2199067
14. "Switching properties of atomic switch controlled by solid state electrochemical reaction"
T. Tamura, T. Hasegawa, K. Terabe, T. Nakayama, T. Sakamoto, H. Sunamura, H. Kawaura, S. Hosoki, and M. Aono
Japanese Journal of Applied Physics **45** (2006) L364–L366.
DOI: 10.1143/JJAP.45.L364
15. "Solid-electrolyte nanometer switch"
N. Banno, T. Sakamoto, N. Iguchi, H. Kawaura, S. Kaeriyama, M. Mizuno, K. Terabe, T. Hasegawa, and M. Aono
IEICE Transactions on Electronics **11** (2006) 1492–1498.
DOI: 10.1093/ietele/e89-c.11.1492
16. "Effect of ion diffusion on switching voltage of solid-electrolyte nanometer switch"
N. Banno, T. Sakamoto, T. Hasegawa, K. Terabe, and M. Aono
Japanese Journal of Applied Physics **45** (2006) 3666–3668.
DOI: 10.1143/JJAP.45.3666

**2007**

17. "Control of local ion transport to create unique functional nanodevices based on ionic conductors"
K. Terabe, T. Hasegawa, C. H. Liang, and M. Aono
Science and Technology of Advanced Materials **8** (2007) 536–542.
DOI: 10.1016/j.stam.2007.08.002
18. "Resistance switching of an individual $Ag_2S$/Ag nanowire heterostructure"
C. H. Liang, K. Terabe, T. Hasegawa, and M. Aono
Nanotechnology **18** (2007) 458202.
DOI: 10.1088/0957-4484/18/48/485202

19. "Anomalous phase transition and ionic conductivity of AgI nanowire grown using porous alumina"
C. H. Liang, K. Terabe, N. Iyi, T. Hasegawa, and M. Aono
Journal of Applied Physics **102** (2007) 124308.
DOI: 10.1063/1.2828141
20. "AgI/Ag heterojunction nanowire: Facile electrochemical synthesis, photoluminescence, and enhanced ionic conductivity"
C. H. Liang, K. Terabe, T. Tsuruoka, M. Osada, T. Hasegawa, and M. Aono
Advanced Functional Materials **17** (2007) 1466–1472.
DOI: 10.1002/adfm.200600590
21. "Material dependence of switching speed of atomic switches made from silver sulfide and from copper sulfide"
T. Tamura, T. Hasegawa, K. Terabe, T. Nakayama, T. Sakamoto, H. Sunamura, H. Kawaura, and M. Aono
Journal of Physics: Conference Series **61** (2007) 1157–1161.
DOI: 10.1088/1742-6596/61/1/229
22. "Electronic transport in $Ta_2O_5$ resistive switch"
T. Sakamoto, K. Lister, N. Banno, T. Hasegawa, K. Terabe, and M. Aono
Applied Physics Letters **90** (2007) 223112.
DOI: 10.1063/1.2777170
23. "A $Ta_2O_5$ solid-electrolyte switch with improved reliability"
T. Sakamoto, N. Banno, N. Iguchi, H. Kawaura, H. Sunamura, S. Fujieda, K. Terabe, T. Hasegawa, and M. Aono
2007 IEEE Symposium on VLSI Technology, 38–39.
DOI: 10.1109/VLSIT.2007.4339718
24. "Nonequilibrium quantum transport properties of a silver atomic switch"
Z. Wang, T. Kadohira, T. Tada, and S. Watanabe
Nano Letters **7** (2007) 2688–2699.
DOI: 10.1021/nl0711054

**2008**

25. "Resistance switching in Anodic oxidized amorphous $TiO_2$ films"
C. H. Ling, K. Terabe, T. Hasegawa, and M. Aono
Applied Physics Express **1** (2008) 064002.
DOI: 10.1143/APEX.1.064002
26. "Structural studies of copper sulfide films: Influence of ambient atmosphere"
M. Kundu, T. Hasegawa, K. Terabe, K. Yamamoto, and M. Aono
Science and Technology of Advanced Materials **9** (2008) 035011.
DOI: 10.1088/1468-6996/9/3/035011
27. "Effect of sulfurization condition on structural and electrical properties of copper sulfide films"
M. Kundu, T. Hasegawa, K. Terabe, and M. Aono
Journal of Applied Physics **103** (2008) 073523.
DOI: 10.1063/1.2903599

28. "A solid electrolyte nanometer switch"
    T. Sakamoto, S. Kaeriyama, M. Mizuno, H. Kawaura, T. Hasegawa, K. Terabe, and M. Aono
    Electrical Engineering in Japan **165** (2008) 68–73.
    DOI: 10.1002/eej.20542
29. "Diffusivity of Cu ions in solid electrolyte and its effect on the performance of nanometer-scale switch"
    N. Banno, T. Sakamoto, N. Iguchi, H. Sunamura, K. Terabe, T. Nakayama, and M. Aono
    IEEE Transaction on Electron Devices **55** (2008) 3283–3287.
    DOI: 10.1109/TED.2008.2004246
30. "On-state reliability of solid-electrolyte switch"
    N. Banno, T. Sakamoto, S. Fujieda, and M. Aono
    2008 IEEE International Reliability Physics Symposium, 707–708.
    DOI: 10.1109/RELPHY.2008.4558999
31. "Excess-silver-induced bridge formation in a silver sulfide atomic switch"
    Z. Wang, T. Gu, T. Tada, and S. Watanabe
    Applied Physics Letters **93** (2008) 152106.
    DOI: 10.1063/1.2963197

**2009**

32. "Nanoionics switching devices: Atomic switches
    T. Hasegawa, K. Terabe, T. Sakamoto, and M. Aono
    MRS Bulletin **34** (2009) 929–934.
    DOI: 10.1557/mrs2009.215
33. "Nonvolatile solid-electrolyte switch embedded into Cu interconnect"
    T. Sakamoto, M. Tada, N. Banno, Y. Tsuji, Y. Saitoh, Y. Yabe, H. Hada, N. Iguchi, and M. Aono
    2009 Symposium on VLSI Technology, 130–131.
34. "Cu-ion diffusivity in $SiO_2$-$Ta_2O_5$ solid electrolyte and its impact of the yield of resistance switching after BEOL processes"
    N. Banno, T. Sakamoto, H. Hada, N. Kasai, N. Iguchi, S. Fujieda, T. Ichihashi, T. Hasegawa, and M. Aono
    2009 IEEE International Reliability Physics Symposium, 359–399.
    DOI: 10.1109/IRPS.2009.5173285
35. "Highly scalable nonvolatile $TiO_x$/$TaSiO_y$ solid-electrolyte crossbar switch integrated in local interconnect for low power reconfigurable logic"
    M. Tada, T. Sakamoto, Y. Tsuji, N. Banno, Y. Saitoh, Y. Yabe, S. Ishida, M. Terai, S. Kotsuji, N. Iguchi, M. Aono, H. Hada, and N. Kasai
    2009 IEEE International Electron Device Meeting, 1–4.
    DOI: 10.1109/IEDM.2009.5424287
36. "First-principles simulations on bulk $Ta_2O_5$ and Cu/$Ta_2O_5$/Pt heterojunction: Electronic structures and transport properties"
    T. Gu, Z. Wang, T. Tada, and S. Watanabe
    Journal of Applied Physics **106** (2009) 103713.
    DOI: 10.1063/1.3260244

37. "Nonstoichiometry-induced carrier modification in gapless type atomic switch device using $Cu_2S$ mixed conductor"
T. Tsuchiya, Y. Oyama, S. Miyoshi, and S. Yamaguchi
Applied Physics Express **2** (2009) 055002.
DOI: 10.1143/APEX.2.055002

**2010**

38. "Toward sub-20nm hybrid nanofabrication by combining the molecular method and electron beam lithography"
C. B. Li, T. Hasegawa, H. Tanaka, H. Miyazaki, S. Osada, K. Tsukagoshi, and M. Aono
Nanotechnology **21** (2010) 495304.
DOI: 10.1088/0957-4484/21/49/495304
39. "Rate-limiting processes determining the switching time of a $Ag_2S$ atomic switch"
A. Nayak, T. Tamura, T. Tsuruoka, K. Terabe, S. Hosaka, T. Hasegawa, and M. Aono
Journal of Physical Chemistry Letters **1** (2010) 601–608.
DOI: 10.1021/jz900375a
40. "Learning abilities achieved by a single solid-state atomic switch"
T. Hasegawa, T. Ohno, K. Terabe, T. Tsuruoka, T. Nakayama, J. K. Gimzewski, and M. Aono
Advanced Materials **22** (2010) 1381–1384.
DOI: 10.1002/adma.200903680
41. "Photo-assisted formation of atomic switch"
T. Hino, H. Tanaka, T. Hasegawa, M. Aono, and T. Ogawa
Small **6** (2010) 1745–1748.
DOI: 10.1002/smll.201000472
42. "Forming and switching mechanisms of a cation-migration-based oxide resistive memory"
T. Tsuruoka, K. Terabe, T. Hasegawa, and M. Aono
Nanotechnology **21** (2010) 425205.
DOI: 10.1088/0957-4484/21/42/425205
43. "The atomic switch"
M. Aono and T. Hasegawa
Proceedings of the IEEE **98** (2010) 2228–2236.
DOI: 10.1109/JPROC.2010.2061830
44. "Off-state and turn-on characteristics of solid electrolyte switch"
Y. Tsuji, T. Sakamoto, N. Banno, H. Hada, and M. Aono
Applied Physics Letters **96** (2010) 023504.
DOI: 10.1063/1.328577
45. "Structural characterization of amorphous $Ta_2O_5$ and $SiO_2$-$Ta_2O_5$ used as solid electrolyte for nonvolatile switches"
N. Banno, T. Sakamoto, N. Iguchi, M. Matsumoto, H. Imai, T. Ichihashi, S. Fujieda, S. Watanabe, S. Yamaguchi, T. Hasegawa, and M. Aono

Applied Physics Letters **97** (2010) 113507.
DOI: 10.1063/1.3488830

46. "Nonvolatile crossbar switch using $TiO_x/TaSiO_y$ solid electrolyte"
M. Tada, T. Sakamoto, N. Banno, M. Aono, H. Hara, and N. Kasai
IEEE Transactions on Electron Devices **57** (2010) 1987–1995.
DOI: 10.1109/TED.2010.2051191
47. "Polymer solid-electrolyte (PSE) switch embedded in 90 nm CMOS with forming-free and 10 nsec programming for low power, nonvolatile programmable logic (NPL)"
M. Tada, T. Sakamoto, K. Okamoto, M. Miyamura, N. Banno, Y. Katou, S. Ishida, N. Iguchi, N. Sakimura, and H. Hada
2010 International Electron Devices Meeting, 16.5.1–16.5.4.
DOI: 10.1109/IEDM.2010.5703376
48. "Reliable solid-electrolyte crossbar switch for programmable logic"
N. Banno, T. Sakamoto, M. Tada, M. Miyamura, Y. Tabe, Y. Saitoh, S. Ishida, K. Okamoto, H. Hada, and N. Kasai
2010 Symposium on VLSI Technology: Digest of Technical Papers (2010) 115–116.
DOI: 10.1109/VLSIT.2010.5556192
49. "Conductive path formation in the $Ta_2O_5$ atomic switch: First-principle analysis"
T. Gu, T. Tada, and S. Watanabe
ACS Nano **4** (2010) 6477–6482.
DOI: 10.1021/nn101410s
50. "Numerical simulation of switching behavior in $Cu/Cu_2S$ nanometer-scale switch"
Y. Okajima, Y. Shibuta, T. Tsuchiya, S. Yamaguchi, and T. Suzuki
Applied Physics Express **3** (2010) 065202.
DOI: 10.1143/APEX.3.065202

**2011**

51. "Short-term plasticity and long-term potentiation in a single $Ag_2S$ inorganic synapse"
T. Ohno, T. Hasegawa, T. Tsuruoka, K. Terabe, J. K. Gimzewski, and M. Aono
Nature Materials **10** (2011) 591–595.
DOI: 10.1038/NMAT3054
52. "Volatile/nonvolatile dual-functional atom transistor"
T. Hasegawa, Y. Itoh, H. Tanaka, T. Hino, T. Tsuruoka, K. Terabe, H. Miyazaki, K. Tsukagoshi, T. Ogawa, S. Yamaguchi, and M. Aono
Applied Physics Express **4** (2011) 015204.
DOI: 10.1143/APEX.4.015204
53. "A polymer-electrolyte-based atomic switch"
S. Wu, T. Tsuruoka, K. Terabe, T. Hasegawa, J. P. Hill, K. Ariga, and M. Aono

Advanced Functional Materials **21** (2011) 93–99.
DOI: 10.1002/adfm.201001520

54. "Memristive operations demonstrated by gap-type atomic switches"
T. Hasegawa, A. Nayak, T. Ohno, K. Terabe, T. Tsuruoka, J. K. Gimzewski, and M. Aono
Applied Physics A **102** (2011) 811–815.
DOI: 10.1007/s00339-011-6317-0
55. "Temperature effects on the switching kinetics of a Cu-$Ta_2O_5$-based atomic switch"
T. Tsuruoka, K. Terabe, T. Hasegawa, and M. Aono
Nanotechnology **22** (2011) 254013.
DOI: 10.1088/0957-4484/22/25/254013
56. "Atomic switches: Atomic-movement controlled nanodevices for new types of computing"
T. Hino, T. Hasegawa, K. Terabe, T. Tsuruoka, A. Nayak, T. Ohno, and M. Aono
Science and Technology of Advanced Materials **12** (2011) 013003.
DOI: 10.1088/1468-6996/12/1/013003
57. "Switching kinetics of a $Cu_2S$ gap-type atomic switch"
A. Nayak, T. Tsuruoka, K. Terabe, T. Hasegawa, and M. Aono
Nanotechnology **22** (2011) 235201.
DOI: 10.1088/0957-4484/22/23/235201
58. "Theoretical investigation of kinetics of a $Cu_2S$-based gap-type atomic switch"
A. Nayak, T. Tsuruoka, K. Terabe, T. Hasegawa, and M. Aono
Applied Physics Letters **98** (2011) 233501.
DOI: 10.1063/1.3597154
59. "Sensory and short-term memory formations observed in an $Ag_2S$ gap-type atomic switch"
T. Ohno, T. Hasegawa, A. Nayak, T. Tsuruoka, J. K. Gimzewski, and M. Aono
Applied Physics Letters **99** (2011) 203108.
DOI: 10.1063/1.3662390
60. "Bulk and surface nucleation processes in $Ag_2S$ conductance switches"
M. Morales-Masis, S. J. van den Molen, T. Hasegawa, and J. M. van Ruitenbeek
Physical Review B **84** (2011) 115310.
DOI: 10.1103/PhysRevB.84.115310
61. "Three-terminal nanometer metal switches utilizing solid electrolytes"
H. Kawaura, T. Sakamoto, N. Banno, S. Kaeriyama, M. Mizuno, K. Terabe, T. Hasegawa, and M. Aono
Electronics and Communications in Japan **94** (2011) 55–61.
DOI: 10.1002/ecj.10214
62. "Polymer solid-electrolyte switch embedded on CMOS for nonvolatile crossbar switch"
M. Tada, K. Okamoto, T. Sakamoto, M. Miyamura, N. Banno, and H. Hada

IEEE Transactions on Electron Devices **58** (2011) 4398–4406.
DOI: 10.1109/TED.2011.2169070

63. "On-state reliability of solid-electrolyte switch under pulsed alternating current stress for programmable logic device"
N. Banno, T. Sakamoto, M. Tada, M. Miyamura, K. Okamoto, H. Hada, and M. Aono
Japanese Journal of Applied Physics **50** (2011) 074201.
DOI: 10.1143/JJAP.50.074201
64. "Conducting mechanism of atom switch with polymer solid-electrolyte"
K. Okamoto, M. Tada, T. Sakamoto, M. Miyamura, N. Banno, N. Iguchi, and H. Hada
2011 International Electron Devices Meeting, 12.2.1–12.2.4.
DOI: 10.1109/IEDM.2011.6131538
65. "Highly reliable, complementary atom switch (CAS) with low programming voltage embedded in Cu BEOL for nonvolatile programmable logic"
M. Tada, T. Sakamoto, M. Miyamura, N. Banno, K. Okamoto, N. Iguchi, and H. Hada
2011 International Electron Devices Meeting, 30.2.1–30.2.4.
DOI: 10.1109/IEDM.2011.6131642
66. "Programmable cell array using rewritable solid-electrolyte switch integrated in 90 nm CMOS"
M. Miyamura, S. Nakaya, M. Tada, T. Sakamoto, K. Okamoto, N. Banno, S. Ishida, K. Ito, H. Hada, N. Sakimura, T. Sugibayashi, and M. Motomura
2011 International Solid-State Circuits Conference, 228–229.
DOI: 10.1109/SSCC.2011.5746296
67. "X-ray absorption, photoemission spectroscopy, and Raman scattering analysis of amorphous tantalum oxide with large extent of oxygen nonstoichiometry"
T. Tsuchiya, H. Imai, S. Miyoshi, P. -A. Glans, J. Guo, and S. Yamaguchi
Physical Chemistry Chemical Physics **13** (2011) 17013–17018.
DOI: 10.1039/clcp21310e

**2012**

68. "Atomic Switch; the atom/ion movement controlled device for beyond von-Neumann computer"
T. Hasegawa, K. Terabe, T. Tsuruoka, and M. Aono
Advanced Materials **24** (2012) 252–267.
DOI: 10.1002/adma.201102597
69. "Effects of moisture on the switching characteristics of oxide-based, gapless-type atomic switches"
T. Tsuruoka, K. Terabe, T. Hasegawa, I. Valov, R. Waser, and M. Aono
Advanced Functional Materials **22** (2012) 70–77.
DOI: 10.1002/adfm.201101846
70. "Controlling the synaptic plasticity of a $Cu_2S$ atomic switch"
A. Nayak, T. Ohno, T. Tsuruoka, K. Terabe, T. Hasegawa, J. K. Gimzewski, and M. Aono

Advanced Functional Materials **22** (2012) 3606–3613.
DOI: 10.1002/adfm.201200640

71. "Flexible resistive switching memory using inkjet printing of a solid polymer electrolyte"
S. R. Mohapatra, T. Tsuruoka, T. Hasegawa, K. Terabe, and M. Aono
AIP Advances **2** (2012) 022144.
DOI: 10.1063/1.4727742
72. "Conductance quantization and synaptic behavior of a $Ta_2O_5$-based atomic switch"
T. Tsuruoka, T. Hasegawa, K. Terabe, and M. Aono
Nanotechnology **23** (2012) 435705.
DOI: 10.1088/0957-4484/23/43/435705
73. "Impacts of temperature and moisture on the resistive switching characteristics of a Cu-$Ta_2O_5$-based atomic switch"
T. Tsuruoka, T. Hasegawa, K. Terabe, and M. Aono
Mater. Res. Soc. Symp. Proc. **1430** (2012).
DOI: 10.1557/opl.2012.901
74. "Flexible polymer atomic switches using ink-jet printing technique"
S. R. Mohapatra, T. Tsuruoka, T. Hasegawa, K. Terabe, and M. Aono
Mater. Res. Soc. Symp. Proc. **1430** (2012).
DOI: 10.1557/opl.2012.1022
75. "Atomically controlled electrochemical nucleation at superionic solid electrolyte surfaces"
I. Valov, I., Sapezanskaia, A. Nayak, T. Tsuruoka, T. Bredow, T. Hasegawa, G. Staikov, M. Aono, and R. Waser
Nature Materials **11** (2012) 530–535.
DOI: 10.1038/NMAT3307
76. "Oxygen migration process in the interfaces during bipolar resistive switching behavior of the $WO_{3-x}$-based nanoionics devices"
R. Yang, K. Terabe, T. Tsuruoka, T. Hasegawa, and M. Aono
Applied Physics Letters **100** (2012) 231603.
DOI: 10.1063/1.4726084
77. "On-demand nanodevice with electrical and neuromorphic multifunction realized by local ion migration"
R. Yang, K. Terabe, G. Liu, T. Tsuruoka, T. Hasegawa, J. K. Gimzewski, and M. Aono
ACS Nano **6** (2012) 9515–9521.
DOI: 10.1021/nn302510e
78. "Improved OFF-state reliability of nonvolatile resistive switch with low programming voltage"
M. Tada, T. Sakamoto, M. Miyamura, N. Banno, K. Okamoto, N. Iguchi, and H. Hada
IEEE Transactions on Electron Devices **59** (2012) 2357–2362.
DOI: 10.1109/TED.2012.2204263

79. "First demonstration of logic mapping on nonvolatile programmable cell using complementary atom switch"
M. Miyamura, M. Tada, T. Sakamoto, N. Banno, K. Okamoto, N. Iguchi, and H. Hada
2012 International Electron Devices Meeting, 10.6.1–10.6.4.
DOI: 10.1109/IEDM.2012.6479020
80. "Improved reliability and switching performance of atom switch by using ternary Cu-alloy and RuTa electrodes"
M. Tada, T. Sakamoto, N. Banno, K. Okamoto, M. Miyamura, N. Iguchi, and H. Hada
2012 International Electron Devices Meeting, 29.8.1–29.8.4.
DOI: 10.1109/IEDM.2012.6479133
81. "Nonvolatile 32x32 crossbar atom switch block integrated on a 65-nm COMS platform"
N. Banno, M. Tada, T. Sakamoto, K. Okamoto, M. Miyamura, N. Iguchi, and H. Hada
2012 Symposium on VLSI Technology, 39–40.
DOI: 10.1109/VLSIT.2012.6242450
82. "Neuromorphic atomic switch networks"
A. V. Avizienis, H. O. Sillin, C. Martin-Olmos, H. H. Shieh, M. Aono, A. Z. Stieg, and J. K. Gimzewski
PLOS ONE **7** (2012) e42772.
DOI: 10.1371/journal.pone.0042772
83. "Emergent criticality in complex tuning B-type atomic switch networks"
A. Z. Stieg, A. V. Avizienis, H. O. Sillin, C. Martin-Olmos, M. Aono, and J. K. Gimzewski
Advanced Materials **24** (2012) 286–293.
DOI: 10.1002/adma.201103053

**2013**

84. "Rate-limiting processes in the fast SET operation of a gapless-type Cu-$Ta_2O_5$ atomic switch"
T. Tsuruoka, T. Hasegawa, I. Valov, R. Waser, and M. Aono
AIP Advances **3** (2013) 032114.
DOI: 10.1063/1.4795140
85. "Volatile and nonvolatile selective switching of a photo-assisted atomic switch"
T. Hino, T. Hasegawa, H. Tanaka, T. Tsuruoka, K. Terabe, T. Ogawa, and M. Aono
Nanotechnology **24** (2013) 384006.
DOI: 10.1088/0957-4484/24/38/384006
86. "Generic relevance of counter charges for cation-based nanoscale resistive switching memories"
S. Tappertzhofen, I. Valov, T. Tsuruoka, T. Hasegawa, R. Waser, and M. Aono
ACS Nano **7** (2013) 6396–6402.
DOI: 10.1021/nn4026614

87. "Nonvolatile three-terminal operation based on oxygen vacancy drift in a Pt/$Ta_2O_{5-x}$/Pt,Pt structure"
Q. Wang, Y. Itoh, T. Hasegawa, T. Tsuruoka, S. Watanabe, S. Yamaguchi, and M. Aono
Applied Physics Letters **102** (2013) 233508.
DOI: 10.1063/1.4811122
88. "Quantized conductance and neuromorphic behavior of a gapless-type Ag-$Ta_2O_5$ atomic switch"
T. Tsuruoka, T. Hasegawa, K. Terabe, and M. Aono
Mater. Res. Soc. Symp. Proc. **1562** (2013).
DOI: 10.1557/opl.2013.719
89. "Synaptic plasticity and memory functions achieved in $WO_{3-x}$-based nanoionics device by using principle of atomic switch operation"
R. Yang, K. Terabe, Y. Yao, T. Tsuruoka, T. Hasegawa, J. K. Gimzewski, and M. Aono
Nanotechnology **24** (2013) 384003.
DOI: 10.1088/0957-4484/24/38/384003
90. "Improved ON-state reliability of atom switch using alloy electrodes"
M. Tada, T. Sakamoto, N. Banno, K. Okamoto, N. Iguchi, H. Hada, and M. Miyamura
IEEE Transactions on Electron Devices **60** (2013) 3534–3540.
DOI: 10.1109/TED.2013.2275188
91. "Bidirectional TaO-diode-selected, complementary atom switch (DCAS) for area-efficient, nonvolatile crossbar switch block"
K. Okamoto, M. Tada, N. Banno, T. Sakamoto, M. Miyamura, N. Iguchi, and H. Hada
2013 Symposium on VLSI Technology, T242–T243.
92. "Room temperature RedOx reaction by oxide ion migration at carbon/Gd-doped $CeO_2$ hetero-interface probed by in-situ hard X-ray photoemission and soft X-ray absorption spectroscopy"
T. Tsuchiya, S. Miyoshi, Y. Yamashita, H. Yoshikawa, K. Terabe, K. Kobayashi, and S. Yamaguchi
Science and Technology of Advanced Materials **14** (2013) 045001.
DOI: 10.1088/1468-6996/14/4/045001
93. "Direct observation of redox state modulation at carbon/amorphous tantalum oxide thin film hetero-interface probed by in-situ hard X-ray photoemission spectroscopy"
T. Tsuchiya, S. Miyoshi, Y. Yamashita, H. Yoshikawa, K. Terabe, K. Kobayashi, and S. Yamaguchi
Solid State Ionics **253** (2013) 110–118.
DOI: 10.1016/j_ssi_2013.09.015
94. "All-solid-state electric-double-layer transistor based on oxide ion migration in Gd-doped $CeO_2$ on $SrTiO_3$ single crystal"
T. Tsuchiya, K. Terabe, and M. Aono
Applied Physics Letters **103** (2013) 073110.
DOI: 10.1063/1.4818736

95. "Morphological transitions from dendrites to nanowires in the electroless deposition of silver"
A. V. Avizienis, C. Martin-Olmos, H. O. Sillin, M. Aono, J. K. Gimzewski, and Adam Z. Stieg
Crystal Growth Design **13** (2013) 465–469.
DOI: 10.1021/cg301692n
96. "A theoretical and experimental study of neuromorphic atomic switch networks for reservoir computing"
H. O. Sillin, R. Aguilera, H. H. Shieh, A. V. Avizienis, M. Aono, A. Z. Stieg, and J. K. Gimzewski
Nanotechnology **24** (2013) 384004.
DOI: 10.1088/0957-4484/24/38/384004

**2014**

97. "Two types of on-sate observed in the operation of redox-based three-terminal device"
Q. Wang, Y. Itoh, T. Tsuruoka, T. Hasegawa, S. Watanabe, S. Yamaguchi, and M. Aono
Key Engineering Material **596** (2014) 111–115.
DOI: 10.4028/www.scientific.net/KEM.596.111
98. "Influence of atmosphere on photo-assisted atomic switch operations"
T. Hino, T. Hasegawa, H. Tanaka, T. Tsuruoka, T. Ogawa, and M. Aono
Key Engineering Material **596** (2014) 116–120.
DOI: 10.4028/www.scientific.net/KEM.596.116
99. "Impact of overshoot current on set operation of atom switch"
T. Sakamoto, M. Tada, M. Miyamura, N. Banno, K. Okamoto, N. Iguchi, and H. Hada
Japanese Journal of Applied Physics **53** (2014) 04ED07.
DOI: 10.7567/JJAP.53.04ED07
100. "Low-power programmable-logic cell arrays using nonvolatile complementary atom switch"
M. Miyamura, T. Sakamoto, M. Tada, N. Banno, K. Okamoto, N. Iguchi, and H. Hada
Fifteenth International Symposium on Quality Electronic Design 2014, 330–334.
DOI: 10.1109/ISQED.2014.6783344
101. "A fast and low-voltage Cu complementary-atom-switch 1Mb array with high-temperature retention"
N. Banno, M. Tada, T. Sakamoto, M. Miyamura, K. Okamoto, N. Iguchi, and H. Hada
2014 Symposium on VLSI Technology, 1–2.
DOI: 10.1109/VLSIT.2014.6894437

102. "Improved switching voltage variation of Cu atom switch for nonvolatile programmable logic"
N. Banno, M. Tada, T. Sakamoto, K. Okamoto, M. Miyamura, N. Iguchi, and H. Hada
IEEE Transactions on Electron Devices **61** (2014) 3827–3832.
DOI: 10.1109/TED.2014.2355830
103. "Three-terminal nonvolatile resistive-change device integrated in Cu-BEOL"
M. Tada, K. Okamoto, N. Banno, T. Sakamoto, and H. Hada
IEEE Transactions on Electron Devices **61** (2014) 505–510.
DOI: 10.1109/TED.2013.2296036
104. "Atom switch technology for low-power nonvolatile logic application"
M. Tada, T. Sakamoto, M. Miyamura, N. Banno, K. Okamoto, and H. Hada
ECS Transactions **61**(6) (2014) 57–64.
DOI: 10.1149/06106.0057ecst
105. "Conduction paths in Cu/amorphous-$Ta_2O_5$/Pt atomic switch: First-principles studies"
B. Xiao, T. Gu, T. Tada, and S. Watanabe
Journal of Applied Physics **115** (2014) 034503.
DOI: 10.11063/1.4861724
106. "In-situ monitoring of oxide ion induced breakdown in amorphous tantalum oxide thin film using acoustic emission measurement"
T. Tsuchiya, K. Ito, S. Miyoshi, M. Enoki, and S. Yamaguchi
Materials Transactions **55** (2014) 1553.
DOI: 10.2320/matertrans.M2014198
107. "Oxygen vacancy effects on an amorphous-$TaO_x$-based resistive switch: A first principles studies"
B. Xiao and S. Watanabe
Nanoscale **6** (2014) 10169–10178.
DOI: 10.1039/C4NR02173H
108. "In situ and non-volatile bandgap tuning of multilayer graphene oxide in an all-solid-state electric double-layer transistor"
T. Tsuchiya, K. Terabe, and M. Aono
Advanced Materials **26** (2014) 1087–1091.
DOI: 10.1002/adma.20134770
109. "Micro x-ray photoemission and Raman spectroscopy studies on bandgap tuning of graphene oxide achieved by solid state ionics devices"
T. Tsuchiya, K. Terabe, and M. Aono
Applied Physics Letters **105** (2014) 183101.
DOI: 10.1063/1.4901103
110. "Self-organized atomic switch networks"
A. Z. Stieg, A. V. Avizienis, H. O. Sillin, C. Martin-Olmos, M. L. Lam, M. Aono, and J. K. Gimzewski
Japanese Journal of Applied Physics **53** (2014) 01AAA02.
DOI: 10.7567/JJAP.53.001AA02

## 2015

111. "Dynamic moderation of an electric field using a $SiO_2$ tunneling layer in $TaO_x$-based ReRAM"
Q. Wang, Y. Itoh, T. Tsuruoka, S. Ohtsuka, T. Shimizu, S. Shingubara, T. Hasegawa, and M. Aono
Physica Status Solidi—Rapid Research Letters **9** (2015) 166–170.
DOI: 10.1002/pssr.201409531
112. "Position detection and observation of a conducting filament hidden under a top electrode in a $Ta_2O_5$-based atomic switch"
A. Nayak, Q. Wang, Y. Itoh, T. Tsuruoka, T. Hasegawa, L. Boodhoo, H. Mizuta, and M. Aono
Nanotechnology **26** (2015) 145702.
DOI: 10.1088/0957-4484/26/14/145702
113. "Redox reactions at Cu,Ag/$Ta_2O_5$ interfaces and the effects of $Ta_2O_5$ film density on the forming process in atomic switch structures"
T. Tsuruoka, I. Valov, S. Tappertzhofen, J. van den Hurk, T. Hasegawa, R. Waser, and M. Aono
Advanced Functional Materials **25** (2015) 6347–6381.
DOI: 10.1002/adfm.201500853
114. "Effects of temperature and ambient pressure on the resistive switching behavior of a polymer-based atomic switch"
S. R. Mohapatra, T. Tsuruoka, K. Krishnan, T. Hasegawa, and M. Aono
Journal of Materials Chemistry C **3** (2015) 5715–5720.
DOI: 10.1039/c5tc00842e
115. "Ultra-low voltage and ultra-low power consumption nonvolatile operation of a three-terminal atomic switch"
Q. Wang, Y. Itoh, T. Tsuruoka, M. Aono, and T. Hasegawa
Advanced Materials **27** (2015) 6029–6033.
DOI: 10.1002/adma.201502678
116. "Nanosecond fast switching processes observed in gapless-type, $Ta_2O_5$-based atomic switches"
T. Tsuruoka, T. Hasegawa, and M. Aono
Mater. Res. Soc. Symp. Proc. **1729** (2015).
DOI: 10.1557/opl.2015.93
117. "Interface structure in Cu/$Ta_2O_5$/Pt resistance switch: A first-principle study"
B. Xiao and S. Watanabe
ACS Applied Materials & Interfaces **7** (2015) 519–525.
DOI: 10.1021/am5066656
118. "Observation of a Ag protrusion on a $Ag_2S$ island using a scanning tunneling microscope"
T. Ohno and T. Hasegawa
Results in Physics **5** (2015) 182–183.
DOI: 10.1016/j.rinp.2015.08.004

119. "Low-power embedded read-only memory using atom switch and silicon-on-thin-buried-oxide transistor"
T. Sakamoto, M. Tada, Y. Tsuji, H. Makiyama, T. Hasegawa, Y. Yamamoto, S. Okanishi, N. Banno, M. Miyamura, K. Okamoto, N. Iguchi, Y. Ogasahara, H. Oda, S. Kamohara, Y. Yamagata, N. Sugii, and H. Hada
Applied Physics Express **8** (2015) 045201.
DOI: 10.7567/APEX.8.045201

120. "Mechanism of OFF-state lifetime improvement in complementary atom switch"
N. Banno, M. Tada, T. Sakamoto, H. Makiyama, K. Okamoto, N. Iguchi, T. Nohisa, and H. Hada
Japanese Journal of Applied Physics **54** (2015) 04DD08.
DOI: 10.7567/JJAP.54.059201

121. "Logic comparable process technology for embedded atom switch in CMOS"
K. Okamoto, M. Tada, N. Banno, N. Iguchi, T. Sakamoto, and H. Hada
Japanese Journal of Applied Physics **54** (2015) 05ED05.
DOI: 10.7567/JJAP.54.05ED05

122. "Cu atom switch with steep time-to-ON-state versus switching voltage using Cu ionization control"
N. Banno, M. Tada, T. Sakamoto, H. Makiyama, K. Okamoto, N. Iguchi, and H. Hada
IEEE Transactions on Electron Devices **62** (2015) 2966–2971.
DOI: 10.1109/TED.2015.2443120

123. "0.39-V, 18.26-μW/MHz SOTB CMOS Microcontroller with embedded atom switch ROM"
T. Sakamoto, Y. Tsuji, M. Tada, H. Miyamaki, T. Hasegawa, Y. Yamamoto, S. Okanishi, K. Maekawa, N. Banno, M. Miyamura, K. Okamoto, N. Iguchi, Y. Ogasahara, H. Oda, S. Kamohara, Y. Yamagata, N. Sugii, and H. Hada
2015 IEEE Symposium in Low-Power and High-Speed Chips, 1–3.
DOI: 10.1109/CoolChips.2015.7158658

124. "Sub-μW standby power, <18μW/DMIPS@25MHz MCU with embedded atom-switch programmable logic and ROM"
Y. Tsuji, X. Bai, M. Miyamura, T. Sakamoto, M. Tada, N. Banno, K. Okamoto, N. Iguchi, N. Sugii, and H. Hada
2015 Symposium on VLSI Circuits, T86–T87.
DOI: 10.1109/VLSIC.2015.7231363

125. "A novel two-varistors (a-Si/SiN/a-Si) selected complementary atom switch (2V-1CAS) for nonvolatile crossbar switch with multiple fan-outs"
N. Banno, M. Tada, K. Okamoto, N. Iguchi, T. Sakamoto, H. Makiyama, Y. Tsuji, H. Hada, H. Ochi, H. Onodera, M. Hashimoto, and T. Sugibayashi
2015 IEEE International Electron Devices Meeting (IDEM), 2.5.1–2.5.4.
DOI: 10.1109/IEDM.2015.7409614

126. "A silicon-on-thin-buried-oxide CMOS microcontroller with embedded atom-switch ROM"
T. Sakamoto, Y. Tsuji, M. Tada, H. Makiyama, T. Hasegawa, Y. Yamamoto, S. Okanishi, K. Maekawa, N. Banno, M. Miyamura, K. Okamoto, N. Iguchi, H. Oda, S. Kamohara, Y. Yamagata, N. Sugii, H. Hada, and Y. Ogasahara
IEEE Micro **35** (2015) 13–23.
DOI: 10.1109/MM.2015.142
127. "Effect of ionic conductivity on response speed of $SrTiO_3$-based all-solid-state electric-double-layer transistor"
T. Tsuchiya, M. Ochi, T. Higuchi, K. Terabe, and M. Aono
ACS Applied Materials & Interfaces **7** (2015) 12254–12260.
DOI: 10.1021/acsami.5b02998
128. "In situ and non-volatile photoluminescence tuning and nanodomain writing demonstrated by all-solid-state devices based on graphene oxide"
T. Tsuchiya, T. Tsuruoka, K. Terabe, and M. Aono
ACS Nano **9** (2015) 2102–2110.
DOI: 10.1021/nn507363g
129. "Atomic switch networks—nanoarchitectonic design of a complex system of natural computing"
E. C. Demis, R. Aguilera, H. O. Sillin, K. Scharnhorst, E. J. Sandouk, M. Aono, A. Z. Stieg, and J. K. Gimzewski
Nanotechnology **26** (2015) 204003.
DOI: 10.1088/0957-4484/26/204003
130. "Resistive switching in a few nanometers thick tantalum oxide film formed by a metal oxidation"
T. Ohno and S. Samukawa
Applied Physics Letters **106** (2015) 173110.
DOI: 10.1063/1.4919724

**2016**

131. "Mechanism for conducting filament growth in self-assembled polymer thin films for redox-based atomic switches"
K. Krishnan, T. Tsuruoka, C. Mannequin, and M. Aono
Advanced Materials **28** (2016) 640–648.
DOI: 10.1002/adma.201504202
132. "Direct observation of anodic dissolution and filament growth behavior in polyethylene-oxide-based atomic switch structures"
K. Krishnan, T. Tsuruoka, and M. Aono
Japanese Journal of Applied Physics **55** (2016) 06GK02.
DOI: 10.7567/JJAP.55.06GK02
133. "Humidity effects on the redox reactions and ionic transport in a $Cu/Ta_2O_5/Pt$ atomic switch structure"
T. Tsuruoka, I. Valov, C. Mannequin, T. Hasegawa, R. Waser, and M. Aono

Japanese Journal of Applied Physics **55** (2016) 06GJ09.
DOI: 10.7567/JJAP.55.06GJ09

134. "Composition of thin $Ta_2O_5$ films deposited by different methods and the effect of humidity on their resistive switching behavior"
C. Mannequin, T. Tsuruoka, T. Hasegawa, and M. Aono
Japanese Journal of Applied Physics **55** (2016) 06GG08.
DOI: 10.7567/JJAP.55.06GG08

135. "Decision maker based on atomic switches"
S. -J. Kim, T. Tsuruoka, T. Hasegawa, M. Aono, K. Terabe, and M. Aono
AMIS Materials Science **3** (2016) 245–259.
DOI: 10.3934/matersci.2016.1.245

136. "Identification and roles of nonstoichiometric oxygen in amorphous $Ta_2O_5$ thin films deposited by electron beam and sputtering processes"
C. Mannequin, T. Tsuruoka, T. Hasegawa, and M. Aono
Applied Surface Science **385** (2016) 426–435.
DOI: 10.1016/j.apsusc.2016.04.099

137. "Kinetic factors determining conducting filament formation in solid polymer electrolyte based planar devices"
K. Krishnan, M. Aono, and T. Tsuruoka
Nanoscale **8** (2016) 13976–13984.
DOI: 10.1039/c6nr00569a

138. "Area-efficient nonvolatile carry chain based on pass-transistor/atom-switch hybrid logic"
X. Bai, Y. Tsuji, T. Sakamoto, A. Morioka, M. Miyamura, M. Tada, N. Banno, K. Okamoto, N. Iguchi, and H. Hada
Japanese Journal of Applied Physics **55** (2016) 04EF01.
DOI: 10.7567/JJAP.55.04EF01

139. "Nanoionic devices enabling a multitude of new features"
K. Terabe, T. Tsuchiya, R. Yang, and M. Aono
Nanoscale **8** (2016) 13873–13879.
DOI: 10.1039/c6nr00956e

140. "In situ tuning of magnetization and magnetoresistance in $Fe_3O_4$ thin film achieved with all-solid-state redox devices"
T. Tsuchiya, K. Terabe, M. Ochi, T. Higuchi, M. Osada, Y. Yamashita, S. Ueda, and M. Aono
ACS Nano **10** (2016) 1655–1661.
DOI: 10.1021/acsnano.5b07374

141. "Comparison of subthreshold swing in $SrTiO_3$-based all-solid-state electric-double-layer transistors wit $Li_4SiO_4$ or Y-stabilized $ZrO_2$ solid electrolyte"
T. Tsuchiya, M. Ochi, T. Higuchi, and K. Terabe
Japanese Journal of Applied Physics **55** (2016) 06GJ03.
DOI: 10.7567/JJAP.55.06GJ03

142. "Nanoionic devices: Interface nanoarchitectonics for physical property tuning and enhancement"
T. Tsuchiya, K. Terabe, R. Yang, and M. Aono

Japanese Journal of Applied Physics **55** (2016) 1102A4.
DOI: 10.7567/JJAP.55.1102A4

143. "$Ag_2S$ atomic switch-based 'tug of war' for decision making"
C. Lutz, T. Hasegawa, and T. Chikyow
Nanoscale **8** (2016) 14031–14036.
DOI: 10.1039/c6nr00690f

144. "Area-efficient nonvolatile carry chain based on pass-transistor/atom-switch hybrid logic"
X. Bai, Y. Ysuji, T. Sakamoto, A. Morioka, M. Miyamura, M. Tada, N. Banno, K. Okamoto, N. Iguchi, and H. Hada
Japanese Journal of Applied Physics **55** (2016) 04EF01.
DOI: 10.7567/JJAP.55.04EF01

145. "Robust Cu atom switch with over-400°C thermally tolerant polymer-solid electrolyte (TT-PSE) for nonvolatile programmable logic"
K. Okamoto, M. Tada, N. Banno, N. Iguchi, H. Hada, T. Sakamoto, M. Miyamura, Y. Tsuji, R. Nebashi, A. Morioka, X. Bao, and T. Sugibayashi
2016 IEEE Symposium on VLSI Technology, 1–2.
DOI: 10.1109/VLSIT.2016.7573403

146. "A 2x logic density programmable logic array using atom switch fully implemented with logic transistors at 40 nm-node and beyond"
Y. Tsuji, X. Bai, A. Morioka, M. Miyamura, R. Nebashi, T. Sakamoto, M. Tada, N. Banno, K. Okamoto, N. Iguchi, H. Hada, and T. Sugibayashi
2016 IEEE Symposium on VLSI Circuits, 1–2.
DOI: 10.1109/VLSITC.2016.7573461

147. "$Ta_2O_5$-based redox memory formed by neutral beam oxidation"
T. Ohno and S. Samukawa
Japanese Journal of Applied Physics **55** (2016) 06GJ01.
DOI: 10.7567/JJAP.55.06GJ01

**2017**

148. "Highly reproducible and regulated conductance quantization in a polymer-based atomic switch"
K. Krishnan, M. Muruganathan, T. Tsuruoka, H. Mizuta, and M. Aono
Advanced Functional Materials **27** (2017) 1605104.
DOI: 10.1002/adfm.201605104

149. "Operating mechanism and resistive switching characteristics of two- and three-terminal atomic switches using thin oxide layers"
T. Tsuruoka, T. Hasegawa, K. Terabe, and M. Aono
Journal of Electroceramics **39** (2017) 143–156.
DOI: 10.1007/s10832-016-0063-9

150. "Quantized conductance operation near a single-atom point contact in a polymer-based atomic switch"
K. Krishnan, M. Muruganathan, T. Tsuruoka, H. Mizuta, and M. Aono
Japanese Journal of Applied Physics **56** (2017) 06GF02.
DOI: 10.7567/JJAP.56.06GF02

151. "P-type polymer-based $Ag_2S$ atomic switch for "tug of war" operation"
C. Lutz, T. Hasegawa, T. Tsuchiya, C. Adelsberger, R. Hayakawa, and T. Chikyo
Japanese Journal of Applied Physics **56** (2017) 06GF03.
DOI: 10.7567/JJAP.56.06GF03
152. "Cu diffusion in amorphous $Ta_2O_5$ studied with a simplified neural network potential"
WW. Li, Y. Ando, and S. Watanabe
Japanese of the Physical Society of Japan **86** (2017) 104004.
DOI: 10.7566/JPSJ.86.104004
153. "Architecture optimization of nanobridge-based field-programmable gate array and its evaluation"
X. Bai, T. Sakamoto, Y. Tsuji, M. Miyamura, A. Morioka, R. Nebashi, M. Tada, N. Banno, K. Okamoto, N. Iguchi, H. Hada, and T. Sughibayashi
Japanese Journal of Applied Physics **56** (2017) 04CF03.
DOI: 10.7567/JJAP.56.04CF03
154. "Nanobridge-based FPGA in high-temperature environments"
M. Miyamura, T. Sakamoto, X. Bai, Y. Tsuji, A. Morioka, R. Nebashi, M. Tada, N. Banno, K. Okamoto, N. Iguchi, H. Hada, T. Sughibayashi, Y. Nagamatsu, S. Ookubo, T. Shirai, F. Sugai, and M. Inaba
IEEE Micro **37** (2017) 32–42.
DOI: 10.1109/MM.2017.3711648
155. "Set/reset switching model of Cu atom switch based on electrolysis"
M. Tada and T. Sakamoto
IEEE Transaction on Electron Devices **64** (2017) 1812–1817.
DOI: 10.1109/TED.2017.2670569
156. "Electrical-pulse-induced resistivity modulation in Pt/$TiO_{2-\delta}$/Pt multilayer device related to nanoionics-based neuromorphic function"
K. Kawamura, T. Tsuchiya, M. Takayanagi, K. Terabe, and T. Higuchi
Japanese Journal of Applied Physics **56** (2017) 06GH01.
DOI: 10.7567/JJAP.56.06GH01
157. "Resonant photoemission and X-ray absorption spectroscopies of lithiated magnetite thin film"
T. Tsuchiya, K. Kawamura, W. Namiki, S. Furuichi, M. Takayanagi, M. Minohara, M. Kobayashi, K. Horiba, H. Kumigashira, K. Terabe, and T. Higuchi
Japanese Journal of Applied Physics **56** (2017) 04GK01.
DOI: 10.7567/JJAP.56.04GK01
158. "Magnetic control of magneto-electrochemical cell and electric double layer transistor"
T. Tsuchiya, M. Imura, Y. Koide, and K. Terabe
Scientific Reports **7** (2017) 10534.
DOI: 10.1038/S41598-017-11114-2

## 2018

159. "Nanoarchitectonics for controlling the number of dopant atoms in solid electrolyte nanodots"
A. Nayak, S. Unayama, S. Tai, T. Tsuruoka, R. Waser, M. Aono, I. Valov, and T. Hasegawa
Advanced Materials **30** (2018) 1703261.
DOI: 10.1002/adma.201703261

160. "Neuromorphic transistor achieved by redox reaction of $WO_3$ thin film"
T. Tsuchiya, M. Jayabalan, K. Kawamura, M. Takayanagi, T. Higuchi, R. Jayavel, and K. Terabe
Japanese Journal of Applied Physics **57** (2018) 04FK01.
DOI: 10.7567/JJAP.57.04fk01

161. "The rate limiting process and its activation energy in the forming process of a $Cu/Ta_2O_5/Pt$ gapless-type atomic switch"
Y. Shigeoka, T. Tsuruoka, and T. Hasegawa
Japanese Journal of Applied Physics **57** (2018) 035202.
DOI: 10.7567/JJAP.57.035202

162. "Thermally stable resistive switching of a polyvinyl alcohol-based atomic switch"
K. Krishnan, M. Aono, and T. Tsuruoka
Journal of Materials Chemistry C **6** (2018) 6460–6464.
DOI: 10.1039/c8tc01809j

163. "Ionic decision-maker created as novel, solid-state devices"
T. Tsuchiya, T. Tsuruoka, S. -J. Kim, K. Terabe, and M. Aono
Science Advances **4** (2018) eaau2057.
DOI: 10.1126/sciadv.aau2057

164. "Development of a molecular gap-type atomic switch and its stochastic operation"
C. Arima, A. Suzuki, A. Kasai, T. Tsuruoka, and T. Hasegawa
Journal of Applied Physics **124** (2018) 152114.
DOI: 10.1063/1.5037657

165. "Oxygen vacancy drift controlled three-terminal ReRAM with a reduction in operating gate bias and gate leakage current"
Q. Wang, Y. Itoh, T. Tsuruoka, M. Aono, H. Deyan, and T. Hasegawa
Solid State Ionics **328** (2018) 30–34.
DOI: 10.1016/j.ssi.2018.11.004

166. "Effects of moisture and redox reactions in VCM and ECM resistive switching memories"
I. Valov and T. Tsuruoka
Journal of Physics D: Applied Physics **51** (2018) 413001.
DOI: 10.1018/1361-6463/aad581

167. "Atomic switch networks as complex adaptive systems"
K. S. Schernhorst, J. P. Carbajal, R. C. Aguilera, E. J. Sandouk, M. Aono, A. Z. Stieg, and J. K. Gimzewski

Japanese Journal of Applied Physics **57** (2018) 03ED02.
DOI: 10.7567/JJAP.57.03ED02

168. "Correlated metal $SrVO_3$ based all-solid-state redox transistors achieved by $Li^+$ and $H^+$ transport"
M. Takayanagi, T. Tsuchiya, W. Namiki, T. Higuchi, and K. Terabe
Journal of Physical Society of Japan **87** (2018) 034802.
DOI: 10.7566/jpsj.87.034802

**2019**

169. "Resistivity control by electrochemical removal of dopant atoms from a nanodot"
W. Hihara, N. Mishima, T. Shima, S. Tai, T. Tsuruoka, I. Valov, and T. Hasegawa
Faraday Discussions **213** (2019) 29.
DOI: 10.1039/c8fd00099a

170. "Conductivity modulation by $CaVO_3$-based all-solid-state redox transistor with ion transport of $Li^+$ and $H^+$"
M. Takayanagi, T. Tsuchiya, W. Namiki, Y. Kitagawa, D. Etoh, D. Nishioka, T. Yamada, T. Higuchi, and K. Terabe
Transactions of the Materials Research Society of Japan **44** (2019) 57.
DOI: 10.14723/tmrsj.44.57

171. "Time-dependent operations in molecular gap atomic switches"
A. Suzuki, T. Tsuchiya, and T. Hasegawa
Physica Status Solidi B **256** (2019) 1900068.
DOI: 10.1002/pssb.201900068

172. "Investigation of Ag and Cu filament formation inside the metal sulfide layer of an atomic switch based on point-contact spectroscopy"
A. Aiba, R. Koizumi, T. Tsuruoka, K. Terabe, K. Tsukagoshi, S. Kaneko, S. Fujii, T. Nishino, and M. Kiguchi
ACS Applied Materials & Interfaces **11** (2019) 27178.
DOI: 10.1021/acsami.9b05523

173. "Significant roles of the polymer matrix in the resistive switching behavior of polymer-based atomic switches"
Karthik Krishnan, Masakazu Aono, Kazuya Terabe, and Tohru Tsuruoka
Journal of Physics D: Applied Physics **52** (2019) 445301.
DOI: 10.1088/1361-6463/ab35bf

The manufacturer's authorised representative in the EU is Springer Nature Customer Service Centre GmbH, Europaplatz 3, 69115 Heidelberg, Germany. If you have any concerns regarding our products, please contact ProductSafety@springernature.com

Printed and bound by CPI Group (UK) Ltd, Croydon, CR0 4YY
25/09/2026
02222901-0002